The World as Idea

In *The World as Idea*, Charles P. Webel presents an intellectual history of one of the most influential concepts known to humanity—that of "the world."

Webel traces the development of "the world" through the past, depicting the history of the world as an intellectual construct from its roots in ancient creation myths of the cosmos, to contemporary speculations about multiverses. He simultaneously offers probing analyses and critiques of "the world as idea" from thinkers ranging from Plato, Aristotle, and St. Augustine in the Greco-Roman period to Kant, Schopenhauer, Nietzsche, Wittgenstein, Merleau-Ponty, and Derrida in modern times. While Webel mainly focuses on Occidental philosophical, theological, and cosmological notions of worldhood and worldliness, he also highlights important non-Western equivalents prominent in Islamic and Asian spiritual traditions. This ensures the book is a unique overview of what we all take for granted in our daily existence, but seldom if ever contemplate—the world as the uniquely meaningful environment for our lives in particular and for life on Earth in general.

The World as Idea will be of great interest to those interested in the concept of "world as idea," scholars in fields ranging from philosophy and history to political and social theory, and students studying philosophy, the history of ideas, and humanities courses, both general and specialized.

Charles P. Webel, Ph.D., is currently a Professor and Guarantor of the School of International Relations at the University of New York in Prague. A five-time Fulbright Scholar and a research graduate of the Psychoanalytic Institute of Northern California, he has studied and taught at Harvard University and the University of California at Berkeley, where he received his Ph.D. He is the author or editor of many books, including *Peace and Conflict Studies* (with David Barash), the standard text in the field, as well as *Terror, Terrorism, and the Human Condition*, and *The Politics of Rationality*.

Routledge Studies in Social and Political Thought

159 Critical Rationalism and the Theory of Society
Critical Rationalism and the Open Society Volume 1
Masoud Mohammadi Alamuti

160 Functionalist Construction Work in Social Science
The Lost Heritage
Peter Sohlberg

161 Critical Theory and New Materialisms
Hartmut Rosa, Christoph Henning and Arthur Bueno

162 Max Weber's Sociology of Civilizations
A Reconstruction
Stephen Kalberg

163 Temporal Regimes
Materiality, Politics, Technology
Felipe Torres

164 Citizenship in a Globalized World
Christine Hobden

165 The World as Idea
A Conceptual History
Charles P. Webel

166 Max Weber and the Path from Political Economy to Economic Sociology
Christopher Adair-Toteff

For a full list of titles in this series, please visit www.routledge.com/series/RSSPT

The World as Idea

A Conceptual History

Charles P. Webel

NEW YORK AND LONDON

First published 2022
by Routledge
605 Third Avenue, New York, NY 10158

and by Routledge
2 Park Square, Milton Park, Abingdon, Oxon OX14 4RN

Routledge is an imprint of the Taylor & Francis Group, an informa business

Library of Congress Cataloging-in-Publication Data
Names: Webel, Charles, author.
Title: The world as idea : a conceptual history / Charles P. Webel.
Description: New York, NY : Routledge, 2021. |
Series: Routledge studies in social and political thought |
Includes bibliographical references and index.
Identifiers: LCCN 2021023019 (print) | LCCN 2021023020 (ebook) |
ISBN 9781138013520 (hardback) | ISBN 9781032115665 (paperback) |
ISBN 9781315795171 (ebook) | ISBN 9781317746720 (adobe pdf) |
ISBN 9781317746706 (mobi) | ISBN 9781317746713 (epub)
Subjects: LCSH: Cosmology. | Existentialism. | Phenomenology.
Classification: LCC BD511 .W43 2021 (print) |
LCC BD511 (ebook) | DDC 113–dc23
LC record available at https://lccn.loc.gov/2021023019
LC ebook record available at https://lccn.loc.gov/2021023020

ISBN: 978-1-138-01352-0 (hbk)
ISBN: 978-1-032-11566-5 (pbk)
ISBN: 978-1-315-79517-1 (ebk)

DOI: 10.4324/9781315795171

Typeset in Times New Roman
by Newgen Publishing UK

Contents

Detailed Contents

Introduction

NASA photo of the Eastern Hemisphere, partially covered with methane gas.
Source: NASA.

The world does and does not exist. Behind this paradoxical assertion lie the facts that "the world" is a linguistic and historical construction, not an ontological given, and therefore "exists" only insofar as meaning-creating organisms frame the boundaries of their being-in-this-world.

"The world" is thus an abstraction, a concept, or idea, but a vital one. Our fate literally depends on it.

We use the term "the world" to provide a limit to our possible experiences and perceptions while living on Earth, and also to distinguish human civilization as a whole from the rest of life, both terrestrial and extra-terrestrial. In this book, *I am using the phrase "the world" to refer to the totality of human artifacts imposed on this planetary orb on which our species has existed.*

The world is normally taken-for-granted. "It" is assumed simply to "be there," undeserving of definition, so self-evident that one doesn't need

DOI: 10.4324/9781315795171-1

explain it.[1] In contemporary geopolitical terms, "the world" is widely assumed to denote an abiding "system" of nation-states. Within this "global system" rules the "human world"—imperiously co-existing with and lording over the "natural world," comprising plants, animals, and insects, as well as an " 'objective' world" largely consisting of inanimate objects, a "worldless" "non-world" from the point of view of some of the most renowned Western philosophers.[2]

Nonetheless, it is useful to distinguish "*the* world" from "*this* world." *The former designates a general context for the possible experiences of sentient creatures as a whole; the latter denotes the specific limits within which your and my lives can be lived and interpreted.*

"The" world is an impersonal factual framework for description and analysis: it is "everything that is the case," as the philosopher Ludwig Wittgenstein described it in the beginning of his book the *Tractatus Logico-Philosophicus.*[3] The world is the scaffolding, the skeletal structure, upon which the flesh, bones, and brains that constitute this world of possible experience has been erected.

"This" world, in contrast, is a subjective and intersubjective boundary for our lived experience. Those two words, this world, denote the personalized horizon that renders "the" world potentially meaningful and value-laden. It is how "I see" "the" world, what it means to me and others, what it's like to be-in-the-world, and how the feelings, desires, and perspectives we have about "the" world are socialized, historicized, politicalized, and individualized. How the words "this world" are used thus depend on their users.

There may or may not be other, possible "worlds" constructed by organisms elsewhere in what we call the "universe" (or "multiverses"). And there are as many different individually- and socially constructed "lifeworlds" on this island Earth as there are symbol-using beings.[4] But the "nature" of "this world"—as symbolized by the text on these pages—is explicable only from the perspective of the person whose existence is bounded by it. It is to this person I briefly turn.

"My" World

"I" denotes the embodied person simultaneously immersed in and reflecting on this, "my," world. My "I" is incarnate, an embodied subjectivity whose brain and perceptual organs literally construct "my" world out of the raw data provided by our senses. "I" designates both collective and individual identities. Our I's, like our eyes, develop through childhood and adolescence through dialogue, social interactions, and psychophysical maturation.

My "I" is the mental agency (not entity) unifying—or in extreme psychological states, disintegrating—what "the" world presents. I am *what* I am because *this* world for me is rendered perspicuous through the body

I call mine. I become *who* I am in *this* and no other world, as long as there is sufficient brainwave activity for me to be capable of conscious awareness of my being-in-this-world.

Without an "outer" world, there is no context for the "inner" world I call my "mind" to develop. Without an inner world, there is no world per se, because there is no sentient subjectivity to construct it. How "the" world in general and "this" world for me are constructed is largely, but not exclusively, a historically developing linguistic and social project. *The world becomes this world for me during the course of my psychophysical life span.*

"My" world is metaphorically and literally how and what I see, feel, and perceive, as well what I do in the world and what others and the world itself do to me. My world is the sum of possible contexts for my states of consciousness and activities. In this sense, it is difficult to imagine my existence without *a* world. But while it is natural to believe that the world will go on without me, it is also hard to conceive of *this* world if I am not in it.

This world is the world for me. The world becomes this world over the course of my life span, during which I acquire impressions, images, feelings, thoughts, and attitudes toward the people and things I directly encounter and those I imagine from exposure to the mass media and other vehicles of communication, representation, and signification—both immediate and virtual. Through this, I develop both my overall sense or image of the world—my *Weltbild* (in German)—as well as my philosophy of the world, my *world-view,* or, following the early 20th-century German philosopher Wilhelm Dilthey, my *Weltanschauung*.[5]

Without me, this world is void. Without *the* world, *this* world would not exist for me, or for anyone else as far as I know. My world is the historically bound psychophysical location I occupy on Earth. As I change locations, my world expands. If I remain stagnant, my world is constricted.

My world is also charged with subjective attributions and emotional attitudes. It can "feel" alive and vibrant, neutral and passive, boring and perilous, meaningless or meaningful—depending on my moods, social relations, and psychophysical health or illness. The meaning(s) of my world may shift, just as the meanings of the word "world" have varied over time. It is to the "meanings" of "the" world that I now turn.

The Origins of "The World"

Lexicologically, "world," in modern English, is a common name—a noun—typically denoting the whole of human civilization, specifically human experience, history, or the human condition in general, as in *world-wide*, i.e., anywhere on Earth.

"World" is a word peculiar to Germanic languages. The modern German equivalent of "the world" is "*die Welt*" (a feminine noun). The corresponding word in Latin is *mundus*, literally "clean, elegant," a translation of the ancient Greek *kosmos*—an "orderly arrangement," or "world," because the ancient Greeks thought that the world was perfectly harmonious and symmetrically ordered.[6]

"World" (*kosmos/mundus*) was also classically used to denote the *material universe*. The Earth was often described as "the center of the world." While the Germanic word "*Welt*" may include a mythological notion of a "domain of Humanity," in contrast with the divine sphere on the one hand and the chthonic sphere of the underworld on the other, the Greco-Latin term expresses a notion of creation as an act of establishing order out of chaos.

Today, "the world" usually refers to the Earth and all the people and things on it, or, more narrowly, to human society and the earthly state of human existence (in contrast with "the next world," presumably to life to come after death), *and is also less often identified with the entire universe of created things.* "Worldly" affairs are *secular* matters (an English adjective derived from the Latin noun *saeculum, meaning* "generation," or "age") and used in Christian Latin to mean "the world" ("of the flesh") as opposed to the Church (or the "world of the spirit").

In common English usage, "*world*" can also be used attributively, denoting what is "global," i.e., "relating to the whole world," and including such usages as "world" (or "global") "community," or "world-canonical" texts. Hence, "this world" distinguishes the entire planet we call the "Earth," or its entire human population, from any particular country or region.

In Chinese culture, the word for (this) "world" is *shi jie*. The word *shi* means 30 years, which indicates how long a person should live, based on premodern Chinese life-expectancies during the Han Dynasty (206 BCE–220 CE), the Qin dynasty (221–206 BCE), and the Three Kingdoms (220–280 CE). Chinese people also use the word: *yi shi*, meaning "one thirty years" to denote a person's life span. The word *shi* is also used to denote a generation, equivalent to the original meaning of 30 years. The word *jie* means boundary or limit. So all together, the words *shi jie*, mean a typical human life span, the limit of one's lifetime on Earth, this world for me.[7]

The Chinese word for "world" comes from Buddhist scriptures, which were translated from Sanskrit texts. In Chinese Buddhism, *shi je* usually refers to the universe. Another relevant word is *yu zhou*, which literally means the universe in modern Chinese. This word is sometimes also used to denote the world. *Zhou* originally meant eaves, but it is generalized to denote space as a whole. *Zhou* means the totality of time. The totality of space and time is the Universe. *Tian xia*, which literally means "under the sky." This word was often used by emperors and generals, and connoted conquest. For the emperor, "everything under the sky is mine."

And in Russian, the commonly used word to denote the "world" as a whole is *mir*, which can also mean peace, part of the universe, kingdom, or a rural community (outdated usage). In addition, the noun *vselennaya* denotes not only universe (and the system of the universe as a whole), but also world, cosmos, or macrocosm. To say "the entire world" in Russian, one would say "*vyes mir*."[8]

In addition, there are compound nouns incorporating "world" in English that incorporate the ordinary use of the term. For example, *world affairs* pertain not just to one place but to the whole world, and *world history* is a field of history that examines events from a global (rather than a national or a regional) perspective. *Earth*, on the other hand, refers to *this* planet as a physical entity, and distinguishes it from other planets and physical objects. By extension, *a world* may refer to *any* planet or heavenly body, especially when it is thought of as inhabited, usually in the context of astronomy, science fiction, or futurology.

There are also multiple "worlds" within and beyond this world as well as particular domains of human experience and activity within "this world," including "the modern world," "the academic world," "the natural world," "the digital world," "the world of work," "the social world," "the political world," and "one's own little world." In political theory, e.g., there are distinguishable realms, or "worlds," of public and private rights and realms. These are, culturally speaking, separable "worlds," but they are always less than the, or this, world.

This division of "the world as a whole" into various mini- and micro-"worlds" also has the function of demarcating particular domains of individual and collective human experience and activity from other life forms, as well as distinguishing objects sharing this planet from humanity and "nature" as a whole, while leaving open the possibility of life on "other worlds."

In a philosophical context, "the world" may refer to both an "external," "outer," or "material" world, as in the whole of the physical universe—at least insofar as humans are capable of perceiving and in some sense knowing it—as well as to an "internal," or "inner," or "mental" world. For sociologists, anthropologists, and other social scientists, there are, of course, also "the social world," "the intersubjective world," "the political world," "the world of finance and business," etc. For many philosophers, philologists, and psychologists, the "mental world," or "ontological" world, has been the subject of rigorous analysis and imaginative leaps since at least what the German classical philologist Bruno Snell called "the discovery of the mind" by the ancient Greeks (and by Asians and other cultures as well).[9]

Furthermore, in a theological context, *world* usually refers to the *material or the profane* sphere, as opposed to the celestial, spiritual, transcendent, or sacred. The "end of the world" refers to scenarios of the final end of human history, often in religious contexts. The fictional or real "end of the world" has been a subject of intense speculation and trepidation for millennia and is connected to what I shall call "the fate of this world."

The Varieties of Fate

Fate, as an English noun, means a predetermined and inevitable outcome, usually adverse. For example, the *fate* of the world is unknown to humans at the present time, but is probably, if things continue the way they are, dire.

In classical mythology, especially in the Homeric epic poems, Fate was one of the goddesses said to control the destiny of human beings. In Latin, *fatum* denoted the prophetic declaration of what must be, an oracular prediction, literally the "thing spoken (by the gods)."[10]

The first known use of *fate* in English as a noun dates from the late 14th and early 15th centuries and meant the power that rules destinies, the agency that predestines events, "supernatural predetermination;" and "destiny personified." The sense of fate as a goddess who determined the course of a human life, often in the sense of bad luck, ill fortune, ruin, a pest, *or a plague* was present in the English language by the end of the 16th century. And the meaning of fate of a "final event" dates from the late 17th century.

Fate as an English verb also dates from the 17th century and means to foreordain, predetermine, and to render inevitable. For example, in the ancient Greek playwright Sophocles' play *Oedipus Rex*, the oracle's prediction *fated* Oedipus to kill his father and marry his mother, and not all his striving could change what would occur.

Synonyms of fate include *destiny, lot, portion, and doom. Destiny* implies something foreordained and often suggests a noble course or end, i.e., their "destiny" was to save their nation. *Lot* and *portion* imply a distribution by fate or destiny, lot suggesting blind chance, e.g., it was her *lot* to die childless. *Portion* refers to the distribution of good and evil, e.g., remorse was his daily *portion. Doom* implies a grim or calamitous fate, e.g., if the rebellion fails, their *doom* is certain.

Some examples of recent uses of *fate* include: "The fate of our species is bound up with those of countless others, with which we share a habitat that we cannot long dominate ....[11] "In recent weeks, a fierce debate has erupted between hospitals and nursing homes over the *fate* of elderly patients."[12] "The announcement followed weeks of speculation over the *fate* of the national vote, which was scheduled for April 22 but has now been pushed to an unspecified future date, depending on how the pandemic develops."[13]

In this context, "the fate of the world" is, at least partially, in our hands ....

Human and Humanity

In everyday speech, the word "human" generally refers to the only extant species of the genus *Homo*—the modern *Homo sapiens*. The term "*Homo sapiens*" was coined by Carl Linnaeus, the Swedish botanist, zoologist, and physician, who is known as "the father of modern taxonomy." In his 18th-century work *Systema Naturae* (*The System of Nature*), Linnaeus formalized binomial nomenclature—in which two terms are used to denote a species of living organism, the first indicating the genus and the second the specific species within the genus—the modern system of naming organisms. The generic name "*Homo*" is an 18th-century derivation from Latin *homō* "man," ultimately "earthly being." The species-name "*sapiens*" means "wise" or "sapient." The Latin word *homo* refers to humans of any gender.[14]

Like "homo," the English term *man* can refer to the species generally (a synonym for *humanity*) as well as to human males. And, the English word "humanity" derives from the Old French noun *humanité*, whose root is Latin substantive *humanitas*, from *humanus*.[15] *Humanity* refers collectively to the human species, and, less often may also denote the moral virtue of humanity (compassion).[16]

The meanings of "human," "humanity," and our "hominid" ancestors have changed somewhat during recent decades due to advances in the discovery and study of the fossil ancestors of modern humans. The previously clear boundary between humans and apes has blurred, resulting in biologists now acknowledging the hominids as encompassing multiple species. The current, 21st-century meaning of "hominid" refers to all the great apes, including gorillas, chimpanzees, and bonobos, as well humans. There is also a distinction between *anatomically modern humans* and *Archaic Homo sapiens*, the earliest fossil members of the species.

Underlying much thinking about our species is the often unquestioned assumption that we, *Homo sapiens*, are not merely distinct from all previous and contemporary beings-in-this world, but are superior to them. Our terrestrial and possibly cosmic "exceptionality" is the premise of what David Wallace-Wells, among others, has called the "anthropic principle."

In cosmology, the anthropic principle means that any consideration of the structure of the universe, the values of the constants of nature, or the laws of nature, assumes the existence of (intelligent) conscious life. Consequently, any data we collect about the universe is filtered by the fact that, in order for it to be observable in the first place, it must be compatible with the conscious and sapient life that observes it. The strong anthropic principle (SAP) states that this is the case because the universe is in some sense destined eventually to have conscious and sapient life emerge within it. Some critics of the SAP argue in favor of a weak anthropic principle (WAP), which states that the universe's ostensible fine tuning is the result of selection bias (specifically, survivorship bias): i.e., only in a universe capable of eventually supporting life will there be living beings capable of observing and reflecting on the matter. Since we are, as far as is believed to be known at present, the only surviving living beings doing this, the human species is "exceptional," i.e., a cosmic "anomaly."

For Wallace-Wells, the anthropic principle

> takes the human anomaly not as a puzzle to explain away but as the centerpiece of a grandly narcissistic view of the universe … because

> only a universe compatible with our sort of conscious life would produce anything capable of contemplating it like this.[17]

But the universe as a whole, and life on Earth in particular—as well as possibly extant or extinct conscious and intelligent extraterrestrial life forms—have existed and evolved for billions of years prior to our existence, and will also likely do so after humanity's extinction. Neither the universe nor the Earth needs us to exist; we need them for our continued existence.

Varieties of Existence

In philosophy and theology, the term "existence" usually refers to the ontological property of *being*. If something "exists," it "is." *That* anything exists constitutes a continuing mystery. *Why* there is something rather than nothing—the "mystery of existence"—is a riddle that has haunted inquiring humans from early recorded history to contemporary evolutionary biology and theoretical astrophysics.[18]

In modern cosmology, what is called "the universe" is said to have *come into existence* about 13.8 billion years ago. Many evolutionary scientists think that people who look like us—anatomically modern *Homo sapiens*—evolved at least 130,000 years ago from ancestors who had remained in Africa, whose brains had reached today's size. But agreement breaks down on the question of when, where, and when early humans became fully human in behavior as well as body, as well as when, and where, human culture born and how these anatomically modern humans began to manifest creative and symbolic thinking.[19] And, of course, although there have been arguments, claims, images, and texts about "God's existence" since the dawn of humanity, there is still no scholarly or scientific consensus regarding the existence of deities in general—including, with the rise of monotheism, the Judeo-Christian-Islamic ("Abrahamic") God in particular.

"Exist" as an English verb dates from about 1600, and is derived from the French *exister*, which comes from the Latin *existere/exsistere* "to step out, stand forth, emerge, appear; exist, be." The English noun term *existence* comes from the Old French noun *existence*, which is derived from the Medieval Latin *existentia*. The term "*existential*," meaning "pertaining to existence," derives from the Late Latin *existentialis/exsistentialis*, and figures prominently in 19th- and 20th-century European philosophy and literature, particularly in Denmark and Germany (in Danish, *eksistentiel*; and in German, *existential*).

In Chinese, the primary word for existence is *zai*. Grammatically, this word can be a verb, a preposition, or a participle. As a verb, *zai* means "to be," "to exist," as well as "to save," as in to save money, to save data, etc. It cannot be used in such expressions as "to save a life." *Zai* also denotes physical being and location, a state of being that does not readily change. As a preposition, *zai* approximates the English "at" and can be used to denote both time and location, but never motion. As a participle, it indicates continuity, equivalent to the English *–ing* (as in sing-*ing*). However, in Chinese, unlike English, there is no differentiation between the continuing present and the continuing past; so, if there is only *zai* + a verb; one does not know if the event is happening now or if it was happening in the past. So, another word or phrase would normally be added to *zai*, like "yesterday," or "three months ago," to resolve that ambiguity. In modern Chinese, the word for existence is *cun zai*. This word is used as root for such words as "existential," "existential crisis," "Existentialism," and so on.

And in Russian, the word for "existence" is *sushchestvovanye*. Russians may also use such expressions as "the meaning of life" (*smysl zhizni*), the "meaning of existence" (*smysl sushchestvovaniya*), and "the meaning of being" (*smysl bytiya*). Perhaps in no other culture is "the meaning of life" so often discussed, possibly due to such historical and cultural factors as Russia's experiences of wars, invasions, and slavery; its literary and religious traditions (especially the novels of Fyodor Dostoyevsky and the poetry of Alexander Pushkin, among others); and, more generally, to the tendency of what is often called "The Russian Soul" (*Russkaya dusha*) to ruminate at length over the dire and ineluctable "fate" (*sud'ba*) of Russia as a nation and Russians as individuals—especially over a bottle of vodka, or three ...

"Existence," in the sense of a human being's individual being-in-this-world, is a central concern of the 19th-century Danish thinker Søren Kierkegaard, for whom "existence," while referring to many things that are, has a special meaning when applied to human life. For Kierkegaard, the most pressing question for each person is the *meaning of his or her own existence*, which arises from their relationships to their individual self, to significant others, and, especially, via a "leap of faith," to God. Twentieth-century Existentialism, both theistic and atheistic, has its roots in the views of Kierkegaard and the late 19th-century German philosopher Friedrich Nietzsche (Existentialism will be discussed in more detail later in this book).

"Existence" also plays a prominent role in ontology, logic, and semantics, dating back to Plato and Aristotle, who developed a comprehensive theory of being (an "ontology"), according to which only individual things, called *substances*, fully have to be, whereas such other things as relations, quantity, time, and place (called the categories) have a derivative kind of being, dependent on individual things. Problems of "being" and "existence" also feature prominently in the metaphysics and ontology of such medieval scholastic philosophers as St. Anselm and St. Thomas Aquinas, as well as in the works of the 17th-century French philosopher and mathematician René Descartes and the 18th-century German philosopher Immanuel Kant.

But the questions "what does the term 'existence' *mean*?" and is "existence" a "property" of individuals at all? came to the fore in 19th- and 20th-century "analytic" philosophy, mathematical logic, and semantics, most notably in the works of the German logician Friedrich Ludwig Gottlob Frege and the English philosopher Bertrand Russell. The contrast between the "meanings of being and existence" in the "analytic" and "continental" traditions of philosophy, will be drawn later in this book.

Planet Earth

The idea of planets has evolved over our history, from the "divine lights" and "wandering stars" of antiquity to the earth-like objects of the scientific age. The concept of a "planet" has expanded to include worlds not only in our solar system, but in hundreds of extrasolar systems. The ambiguities inherent in defining planets have led to much scientific controversy.

The term *planet* is ancient, with ties to natural philosophy, history, astrology, science, mythology, and religion. Five planets in our solar system are visible to the naked eye—Mercury, Venus, Mars, Jupiter, and Saturn. These were regarded by many early cultures as divine, or as emissaries of deities. As scientific knowledge advanced, human perception of the planets changed, incorporating a number of disparate objects. Today, a planet is generally defined an astronomical body orbiting a star or stellar remnant that is massive enough to be rounded by its own gravity, is not massive enough to cause thermonuclear fusion, and has cleared its neighboring region from the rock-type objects formed in the early solar system from collisions with other objects in the solar system, i.e., "planetesimals."

The modern English word "planet" comes from the Middle English noun *planete* and from the Old French *planete*, deriving from the Latin *planeta*, *planetes*, the Roman equivalent of the Ancient Greek *planḗtēs*, meaning "wanderer". The term *planet* originally meant any star that wandered across the sky, and generally included comets as well as the Sun and Moon. In ancient Greece, China, Babylon, and in virtually all premodern civilizations, it was almost universally believed that Earth was the center of the universe and that all the "planets" circled Earth. The reason for this perception was that stars and planets appeared to revolve around Earth each day, as well as for the apparently common-sense perceptions that Earth was solid and stable and that it was not moving but at rest. With the Copernican revolution of the early 16th century, the Earth was recognized as a planet, not the center of the cosmos, and was observed to be a fundamentally different celestial orb from the Sun—now considered a relatively minor star around which the Earth and the other planets in our solar system revolve.

There are continuing debates in astronomy as to which large objects should be designated as "planets." The first asteroids were also thought to be planets, but were reclassified when it was realized that there were a great many of them, crossing each other's orbits, in a zone where only a single planet had been expected. Likewise, Pluto was found where an outer planet had been expected, but doubts were raised when it turned out to cross Neptune's orbit and to be much smaller than the expectation for planetary size required.

In 2006, the International Astronomical Union adopted a resolution defining planets within our solar system. This followed their announcement the previous year regarding the status of Eris, an object in the same region of our solar system as Pluto but thought to be about 27% more massive than Pluto, created the desire for an official definition of a planet. This definition is controversial because it excludes many objects of planetary mass. Although eight of the planetary bodies discovered before 1950 remain "planets" under the current definition, some celestial bodies, such as Ceres, Pallas, Juno, and Vesta (each an object in the solar asteroid belt), and Pluto, which were once considered *planets* by the scientific community, are no longer viewed as planets under the current definition. The number of IAS-designated planets dropped to the eight significantly larger bodies that had cleared their orbit (Mercury, Venus, Earth, Mars, Jupiter, Saturn, Uranus, and Neptune), and a new class of solar-orbiting large objects was created, initially containing three objects—Ceres, Pluto, and Eris.

Importantly, there is also the continuing search for what astronomers call extrasolar or exoplanets, i.e., for planets outside our solar system. This coincides with the search for life outside Earth.[20] As Sara Seager, a professor of planetary science and physics at MIT, has said, "If we can identify another Earth-like planet, it comes full circle, from thinking that everything revolves around our planet to knowing that there are lots of other Earths out there."[21]

Our Earth

Our home, the ground of our lifeworld, is called the Earth, which has been considered a planet since the early 14th century.[22] Earth is the third planet from the Sun and the only astronomical object known to harbor life. According to radiometric dating and other evidence, Earth was formed over 4.5 billion years ago. Earth's gravity interacts with other objects in space, especially the Sun and the Moon, which is Earth's only natural satellite. Earth's axis of rotation is tilted with respect to its orbital plane, producing seasons on Earth. Earth is the densest planet in the solar system and the largest and most massive of the four rocky planets.

Unlike the other planets in our solar system, Earth does not directly share a name with an ancient Roman deity. The modern English word *Earth* derives, via Middle English, from an Old English noun most often spelled *eorðe*. Earth has cognates in every Germanic language. The ancient German noun *eorðe* (in modern German, the feminine noun *die Erde*) was used to translate the many senses of the Latin substantive *terra* and the ancient Greek noun *Gē*, meaning the ground, its soil, dry land, the human world, the surface of the world (including the sea), and the globe itself. From *Earth* comes the adjective *earthly*; from the Latin *terra* comes *terrestria*l. As with the Roman *terra* and the ancient Greek *Gaia*, Earth may have been a personified goddess in Germanic paganism: late Norse mythology, e.g., included Jörð ("Earth"), a giantess often portrayed as the mother of the god Thor.

Originally, *earth* was written in lowercase, and from early Middle English, its sense as "the globe" was expressed as *the earth*. Occasionally, the name *Terra* is used in scientific writing and in science fiction to distinguish our inhabited planet from others, while in poetry *Tellus* has been used to denote personification of the Earth. The Greek poetic name *Gaia*, who for the ancient Greeks, was the ancestral mother of all life—the primal "Mother Earth" goddess—has come into common usage due to the "Gaia theory" or the "Gaia hypothesis."

The Gaia theory posits that the Earth is a self-regulating complex system involving the biosphere, the atmosphere, the hydrosphere, and the pedosphere, tightly interacting as an evolving system. The theory claims that this system as a whole, called Gaia, requires a physical and chemical environment optimal for contemporary life. The originality of the Gaia theory relies on the assessment that such a homeostatic balance is actively pursued with the goal of keeping the optimal conditions for life, even when terrestrial or external events menace them.

The Gaia hypothesis was formulated by the chemist James Lovelock and co-developed by the microbiologist Lynn Margulis in the 1970s. Initially received with hostility by the scientific community, it is now examined in geophysiology and Earth system science, and some of its principles have been adopted in such fields as biogeochemistry and systems ecology. This hypothesis has also inspired analogies and various interpretations in the social sciences, politics, and religion, as well as an ecological movement. The assumption behind this theory is that life on Earth is an evolving, interconnected whole.

Life on Earth

Within the first billion years of Earth's history, life appeared in the oceans and began to affect Earth's atmosphere and surface, leading to the proliferation of anaerobic (living without oxygen) and, later, aerobic (living with oxygen) organisms. Some geological evidence indicates that life may have arisen as early as 4.1 billion years ago. Since then, the combination of Earth's distance from the Sun, physical properties, and geological history have allowed life to evolve and thrive.

In the history of life on Earth, biodiversity has gone through long periods of expansion, occasionally punctuated by mass extinctions. While estimates of the number of species on Earth today vary widely; most species have not yet been described. According to a 2018 UN Report, there are an estimated 8 million plant and animal species on Earth, including 5.5 million insect species.[23]

Over 99% of all species that ever lived on Earth are extinct, and our planet now faces a global extinction crisis never witnessed by humankind.

The UN Report finds that around 1 million animal and plant species are now threatened with extinction, many within decades, more than ever before in human history. And, due mostly to anthropogenic (human-generated) global warming, an estimated 5% of species are at risk of extinction from a 2°C increase in warming, rising to 16% of species at risk for extinction at 4.3°C warming. Even with a global warming of 1.5–2°C, the majority of terrestrial species are projected to shrink profoundly. The UN Report concludes that current global response is insufficient, and that "transformative changes" are needed to restore and protect nature, and also that opposition to such changes by vested interests can be overcome for the public good and for species-preservation as well. Human population increase is also a key contributor to the unprecedented increase in species extinction.

Humanity on Earth

It is estimated that only about one-eighth of Earth's surface is suitable for human habitation. About three-quarters of Earth's surface is covered by oceans, and approximately one-quarter is land. Half of that land area consists of deserts, high mountains, or other virtually uninhabitable terrains, leaving about 12.5% of the planet appropriate for human settlement.

Almost 8 billion humans live on Earth and depend on its biosphere and natural resources for their survival. It took over 200,000 years of human history for the world's population to reach 1 billion, and only 200 years more to reach 7 billion. The number of humans on Earth is expected to reach 8 billion in the early 2020s and may exceed 10 billion by 2050. Most of the population growth is expected to take place in developing nations. Human population density varies widely around the world, but a majority live in Asia, with over 2.5 billion people in two countries alone, China and India. A growing majority of the world's population lives in urban, rather than rural, areas. Almost 70% of the land mass of the world is in the northern hemisphere, which is where about 90% of humans live. Politically, the world has about 200 sovereign states.

Independent nations claim the planet's entire land surface, except for some parts of Antarctica, a few land parcels along the Danube river's western bank, and an unclaimed area of land between Egypt and Sudan. As of early 2021, there were 193 member states of the United Nations, plus two non-member observer states (the Holy See and Palestine), as well as 72 dependent territories and states with limited recognition. Our world has never had a sovereign government with authority over the entire globe, although some nation-states and empires have striven for world domination and failed.

The United Nations, which was created in the aftermath of World War II, is a worldwide intergovernmental organization that was designed with the goal of intervening in the disputes between nations, thereby avoiding armed conflicts, especially between nuclear powers. The UN serves primarily as a forum for international diplomacy and international law. When the consensus of the membership permits, it provides a mechanism for armed intervention. But due to the veto power wielded by the five permanent members of the UN Security Council, it is often difficult to achieve unanimous consent for resolutions opposed for political reasons by the United States, the United Kingdom, China, France, or Russia.

My Perspective on The World

Scientists have described the world, philosophers have interpreted it, revolutionaries have tried to change it, and we all continue to live in it as best we can. The task for humanity as a whole in this century is to do all of this in a way that preserves and transforms both the best in our species and in the planet we have colonized.

In this book, the fate of this world will be approached … existentially. That is to say that while the chapters that follow will present a *multidisciplinary description* of how and why this world is becoming increasingly imperiled and what might be done to address this historically unprecedented peril, my *analysis and recommendations* will largely be based on the *existential-ontological meanings* of potential world-annihilation for us as individual persons, for humanity as a whole, and for life on Earth.

The Fate of this World is my attempt to provide an interpretive analysis of this world in order to abet revolutionary changes in the way that humans inhabit Earth by understanding how and why this world has become what it is today. *The World as Idea*: *A Conceptual History* is the first volume of the project, and with due reference to Arthur Schopenhauer's *The World as Will and Idea*, or *Representation* (German: *Die Welt as Wille and Vorstellung*), provides an intellectual history of the world as idea from ancient to postmodern times. In the second volume, *The Reality of the World*, I will seek not merely to understand but also to motivate responsible nonviolent change of the social, political, and economic dimensions of our worldly existence. This is a very ambitious task, but an indispensable one, if we are to bequeath to our descendants a world that is really worth living in.

Notes

1 For example, see the recent popular book *The World: A Brief Introduction* by Richard Haass (New York: Penguin Press, 2020), which omits any clarification of what "The World" means while spending hundreds of pages presenting the author's overview of global geopolitics. Noam Chomsky's recent political essays, while coming from a very different perspective, also take for granted "the world" as self-evident.

2 See Jacques Derrida's discussion of this point in his "The 'World' of the Enlightenment to Come," as summarized in Sean Gaston, *The Concept of World from Kant to Derrida* (London and New York: Rowman & Littlefield, 2013), 115.

3 Ludwig Wittgenstein, *Tractatus Logico-Philosophicus*, 1-2.063, in *Major Works, Ludwig Wittgenstein* (New York: Harper Perennial, 2009), 5–9.
4 For the "lifeworld" (*Lebenswelt* in German), see Edmund Husserl. *The Crisis of the European Sciences*, trans. David Carr (Evanston, IL: Northwestern University Press, 1970), 108–9 and *passim*. Alfred Schutz and Thomas Luckman. *The Structures of the Life-World. (Strukturen der Lebenswelt.)*, trans. Richard M. Zaner and H. Tristram Engelhardt, Jr. (Evanston, IL: Northwestern University Press: 1973); and Jürgen Habermas. *The Theory of Communicative Action,* two volumes, trans. Thomas McCarthy (Boston: Beacon Press, 1989), *passim*.
5 See David. K. Naugle, *Worldview: The History of a Concept* (Wm. B. Eerdmans Publishing: Grand Rapids, MI, 2002) especially for a discussion of the use of the term "*Weltbild*" in Wiittgenstein's writings. For Dilthey's "*Weltanschauung*," see: www.encyclopedia.com/philosophy-and-religion/philosophy/philosophy-terms-and-concepts/worldview-philosophy; also see: https://en.wikipedia.org/wiki/Worldview/.
6 Both the English noun "world" and the German "*die Welt*" stem from the Old High German *weralt*—a compound of "*wer*," or "man," and "*alt*," or "*andeld*," meaning "past," "age," or "old," as in the "Age of Man," leading to the Old English *wer, meaning "man"* (still in use, as in *werewolf*). See Gerhard Köbler, *Althochdeutsches Wörterbuch*, 6th Edition, 2014, www.koeblergerhard.de/ahdwbhin.html.
7 This summary of Chinese usages of "world" and its equivalents in Mandarin Chinese has been informed primarily by my former student Russell Lo Su, who is from China and is now completing his PhD in Clinical Psychology at Charles University in Prague, Czech Republic.
8 would like to thank two former University of New York students, Natalia Khozyainova and Ekaterina Starodumova, for providing assistance with Russian/English translations.
9 Bruno Snell. *The Discovery of the Mind; The Greek Origins of European Though* (Oxford: Blackwell, 1953).
10 The Latin meaning of *fatum* as the "sentence of the Gods" is derived from the ancient Greek (*theophaton)* and is related to the ancient Greek noun (*moira*), meaning a lot, or portion, personified as a goddess.
11 John Gray, *Times Literary Supplement*, September 11, 1992.
12 *Los Angeles Times*, "L.A. County death toll nears 80 as number of coronavirus cases skyrockets past 4,000," April 2, 2020.
13 Ann M. Simmons, *Wall Street Journal*, "Coronavirus Forces Putin to Delay Vote That Could Keep Him in Power," March 25, 2020.
14 For a comprehensive but somewhat idiosyncratic history of the evolution of *Homo sapiens*, see Yuval Noah Harari, *Sapiens: A Brief History of Humankind* (New York: Harper), 2015.
15 The English adjective *human* is a Middle English loanword from Old French *humain*, ultimately from the Latin *hūmānus*, the adjective form of *homō* "man." The word's use as a noun (with a plural: *humans*) dates to the 16th century.
16 See Jonathan Glover. *Humanity: A Moral History of the Twentieth Century* (New Haven: Yale University Press, 2000).
17 David Wallace-Wells, *The Uninhabitable Earth Life after Warming* (New York: Tim Duggan Books, 2019,) 225 and *passim*. See also: https://en.wikipedia.org/wiki/Anthropic_principle.

18 See Jim Holt, *Why Does the World Exist?* (New York: Liveright, 2012), 5 and *passim*.

19 John Noble Wilford, "When Humans Became Human," *The New York Times:* February 26, 2002: www.nytimes.com/2002/02/26/science/when-humans-became-human.html. For a popular account of where and when things arose, see David Christian, *Origin Story: A Big History of Everything* (London: Allen Lane, 2018).

20 Recently, a team of astrophysicists announced the possible discovery of phosphine in the upper atmosphere of Venus, which has long thought to be inhospitable to life. However, according to the Harvard astrochemist Clara Sousa-Silva, who is investigating if phosphine might be a biosignature for life on Venus:

> Life is very resilient and very resourceful on Earth and there's no reason to think that's some special characteristic of life on Earth rather than of life itself. We have ignored Venus because Venus is quite horrid to us. When we sent probes, they melted dramatically so we didn't feel particularly welcome. It seems easier to imagine a place like Mars as habitable, even though actually there's so little atmosphere and so little protection from the sun's radiation that it's really not an easily habitable surface. Mars is mostly uninhabitable, like Venus, just in a much quieter way. Mars will kill you, but it doesn't melt you, so it feels more habitable, though I have no loyalty to either planet as a place to find life. This is hopefully going to help us think of habitability in a less anthropocentric way—or at least a less terra-centric way—and to think of habitability not just as a rocky planet with liquid water on the surface, but to think of subterranean habitats, moons of gas giants—something people already consider—and envelopes of an atmosphere as potentially habitable places in an otherwise uninhabitable planet.

(Full report available at: https://news.harvard.edu/gazette/story/2021/01/astrochemist-brings-search-for-extraterrestrial-life-toharvard/?utm_source=SilverpopMailing&utm_medium=email&utm_campaign=Daily%20Gazette%2020210105%20(1))

21 Sara Seager, as quoted on: https://exoplanets.nasa.gov/what-is-an-exoplanet/how-do-we-find-habitable-planets/.

22 For a popular history of the Earth and how it has shaped the human race, see Lewis Dartnell, *Origins: How the Earth Shaped Human History* (London: Vintage, 2020). Also see: www.etymonline.com/word/earth.

23 The United Nations Report is available at: www.un.org/sustainabledevelopment/blog/2019/05/nature-decline-unprecedented-report/; and https://ipbes.net/global-assessment.

1 The World and Its History

The "world" denotes in large part the ways in which this English-language noun, and its equivalents in other languages, have been used. It also signifies a wide range of everyday and academic expressions, ranging from "world history" to the "world of ideas," as well as the terrestrial and political habitat within which humans and all extant species exist and perish. The "ancient" and "medieval" "worlds" were constructed and transformed by the thinkers and actors who will be discussed in the chapters to come.

While there are numerous "histories of the world," many written by stellar but conventional historians, my orientation is more theoretical and multidisciplinary than most. Accordingly, in this book, I present what might be deemed an *intellectual history of the idea of the world,* beginning with the "ancient world," in which "science," "philosophy," and "politics" as we now know them arose several thousand years ago, and concluding with some prominent 20th-century and early 21st-century accounts of this world in which we now live.

My principal focus in this book will be on the world as a scientific and cosmological *object* of speculation, analysis, and observation on the one hand, and as a philosophical–theological *concept and collective construction* on the other hand. In the chapters to come, influential social, psychological, political, and ecological dimensions of, and threats to, "this world as we know it" will also be explored.

The Historical Periodization of the World

Ancient, Medieval, Modern, and Postmodern Worlds

The historical periodization of the world is usually taken for granted, and, therefore, seldom analyzed. For those of us brought up in an Occidental culture and educated in Western schools, the history of the world may seem to be a self-evident "progression" from "ancient" to "medieval," "modern," and "postmodern times." The past may therefore appear unintelligible unless it has been subdivided into such distinct units of time. However, apparently distinctive and widely accepted period-demarcations can become intellectual straitjackets that limit the way we "see the world"—and perceive the

DOI: 10.4324/9781315795171-2

beginning, middle, and ending of things—to what we have read in conventional texts and been exposed to in popular and digital cultures. The power of periodization is nowhere more evident than in the way in which most conventional Western historians, and historians of science and ideas in particular, have usually employed a tripartite mode of periodization, usually dividing history into ancient (c. 6000 BCE–500 CE), medieval (c. 500 CE–1500), and modern periods (c. 1500–the present), occasionally adding a "postmodern era" (since, roughly, the 1970s).[1]

For the past half-century or so, this convenient framework for organizing huge amounts of data has increasingly been called into question by writers and activists who view such periodization as simplistic, parochial, and often demeaning of the populations usually omitted from such historical accounts, namely intellectual and political dissidents, working people, women, people of color, sexual minorities, and non-Western cultures. While I agree with this critique, I also realize that I do not have the competence or space to cover these neglected and often marginalized perspectives. Accordingly, the historical and intellectual worldview manifested in this text tends to reflect the explicit and tacit assumptions of the Occidental cultures in which I have lived.

The Cosmos and the World

The words "cosmos" and "world" have numerous meanings. In some contexts, they refer to everything that makes up "external reality," or the physical universe as a whole. In others, they can have an ontological sense, denoting "being" as such, or everything that "is." While clarifying the concepts of cosmos and world has been among the basic tasks of major Occidental and Oriental intellectual and scientific traditions, this endeavor has been the subject of continuous debate. The question of what the world and cosmos are and mean has not been settled.

Philosophy as a whole, and, in particular, what came to be later known as natural philosophy (or physical science and the philosophy of science), are widely believed to have begun as speculations about the nature and origins of the cosmos in general and this world in particular, primarily in ancient Greek-speaking settlements. The 20th-century German existential philosopher Martin Heidegger posed two questions about this origin of philosophy: (a) "How does philosophy arise from the Greek residence in the midst of phenomena?" and (b) "From where does philosophy receive its first impetus, which sets it upon its way?"[2] His answer to these questions was naturalistic. According to Heidegger, philosophy was, for the Greeks, an outcome of their encounter with the "abundance" of natural phenomena present in their "world." For example, the Presocratic thinker Thales of Miletus (c. 624/623–c. 548/545 BCE, from Asia Minor, now modern Turkey) was said to have been so struck by the over-abundance of the world of stars that he was compelled to direct his gaze towards the heavens alone, thus giving rise to philosophy, astronomy, and cosmogony.

What the 18th-century German philosopher Immanuel Kant called the "awe-inspiring starry firmament above," therefore, provided the "overabundant presence" for both the ancient Greeks' theories of being (ontology) and for how the *kosmos* (Greek: the "harmonious order of things") had *come into* being (cosmogony).[3] The earliest recorded Western speculations about the "nature of things"—the search for the common element, stuff, or substance (*phusis*) from which everything in the universe (*kosmos*) was made—were fragmentary ruminations by Greek-speaking philosopher-cosmogonists.

Cosmogony and Cosmology

Cosmogony is any explanatory model concerning the origin of the universe as a whole and of the individual bodies that compose it.[4] Since cosmogony attempts to deal with creation, cosmogonies of the past have been a part of religion, mythology, and natural philosophy, as well as astronomy and theoretical astrophysics.

Cosmogony is often distinguished from *cosmology*, which studies the universe at large—a description of its existence as a whole as well as its individual components and physical laws—but does *not* inquire directly into the *source of its origins*. In contrast, *cosmogony* refers to the study of the origin of particular astrophysical objects or systems, and is most commonly used in reference to the *origin of the universe*, the Solar System, or the Earth–Moon system. Yet, for many theologians and philosophers, there is no hard-and-fast distinction between cosmogony and cosmology. For instance, the "cosmological argument" made by theologians and philosophers of religion regarding the existence of God is often an appeal to cosmogonical rather than cosmological ideas, as are theological or "supernatural" claims for God or some other eternal being(s) as the creator(s) of the universe.

An attempt to explain the origin of the solar system by natural rather than supernatural or theological processes was first made by Emanuel Swedenborg and Immanuel Kant in the mid-1700s. For many contemporary astrophysicists, there remains the working distinction between cosmological and cosmogonical ideas. Questions regarding *why* the universe behaves in such a way have often been described by physicists and cosmologists as being "*extra*-scientific" (i.e., metaphysical, ontological, supernatural, or theological), although sometimes scientists make speculations or posit theories that include extrapolations of scientific theories to philosophical or religious ideas.

The prevalent cosmological model of the early development of the universe is the Big Bang theory. According to this theory, the universe came into being at a single point, a "singularity," about 13.8 billion years ago, and when the singularity of the universe started to expand, the Big Bang occurred, hence the beginning of this universe. An alternate theory, the Steady State Theory, holds that the universe had no beginning and that the universe is much the same now as it always has been.

Stephen Hawking, among others, has argued that "time" did not exist when it emerged along with the universe. This implies that the universe does *not* have an origin story. For these theorists, it is unclear whether such astrophysical properties as space and time emerged with the singularity and the universe as it is known. Other proposed cosmogonical scenarios include string theory, M ("magic," "mystery," or "membrane") theory, cosmic inflation, and the "ekpyrotic" universe.[5] Some of these proposed cosmogonies, like string theory, are compatible; while others are not.[6]

Some Asian Theories of the Universe and World

The universe (or cosmos) and the world have also figured prominently in non-Western cultures and spiritual traditions as well, especially in Asia. In Hinduism, for example, the *Rigveda,* written in India from the 15th–12th centuries BCE, describes a cyclical or oscillating universe in which a "cosmic egg," or *Brahmanda*, containing the whole universe (including the Sun, Moon, the planets, and space), expands out of a single concentrated point called a *Bindu* before subsequently collapsing again. The universe cycles infinitely between expansion and total collapse. This is eerily similar to several contemporary astrophysical cosmogonical theories of the universe.

In Buddhism, the world denotes society, as distinct from the monastery. It refers to the material world and to worldly gain, including wealth, reputation, jobs, and war. Human access to the spiritual world and the path to enlightenment requires changes in what we could call the psychological realm.

Myths of the World

Cosmogony is not only a focus of scientific theories, but also has connections to the humanities, more specifically to creation myths. The term "myth" often refers to stories that are fictional and purely for entertainment. However, myths help provide insight into the theologies, practices, traditions, social organization, and political structures of many cultures.

The word *myth* comes from Ancient Greek *mȳthos*, meaning speech, narrative, conversation, narrative, story, tale, word, fiction, and plot. This Greek word began to be used in English (and was adapted into other European languages) in the early 19th century, in a much narrower sense, as a scholarly term for a traditional story, typically involving supernatural beings or forces, which embodies and provides an explanation, etiology, or justification for something such as the early history of a society, a religious belief or ritual, or a natural phenomenon. Plato used *mythología* as a general term for "fiction" or "story-telling" of any kind.

Another 20th-century German philosopher, Ernst Cassirer, wrote extensively about the symbolic and political uses and abuses of myths, which he regarded as form of "primitive" (preliterate) thought, an intuition, and a life-form, as well as a force in modern politics, particularly by such state dictatorships as Nazi Germany. Cassirer regarded modern totalitarian societies as attempts to establish the state on a mythical rather than a rational base, and that political myths follow the same logic as do premodern myths; their use in modern society essentially represents a regression to non-rational modes of thought. The rational ordering of society, which has been the subject of political philosophy since Plato, according to Cassirer, is most strongly threatened during times of social and economic stress (such as the one we are experiencing today). When reason fails, it is always possible to resort to the earlier mode of mythical thought. "For myth," Cassirer stated, "has not been really vanquished and subjugated. It is always there, lurking in the dark and waiting for its hour and opportunity."[7] The hour and opportunity for such regression is often seized by authoritarian and protofascist leaders and their supporters.

Myths of Life, Death, and Their Meanings

Mythology, like its religious, scientific, and philosophical successors, tries to answer some of most difficult and basic questions of human existence: Who am I? Where did the universe, this world, and I come from? Why am I here? Where am I going? In many preliterate and prehistorical cultures, the *perceived meaning or the symbolic significance* of the stories—transmitted orally or in the visual arts before the developments of writing and widespread literacy—was most important, *not the literal truth* of the details of a certain version of a tale.

The oldest purported myth is a Babylonian tale whose narrative may be interpreted as providing accounts of the inevitability of death and the individual's attempt to find meaning in life. This myth, called the *Epic of Gilgamesh*, is also considered the earliest surviving great work of literature and the second oldest religious document, after the *Pyramid Texts*, which are hieroglyphs on the walls of royal pyramids, dating back to the Old Kingdom of ancient Egypt from c. 2400 to 2300 BCE.[8]

The *Gilgamesh Epic* was orally transmitted in the late third millennium BCE and written down on clay tablets in Babylonia about 1800 BCE. It developed in Mesopotamia from Sumerian poems relating to the historical Gilgamesh, king of Uruk, who was later elevated to the status of a demigod. In the story, Gilgamesh is a proud king who is so haughty that the gods feel he needs a lesson in humility. They groom the wild man Enkidu as a worthy opponent to the king and the two fight but, when neither can get the best of the other, they become best friends. Enkidu is later killed by the gods for affronting them and Gilgamesh, grief-stricken, embarks on a quest for the meaning of life as represented by the concept of immortality. Although he fails to win eternal life, his journey enriches him and he returns to his kingdom a wiser and better man and king.

What one today calls "mythology" frequently constituted the religions of many ancient cultures. Such tales as the aforementioned *Gilgamesh*

Epic, which collectively make up the corpus of ancient mythology, served the same purposes for the people of preliterate times as the stories from accepted or "canonical" scriptures do for people of faith today: they explained, comforted, and directed an audience and, further, provided a sense of unity, cohesion, meaning, purpose, and protection to a community of like-minded believers. Among these stories, creation tales have served a vital social and psychological function.

Myths of Creation

Before cosmogony came to the fore in early Greek (the centuries before the 5th century BCE, or "Presocratic") scientific theories, creation myths were used to provide explanations for the origin story of the universe.

Mythological cosmologies often depict the world as centered around an *axis mundi*—a symbol representing the world axis and center of the world, where the heaven connects with the earth and delimited by a boundary, such as a world ocean or a world serpent. Creation, or cosmogonic myths, explain the creation of the universe or the cosmos by a supreme being, or beings; the process of metamorphosis; the copulation of female and male deities; through chaos; via the cosmic egg (also known as the world egg), or from a great sea.

For instance, the *Eridu Genesis*, possibly the oldest known creation myth—on Sumerian tablets dating back to the third millennium BCE—depicts the creation of the world, the founding of cities, and a great flood. In this myth, the universe was created from the primeval sea and then the gods were created, and men were later formed to tend to nature.

While myths regarding the creation of the world vary, there are also some striking similarities. The creation story as related in the biblical Book of Genesis, for example, where God speaks existence into creation, is quite similar to creation stories from ancient Sumeria, Babylonia, Egypt, Phoenicia, Northern Europe, and China, as well as in the Vedic and Hindu cosmologies of ancient India.

In the Babylonian creation story *Enûma Eliš*—probably dating to the late second millennium BCE, the time of Hammurabi—the universe was in a formless state and is described as a watery *chaos*. From it emerged two primary gods, the male Apsu and female Tiamat, and a third deity who is the *maker* Mummu and has the power to give birth, to procreate.[9]

Norse and early Germanic mythology describe *Ginnungagap* as the primordial abyss, the "nothingness" from which sprang the first living creatures, including the giant Ymir, whose body eventually became the world, and whose blood became the seas.[10] In this myth, the bottomless gap was all there was prior to the creation of the cosmos, and the cosmos will collapse into it once again during *Ragnarok*, the "Twilight or Demise of the Gods" and the immersion of the world in water. These early legends were later compiled in the 13th century CE Norse tales called the *Poetic Edda*, which would serve as a basis for the four-opera *Ring Cycle* of the 19th century by the German composer Richard Wagner.

The total silence and darkness prior to the creation of the universe has close counterparts in other creation stories, most notably, perhaps, in the first chapter of the book of *Genesis*. This book is in what has come to be called the *Old Testament*—the Hebrew Bible as interpreted among the various branches of Christianity—which describes the state of the universe prior to the intervention of *Elohi*m, the ancient Jewish deity, usually referred to simply as God:

> In the beginning God created the heavens and the earth. Now the earth was formless and void, and darkness was over the surface of the deep. And the Spirit of God was hovering over the surface of the waters. And God said, "Let there be light," and there was light.[11]

The extreme contrast between the divinely ordered cosmos on the one hand—the "light" illuminating the universe—and the lawless chaotic "darkness" from which the universe and everything in it arose and to which they may return, is perhaps one of the most common themes in religion and in cosmogonies more generally. It also presages the binary oppositions between "Being" and "Nothingness," "existence" and "non-existence," and "meaning" and "meaninglessness" present in the texts of many European philosophers in the 19th and 20th centuries, particularly the Existentialists.

In contrast with Judeo-Christianity, there is no single story of creation in Hinduism, and Ancient Hindu creation tales are derived from various sources, including the *Vedas*, the *Brahmanas*, and the *Puranas*, which are believed to have been composed in Sanskrit between 1500 BCE and 500 BCE in northern India. Some are philosophical, based on concepts, and others are narratives, founded on orally transmitted sagas.

The *Rigveda* mentions the *Hiranyagarbha* ("*hiranya*" = "golden," or "radiant," and "*garbha*" = "filled /womb") as the source of the creation of the Universe, similar to the world egg motif found in the creation myths of many other civilizations. It also contains a story in which the creation of the world stems from the dismemberment of a cosmic being (the *Purusha*) who is sacrificed by the gods. In the *Rigveda,* the gods came into being after the world's creation, and it is unknown when the world first came into being. In the later Puranic texts, the chief god, Brahma, is described as performing the act of "creation," of "propagating life within the universe." Brahma is a part of the trinity of gods that includes Vishnu and Shiva, who are responsible for "preservation" and "destruction" (of the universe), respectively. Many Hindu texts also mention the cycle of creation and destruction. The *Shatapatha Brahmana*, for example, states that humanity descends from Manu, the only man who survived a great deluge after being warned by the gods. This legend is comparable to the other flood legends, such as the story of the Noah's Ark mentioned in the *Bible* and the *Quran*.[12]

This story of the "Great Flood" can be found in the mythology of virtually every culture on Earth but takes its biblical form from the *Atrahasis* ("the good, wise man") epic, originating during the 18th century BCE in Akkadia (Mesopotamia). The figure of the Dying and Reviving God (a deity who dies for the good of, or to redeem the sins of, his people, goes down into the Earth, and rises again to life) can be traced back to ancient Sumeria in the *Gilgamesh* epic; the poem *The Descent of Inanna*; the Egyptian myth of Osiris; the Greek myths involving Dionysus, Adonis, and Persephone; the Ugaritic *Baal Cycle*; the Hindu deity Krishna (the eighth avatar of the god Vishnu), as well as, of course, to the most famous of Western resurrected figures, Jesus Christ.

Perhaps the most influential Occidental creation tales prior to those of Judeo-Christianity arose in the Greek-speaking world and were primarily represented on pottery, until the oral recitations of the epic poems attributed to Homer (the *Iliad* and *Odyssey*) and to Hesiod, who, in the late 8th century BCE, in the *Theogony* ("The genealogy or birth of the gods") provided a protophilosophical account of the beginning of things.[13]

Originally, according to Hesiod, there was Chaos, a yawning nothingness. Out of the void emerged *Gaia* (the Earth) and some other primary divine beings, *Eros* (Love), the *Erebus* (a Greek deity personifying darkness), and the *Tartarus*—the deep abyss that is used as a dungeon of torment and suffering for the wicked and as the prison for the Titans, also the place where, according to Plato's *Gorgias*, souls are judged after death and where the wicked received divine punishment.[14]

In Greek mythology, the Earth was viewed as a flat disk afloat on the river of Ocean. The Sun (*Helios*) traversed the heavens like a charioteer and sailed around the Earth at night in a golden bowl. Natural fissures were regarded as entrances to the subterranean house of Hades—i.e., the home of the dead. And so before the rise of their science and philosophy between the 7th and 4th centuries BCE, for the ancient Greeks there were three planes of existence—the heavens, the Earth, and the underworld—a cosmology similar to those of later, Christian visions of creation and the universe, especially Dante's in *The Divine Comedy,* approximately 2 millennia later.

Early Greek mythology and poetry also represented Nature as a whole (*phusis*), or what we might call the "natural world," in the way in which we might regard rivers, rocks, trees, or animals as parts of a totality that usually does not include ourselves. The natural world, especially in some Homeric depictions, is an alternative to—rather than a carrier of—human significance. In later Greek texts, as mythology and poetry faded and natural philosophy emerged, the natural world was not merely a background framing human action but instead became the primary object of attention.

Mythos, Nous, and Logos

As previously noted, the word *mythos* appears frequently in the works of Hesiod, Homer, and other poets of the Presocratic era, as did the enduring

Ancient Greek terms *logos* and *noos/nous*, which became central concepts in Western philosophy, psychology, rhetoric, and religion."[15]

Logos, for example, became a technical term in Western philosophy beginning with Heraclitus (c. 535–c. 475 BCE), who used the word to denote a principle of order and knowledge. *Mythos* expressed for the ancient Greeks whatever can be delivered in the form of words, in contrast with *ergon*, a Greek term for action, deed, and work. *Noos/nous* later became the linguistic equivalent of "mind." *Importantly, mythos lacks an explicit distinction between true or false narratives, unlike the logical (from logos) arguments emanating from rational minds (from nous) and providing good reasons and sound evidence to support the truth of their claims.*

Incarnate in the earliest known forms of Greek "mind" *(nous)* and "reason" (*logos*) were the physical activities of seeing and speaking. Hence, *nous* (*noos* in the Homeric epics) contained both an intellectual and a volitional element; it was a planning agency closely linked to a kind of visual imagination and was thought to be implanted in the chest. As an organic function, Homeric *noos* was portrayed as the immediate cause of violent or unruly emotions as well as the type of multifaceted resourcefulness characteristic of Ulysses. In addition, the connotation of sight or vision implied a glimpse of a *more real world beneath the outward, often deceptive appearance of stability—an implication that would become explicit in Platonic dualisms, especially between the permanence of the intellectual worlds of the Forms and the impermanence of the world of the senses.* The epic hero of *noos* was a person of strong intuitions and keen insight, whose perspicacity facilitated the disclosure of the truth concealed by the deceits of other humans and the surface lawlessness of nature.[16]

The Homeric mythological image of the "mind" (noos) is far from the Platonic and Aristotelian mental faculties, much less the "cosmic mind" of the Stoics. Some Presocratic thinkers would identify *nous* with God or the *kosmos* (Orpheus legend), *nostra scientia intelligentia* ("our scientific intelligence," Demokritos) or a *Mentum divinum* ("divine Mind," Anaxagoras).[17] But Homeric heroes were not abstract speculators. In them, reason was as visceral and substantive as it was cognitive.

The history of *noos-nous* is also closely tied to the Greeks' usages of the words wisdom (*sofia*), willpower (*thumos*), soul (*psuke*), and technique (*techne*). These words, which in slightly altered forms and with considerably expanded connotations have come to form a considerable part of the conceptual vocabulary of the Western intellectual tradition, were originally connected with bodily organs and physical perception. *More than two millennia later, human mental activity is again closely associated with, and often causally linked to physical processes, namely brain activities.*

Ancient Greek and Roman philosophers used the term *logos* in different ways. The sophists used the term to mean discourse. In his *Rhetoric*, Aristotle used the term to refer to *logos* as "*reasoned* discourse," or "the argument" made to persuade an audience at the intellectual level, and he considered it one of the three modes of persuasion alongside *ethos* (an appeal to the audience based on the character, the *ethos*, of the speaker) and *pathos* (an appeal to the audience's experiences and emotions, their *pathea,* especially fear, suffering, sympathy, and empathy). Such ancient skeptical philosophers as Pyrrho and Sextus Empiricus (4th–2nd centuries BCE) used *logos* to refer to dogmatic accounts of non-evident matters. Early Stoic thinkers, incorporating the ideas of Zeno of Citium (4th–3rd centuries BCE), the founder of Stoicism, spoke of the *logos spermatikos*—the generative principle of the Universe, denoting universal Reason, the active rational and spiritual force that permeates nature, and living in accordance with reason was the purpose of human life.[18] The later Stoics' deification of l*ogos*, especially by Seneca and Epictetus during Roman imperial times, foreshadowed related concepts in Neoplatonism and early Christianity. Within Hellenistic Judaism, Philo (c. 20 BCE–c. 50 CE) adopted the Greek word *logos* into Jewish philosophy and distinguished between *logos prophorikos* ("the uttered word") and the *logos endiathetos* ("the word remaining within").[19] For Philo, *logos* was the intermediary between God and the cosmos, being both the agent of creation and the agent through which the human mind can apprehend and comprehend God.

Mythos and Logos in Presocratic Greek Thought

Classical Greek mythology provides narratives to make sense of physical and social reality, to conceive of the universe—the cosmos—as a whole, and this universalizing impulse was fundamental for the first Occidental efforts at speculative theorizing. Myth made something like *cosmological* speculation possible. The earliest recorded Western theoretical speculations were kindled by the experiences of amazement and curiosity: Plato and Aristotle were later to declare that philosophy begins in wonder—about the nature (*phusis*) of things and the conviction that cosmic reality forms a beautiful and harmonious whole.

Presocratic philosophy is ancient Greek philosophy before Socrates and his contemporaries. Aristotle referred to the Presocratics as *physiologoi*—or what we might call in English "natural philosophers"—and *phusikoi* (from *phusis*, meaning "substance" or "universal element," from which the English word "physics" is derived), to differentiate them from the earlier *theologoi* (theologians), or *mythologoi* (storytellers and bards), who appealed to the gods to provide explanations for things and events. Their inquiries attempted to describe and explain the workings of the natural world and the essence of Being itself, as well as the nature of human society, ethics, and religion, by seeking causal explanations based on sense-derived principles rather than the actions of supernatural gods. In their discourses, *they introduced to the West the notion of this world as a cosmos, an ordered arrangement that could be understood via rational inquiry.*

Starting in the early the 7th century BCE, two types of discourse emerged that over the centuries became more "realistic" alternatives to the mythic poetry of Hesiod and Homer—history (especially by Thucydides), and philosophy. While incorporating much of the theoretical vocabulary found in classical mythology, the Presocratic philosophers tended to reject mythological stories in favor of more rational causal explanations for what occurs. They asked such questions about "the essence" or "nature" (*phusis*) "of things," as:

> From where does everything come?
> From what is everything created?
> How do we explain the plurality of things found in nature?
> How might we describe nature mathematically?

Some Presocratic thinkers, especially Zeno, also concentrated on defining problems by posing paradoxes that became the basis for later mathematical, scientific, and logical inquiries.

Significant Presocratic philosophers include Thales of Miletus (founder of the "Milesian School"), Anaximander (an "Ionian"), Heraclitus, Parmenides, Empedocles, Zeno of Elea (first of the "Eleatics"), Democritus, and Anaxagoras, among others. None of their texts have survived in complete form. All that is available are quotations and textual fragments, as well as testimonies by later philosophers and historians. The knowledge we have of them derives from accounts or textual summaries of their fragmentary writings—known as doxography—by later writers (especially Aristotle, Plutarch, and Diogenes Laërtius), as well by those of some early theologians (particularly Clement of Alexandria and Hippolytus of Rome).[20]

In Presocratic thinking, *archê* (from the ancient Greek, *arkhé*) is the prime element or first principle of all things, the permanent substance from which everything in the universe comes to be, and into which everything will finally resolve. In Presocratic and classical Greek philosophical discourse, *archê* meant the "beginning," the "origin," and the "source of action" ("from the beginning," or, "the original argument"—the *logos*). By extension, it also denoted the "first place," "the power," the "method of government," the "empire" or "realm," the "authorities," and a "command." This first principle or element corresponds to the "ultimate underlying substance" (*phusis*) and the "ultimate principle." Aristotle emphasized the meaning of *archê* as the element or principle of a thing, which, although intangible, provides the conditions for the possibility of that thing's nature and development.

Thus, the Presocratics and classical Greek-speaking natural and political philosophers were striving to uncover the ultimate "essence," "nature," or "stuff" from which everything visible and invisible arises and is composed, and that persists after the things it has given rise to have decomposed. *They attempted to explain this world in particular and the cosmos as a whole. They did so by focusing on the description of the growth and emergence of everything that is, that exists, from a beginning, from the archê. They believed that the world arose from a primal unity, and that this substance was the permanent base of all that is.*

Thales (7th–6th century BCE), considered the first major Greek-speaking philosopher, claimed that the first principle (*archê*) of all things is water. His Milesian successor Anaximenes (585–525 BCE) considered this principle to be air, conceiving it as modified, by thickening and thinning, into fire, wind, clouds, water, and earth. Anaximander (also the 6th century BCE), was the first philosopher who used the term *archê* for which writers commencing with Aristotle call the "substratum," and asserted that the beginning, or first principle, of everything that exists is a limitless entity (*apeiron*, "without limit") subject to neither age nor decay, and from which all things are born and later end. Anaximander also claimed that the *apeiron* is a divine and perpetual substance; everything is generated from the *apeiron*, and must return there according to necessity or destiny (Greek: *anánkē*). *These Presocratic thinkers also had a conception of the nature of the world in which the Earth below its surface stretches down indefinitely* and has its roots on or above what they called *Tartarus*, the lower part of the underworld. The sources and limits of the earth, the sea, the sky, *Tartarus*, and all things are located in a great windy gap, which seems to be infinite, and is a later specification of "chaos."

Later philosophers rejected many of the answers the early Milesian philosophers provided, but continued to place importance on their questions. Furthermore, the cosmologies proposed by them have been updated by later developments in science.

Heraclitus

Presocratic philosophers were interested in the natural world and how it is created and ordered. Heraclitus (544–480 BCE) from Ephesus, a Greek-speaking city in Ionia (coastal Turkey), however, went beyond his Milesian predecessors and examined the nature of philosophical inquiry itself, in addition to observing and explaining natural and physical processes. He wanted to figure out the proper methods of understanding human knowledge and the ways humans fit into the world. This was different from the natural philosophy that had been done by previous philosophers *as it questioned how the cosmos operates as well as the human activity within the universe.*

Heraclitus posited that all things in nature are in a *state of perpetual flux*, everything is constantly changing. But underlying this constant flux in the natural and human worlds is a universal logical structure or pattern, the *logos.* To Heraclitus, fire, one of the four classical elements (along with earth, air, and water), catalyzes this eternal pattern. From fire, all things originate, and return to it again, in a process of eternal cycles.

Xenophanes

Eleatics philosophers (from Elea, in contemporary Sicily) emphasized the doctrine of "the One," a philosophical doctrine sometimes called

"Monism," in contrast with their "pluralist" predecessors (like Heraclitus). For example, Xenophanes (570–470 BCE) declared that there is a "greatest god," an eternal unity, permeating the universe, and governing it by his thought. According to Xenophanes, "god shakes all things," "all things are from the earth and to the earth all things come in the end," and that "all things which come into being and grow, are earth and water." He thus seemed to believe in a two-substance (earth and water) principle of all things (*archê*). For him, human life also alternated between extinction and regeneration.[21]

Xenophanes removed natural phenomena from all vestiges of religious or spiritual significance. His demythologized account of nature and the gods was in sharp contrast with traditional Greek mythology and religion. Xenophanes also believed that there are an *infinite number of worlds, not overlapping in time, a concept that has loose parallels in contemporary ideas of many worlds* (in quantum dynamics theories) and multiple universes (aka "multiverses" in theoretical astrophysics and string theories).[22]

Xenophanes also examined the conditions under which human beings can achieve knowledge. His *distinction between knowledge and true opinion became an axiom of ancient Greek accounts of knowledge, especially those of Socrates and Plato, and is a fundamental assumption of most contemporary epistemology* (theory of knowledge). He posed the challenge that, given the *severely limited character of human experience, how is it possible plausibly to claim to have discovered the truth about matters lying beyond anyone's capacity to observe the truth about the world first-hand?*

Parmenides

Parmenides (510–440 BCE), Xenophanes' Eleatic successor, is often considered the first philosopher to inquire into the nature of existence—of Being, or "what is." He is also regarded the "Father of Metaphysics"—literally "beyond physics" (*phusis*), the deep "nature" or "substance" from which everything is said to have arisen—and the founder of Western ontology, the study of "Being."

It is likely that Parmenides composed only a single work. This was a cosmological poem now called "On Nature," of which only fragments have survived. In this poem, a "goddess" declares: "It is necessary to say and to think that What Is; for it is to be, but nothing it is not." "The thought of mortals who have supposed that it is and is not the same and not the same" involves an intermingling of being and not-being altogether different from what one sees. The goddess then begins her account of "true reality:" "What Is" (*to eon*) "is ungenerated and deathless/whole and uniform, and still and perfect."[23]

One interpretation of Parmenides' work is that he argued that the everyday perception of the "reality" of the physical world is mistaken, and that the reality of the world is "One Being" an unchanging, ungenerated, indestructible whole. "The One," "what is,"—unchanging Being—is alone true and capable of being conceived, and multitude and change are an appearance without reality. *Therefore, the world as perceived by the senses is unreal. Only what is disclosed by reason can be real, and beliefs derived from sense experience are thereby rejected as altogether deceptive.*

Based on this *fundamental ontological dualism, according to Parmenides, the cosmos itself is divided between two basic principles, the light and the night. Aristotle interpreted Parmenides' cosmological thesis that what is, is one* (*hen to on,* in Ancient Greek), *and is not subject to generation and change as belonging, not to natural philosophy, but to what Aristotle called "first philosophy," or metaphysics.*[24] Parmenides' vision of the relation between What Is and the developed cosmos, as coterminous but not consubstantial, is comparable to Xenophanes' conception of the relation between his one greatest god and the cosmos, as well as Empedocles' conception of the divinity.

Parmenides' doctrine was defended by Zeno of Elea (490–430 BC) in an argument against common opinion, which, according to Zeno sees in physical things multitude, becoming, and change. Zeno's legendary paradoxes, much debated by later philosophers and mathematicians, try to show that the assumption that there is any change or multiplicity leads to contradictions. Plato, in his *Dialogue* "Parmenides," while somewhat critical of Eleatic metaphysics, would adopt Parmenides' conception of *there being "Two Worlds," one eternal, true, unchanging, and grasped only intellectually—the world of the Forms, or Ideas—and the other temporary, erroneous, inconstant, and apprehended by our senses—the physical world as we see it.*

Pythagoras and the Pythagoreans

Pythagoras (c. 570–490 BCE) was, according to the Greek philosopher–historian Plutarch (*c.* 45–120 CE), reported to have "named the compass of the whole a Cosmos, because of the order which is in it."[25] *The idea that everything that exists and can be known forms a single orderly and comprehensive system, in which every event is linked with every other by causal necessity, is something many of us take for granted and might attribute to "the world" or "the universe."* The notion of a unified, harmonious cosmos that, like the harmonies of well-tuned musical strings, is organized according to strict mathematical proportions, has also been attributed to Pythagoras and his Pythagorean followers, as, indeed, has the claim that Pythagoras was the "inventor of philosophy," and was "the master philosopher, from whom all that was true in the Greek philosophical tradition derived."[26]

It is, however, difficult to say what Pythagoras himself believed because, like Socrates, if he wrote anything at all, nothing has survived. And, moreover, it is also debatable if the first detailed accounts of Pythagoras, mostly written centuries after his death, were accurate or forged. And neither Plato, often considered a metaphysical successor of Pythagoras, nor Aristotle, appeared to have shared the view that Pythagoras was philosophically or mathematically significant. Instead, Aristotle found evidence for Pythagoras only as a "wonder-worker" and founder of a way of life—including religious rituals, moral discipline, and what would later be called vegetarianism—even of giving birth to what we might today call hippie communities! For Plato and Aristotle, Pythagoras did not belong to the succession of Greek-speaking thinkers starting with Thales, who were attempting to explain the basic principles of the natural world and the cosmos as a whole. Plato's only reference to Pythagoras (*Republic*, 600a) also treated him as the founder of a way of life, and, when Plato traced the history of philosophy prior to his time in the *Sophist* (242c–e), there is no allusion to Pythagoras. So, even if he did not invent the word, what can we say about the "philosophy" of Pythagoras?

For many classical thinkers and scholars, Pythagoras was known as an expert on the fate of our soul, or "psyche" (Ancient Greek: *psukhḗ*, Latin *psychē*, from which the word "psychology," as the "study of the soul," is derived) after death, ostensibly the belief that we would never really die but instead might go, after the death of our bodies, to a place where we would eternally possess all good things, or even that the individual human soul transmigrates into other animals before returning to human form millennia later This is sometimes called the doctrine of "metempsychosis," and the idea of the "transmigration of souls" explicitly appears in some works of Empedocles and Plato (as well as in Greek Orphic religious cults and in ancient Hinduism).

In antiquity, Pythagoras was credited by some ancient thinkers with many mathematical and scientific discoveries, including what have come to be called the Pythagorean theorem in geometry, Pythagorean tuning in music, the five regular solids, the Theory of Proportions, the sphericity of the Earth, the identity of the morning and evening stars as the planet Venus, and the (astronomical) "music of the spheres."[27]

In addition, Pythagoras was thought to have believed, in addition to the immortality of the soul, that, after certain periods of time, the things that have happened once happen again, and nothing is absolutely new. This doctrine of "eternal recurrence," however, has usually been ascribed to the Pythagoreans rather than to Pythagoras himself. The initial doctrine of transmigration seems to have been extended to include the idea that *we and indeed the whole world will be reborn into lives that are exactly the same as those we are living and have already lived. The late 19th-century philosopher Friedrich Nietzsche would reintroduce the idea of "eternal recurrence" to modern readers.*

Pythagoras' vision of the cosmos is said to have embodied mathematical relationships and combined them with moral ideas tied to the fate of the soul. Where we might hope to go, if we have lived a good life, is, in this view, the sun and the moon. For Pythagoras, apparently, the planets were also agents of vengeance for wrongdoing. This resembles some of the myths appearing at the end of such Platonic dialogues as the *Phaedo*, *Gorgias*, and *Republic*, where cosmology has a primarily moral purpose.

There is considerable controversy concerning Pythagoras' role as a scientist and mathematician, with some contemporary scholars claiming he was not the originator of some of the most important doctrines ascribed to him, and others dissenting from this position. Clearly, decisions about sources are crucial in addressing the question of whether Pythagoras was a noted philosopher, mathematician, and scientist. The consensus among classical scholars today is, nonetheless, that Pythagoras was neither a distinguished mathematician nor a scientist.

This section on Pythagoras and the Pythagoreans *should serve as a cautionary tale. We rely frequently on the testimony of authoritative others for framing our own beliefs about the past, what happens in the world today, and the nature and fate of the cosmos itself. But often these narratives are, at best, fragmentary and unreliable. At worst they are untrustworthy and even false. How to distinguish what is "real" and what is "fake" is a challenge this world presently confronts, but its sources are ancient.*

Empedocles

Empedocles (c. 494–434 BCE?), also from Elea (in Campania, the south of Italy), posited in his main work, the poem "On Nature," apparently for the first time, a plurality of elementary substances—i.e., the four classical elements, earth, water, air, and fire—as well as two universal competing forces he called "love" and "strife." The cosmos as a whole and the world in particular are generated and ruled by love, as the cause of their union, and strife as the cause of their separation.[28]

Empedocles' cosmogonical vision pictures love and strife in eternal contention, waxing and waning in different combinations, and generated from an originating sphere. The cosmos and all creatures in it come to be and pass away. Some portions of love and strife become intermingled. Love asserts its influence by forming the cosmos (consisting of a world-order with continental land-masses, oceans, rivers, winds, sun, moon, seasons, planets, stars, etc.). From the mixture of love and strife arise the various forms of animal life (zoogony). Ultimately, both animals and the cosmos itself perish; the sphere is restored and the cosmos ends. His cosmic vision stands in contrast to Parmenides' rejection of change while also embracing religious injunctions, vegetarianism, and magical practices. As a result, Empedocles's theories combined elements of both *mythos* and *logos*, religion and science.

Anaxagoras

Anaxagoras (500–428 BCE), originally from a Greek-speaking city in Ionian Asia Minor, was the first of the major Presocratic philosophers to live in Athens. As with other Presocratics, Anaxagoras' work survives only in fragments quoted by later philosophers and commentators (especially in Diels-Kranz), as well as comments in many ancient sources about his views.

Anaxagoras believed that the original state of the cosmos was a primordial mixture of all its ingredients, which existed in infinitesimally small fragments. This mixture was not entirely uniform and varied from place to place. At some point in time, this mixture was set in motion by a universal mind, called *nous* by Anaxagoras, and the whirling motion shifted and separated out the ingredients, ultimately producing the cosmos of separate material objects, all with different properties, that we see today.

According to Anaxagoras, animals, plants, human beings, the heavenly bodies, and so on, are *natural constructs*. They are *constructs* because they depend for their existence and character on the ingredients from which they are constructed. Yet, they are *natural* because their construction occurs as one of the processes of nature. Unlike human-made artifacts (which are similarly constructs of ingredients), they are not predetermined to fulfill some purpose.

Anaxagoras also claimed that *nous* is completely different from the ingredients that constituted the original mixture; *nous* is present in some things, but it is not an ingredient as flesh and blood are ingredients in a dog. Furthermore, *nous* is not only first cause of motion but is also the preserver of order in the cosmos, as it maintains the rotations that govern all the natural processes. For Anaxagoras, *nous is the "finest, purest" and most powerful thing in the cosmos*; and "has control over all things that have soul, both the larger and the smaller" (DK, 59 B12). *Just as we control our bodies by our thoughts, so the cosmos is controlled by nous.* However, Anaxagoras did not identify *nous* with a divine principle or god, but rather suggested that *nous is a non-corporeal entity that can pervade and control a body, or even the whole cosmos without being a material part of it.*[29] Accordingly, according to Anaxagoras, since all *nous* is alike, *cosmic nous and our individual minds share the same nature.*

Anaxagoras provided a complete account of the universe: of the heavens, the Earth, and also of geological and meteorological phenomena. Further, *he claimed that the cosmic rotary motion could produce other worlds like our own*. Modern commentators have suggested that *the "other worlds" are on the moon and/or other planets, elsewhere on the face of our Earth, or even contained within ourselves (and all other things), so that there are potentially infinite worlds within worlds.*

Anaxagoras is also credited with having discovered the causes of eclipses—the interposition of other "unseen" celestial bodies between the Earth and the Sun, or between the Earth and the Moon. He also was reported to have provided explanations for the light of the Milky Way, the formation of comets, the solstices, and the composition of the moon and stars.

Anaxagoras acquired the nickname "Mr. Mind" (DK 59 A1); his view that the cosmos is controlled by *nous* first attracted and then disappointed Socrates (Plato, *Phaedo*, 97b). Plato partially adopted some of Anaxagoras' language and, like Anaxagoras's *nous*, Platonic forms are "themselves by themselves" in being self-explanatory. Aristotle, although impatient with the gaps in Anaxagoras' account of *nous*, expressed admiration for his recognition that mind has a role to play in guiding the cosmos, and he treated Anaxagoras's explanation of eclipses as a model of scientific explanation.

More generally, *Anaxagoras's focus on a universal mental agency that sets the cosmos in motion, found its way, in altered forms, into medieval Christianity and, much later, modern idealism, particularly the philosophies of Hegel and Schelling in the early 19th century.*

Atomistic Materialism: Leucippus, Democritus, Epicurus, and Lucretius

The first explicitly materialistic Occidental accounts of the cosmos and the human psyche were created by Leucippus (5th century BCE) and his pupil Democritus (460–370 BCE), and later also included Epicurus and Lucretius.[30] This was the doctrine of *atoms* (from the Greek "*atomos*," meaning "uncuttable")—small bodies infinite in number, indivisible, and imperishable, qualitatively similar, but distinguished by their shapes. Moving eternally through the infinite void, they collide and unite, thus generating objects which differ in accord with the varieties, in number, size, shape, and arrangement, of the atoms which compose them. *All of reality and all the objects in the universe are composed of different arrangements of these eternal atoms and an infinite void, in which they form different combinations and shapes.*

Since Leucippus's exact dates are not recorded and he is often mentioned in conjunction with Democritus, it is difficult to determine which contributions to atomism come from Democritus and which come from Leucippus, or even if he really existed. The title most attributed to Leucippus is the lost work *Megas Diakosmos* (*Big World-System*, but this title was also attributed to Democritus, whose companion work was *Micros Diakosmos, Little World-System*).

Leucippus is believed to have been an Ionian Greek and a contemporary of Zeno of Elea and Empedocles, and to have belonged to the same Ionian School of naturalistic philosophy as Thales, Anaximander, and Anaximenes. While causality was implicit in the philosophies of Thales and Heraclitus, Leucippus is considered the first to explain that all things happen due to "necessity." i.e., according to their "nature."

Democritus, known in antiquity as the "laughing philosopher" because of his emphasis on the value of "cheerfulness," elaborated the atomistic ideas originated by his teacher Leucippus into a *fully materialist account of the natural world.*[31] According to ancient reports, Democritus was born about 460 BCE (thus, he was a younger contemporary of Socrates) and was a citizen of Abdera, although some reports mention Miletus. As well as his associate or teacher Leucippus, Democritus is said to have known Anaxagoras. The work of Democritus has survived only in often unreliable secondhand reports, the vast majority of which refer either to both Leucippus, or to Democritus alone; the developed atomist system is often regarded as essentially that of Democritus.

These ancient atomists theorized that *the two fundamental and opposing constituents of the natural world are indivisible bodies—atoms—and the void. Changes in the world of macroscopic objects are caused by rearrangements of the atomic clusters.* Atoms can differ in size, shape, order, and position (the way they are turned); they move about in the void, and—depending on their shape—some can temporarily bond with one another by means of tiny hooks and barbs on their surfaces. While the atoms are eternal, the objects compounded out of them are not. *Clusters of atoms moving in the infinite void come to form kosmoi, or worlds,* as a result of a circular motion that gathers atoms up into a whirl, creating clusters within it; *these kosmoi are impermanent. Our world and the species within it have arisen from the collision of atoms moving about in such a whirl, and will likewise disintegrate in time.*

The Earth is depicted by atomists as a flat cylindrical drum at the center of our cosmos. Species are not regarded as permanent abstract forms, but as the result of chance combinations of atoms. In common with other early ancient theorists of living things, including Aristotle, Democritus seems to have used the term *psychê* to refer to that distinctive feature of living things that accounts for their ability to perform their life-functions. Democritus seems to have considered thought also to be caused by physical movements of atoms. This is sometimes taken as evidence that Democritus denied the survival of a personal soul after death.

Democritus' atomism was revived in the early Hellenistic period, and an atomist school founded in Athens about 306, by Epicurus (341–270 BCE), who lived most of his life in that city.[32] The Epicureans formed a closed community, and promoted a philosophy of a simple, pleasant life lived with friends. But Epicurus' view of the motion of atoms also differs from Democritus.' Rather than talking of a motion towards the center of a given cosmos, possibly created by the cosmic vortex, Epicurus ascribed to atoms an innate tendency to downward motion through the infinite cosmos.

Epicurus believed that the natural goal of a human life was happiness, resulting from absence of physical pain and mental disturbance. *Based on atomistic materialism, Epicurus provided a naturalistic account of evolution, from the formation of the world to the emergence of human societies.* He claimed that, on the basis of a radical materialism which dispensed with transcendent entities such as the Platonic Ideas or Forms (to be discussed later in this chapter), he could disprove the possibility of the soul's survival after death, and hence the prospect of punishment in the afterlife. He regarded the unacknowledged fear of death and punishment as the primary cause of anxiety among human beings, and anxiety in turn as the source of extreme and irrational desires. The elimination of the fears and corresponding desires would leave people free to pursue the pleasures, both physical and mental, to which they are naturally drawn, and to enjoy the peace of mind that should follow the absence of vain desires.

Epicurus was aware that deeply ingrained habits of thought are not easily corrected, and thus he proposed various exercises to assist novice learners. His system included advice on the proper attitude toward politics (avoid it where possible) and the gods (do not imagine that they concern themselves about our behavior), the role of sex (dubious), marriage (also dubious), and friendship (essential), reflections on the nature of various meteorological and planetary phenomena, about which it was best to keep an open mind in the absence of verification, and explanations of such processes as gravity (i.e., the tendency of objects to fall to the surface of the earth) and magnetism.

Having established the atomic basis of the world, Epicurus explained the nature of the *psyche*, which also consists of atoms and the void. Epicurus maintained that "soul atoms" are fine and are distributed throughout the body, and it is by means of them that we have sensations (*aisthêseis*) and the experiences of pain and pleasure, which Epicurus calls *pathê* (a term used by Aristotle and others instead to signify emotions).

According to Epicurus, a body without soul atoms is unconscious and inert, and when the atoms of the body are disarranged so that it can no longer support conscious life, the soul atoms are scattered and no longer retain the capacity for sensation. There is also a part of the human soul that is concentrated in the chest, and is the seat of the higher intellectual functions. For Epicurus, it was in the rational part of the soul that errors of judgment occur. Sensations, like pain and pleasure, are not susceptible to errors because they are functions of the non-rational part of the soul, which, unlike its rational part, does not modify a perception by the addition of opinion or belief. *The distinction between "rational" and "irrational" parts of the soul is also an essential component of such other ancient philosophers as Plato and Aristotle, and also finds its way, though expressed in different terms, into the psychological theories of such modern thinkers as Sigmund Freud.*

For Epicurus, the corporeal nature of the soul has two major consequences. First, the soul does not survive the death of the body. Hence, there can be no punishment after death, nor any regrets for the life that has been lost. Second, the soul is responsive to physical impressions, and so there are no purely mental events, in the sense of disembodied states or objects of pure consciousness separate from embodiment. Sensations of pleasure and pain, rather than abstract moral principles of goodness or badness, are, accordingly, the fundamental guides to what is good and bad. The function of the human mind is not to seek "higher things," like abstract ideals, but to maximize pleasure and minimize pain. According to Epicurus, although human beings, like everything else, are composed of atoms that move according to their fixed laws, *our actions are not wholly predetermined. The little freedom we enjoy in such a mechanistic universe is the existence of a certain randomness in the motion of atoms.*

Titus Lucretius Carus (c. 99–55 BCE), known simply as Lucretius, was a Roman poet and philosopher. His only known work is the Latin philosophical poem *De rerum natura* ("On the Nature of Things"), focusing on the principles of Epicureanism.[33] In *De rerum natura*, Lucretius set out the fundamental ontology of Epicurean atomism, which is based on the premise that nothing comes into being out of nothing or perishes into nothing. Time and historical facts were claimed by Lucretius to be dependent on the presently existing world, and thus not independently existing entities. *The universe is infinite, consisting of infinitely extended space and an infinite number of atoms. Our world as formed around a spherical earth is not itself located at the universe's center, in contrast with a Platonic vision that privileging our own world as unique.*

For Lucretius, the soul consisted of two parts. The "spirit" (*anima*) is spread throughout the body, while the "mind" (*animus*) is located in the chest. These two aspects of the soul can be corporeal, Lucretius argued, and its sensitivity and mobility are explicable by the special combination of atoms that constitute it, namely the types of atoms constitutive of air, wind, and fire, along with a fourth, type unique to soul. *Lucretian "mind" and "spirit" might be understood as the ancient equivalents of our brains and nervous systems, respectively.*

According to Lucretius, the soul, since it is atomically constituted, must like every atomic compound be destined for eventual dissolution. Once the body dies, there is nothing to hold the soul together, and its atoms will disperse. There is, therefore, contrary to most religious traditions, no survival after death, no reincarnation, and no punishment in Hades. Since death is simply annihilation, to fear, while one is alive, a future state of death, Lucretius argued, is to make the mistake of supposing oneself present to regret and bemoan one's own non-existence. Lucretius concludes *that being dead will be no worse (or no better) than it was prior to our birth in this world.*

Our world is no more than a transient amalgam of atoms, according to Lucretius. Besides, he argued, the world is an environment too hostile to human beings for anyone rationally to argue that it has been created especially for our species.

Lucretius became a key influence on the emergence of early modern atomism in the 17th century—due to Pierre Gassendi's construction of an atomistic system that was based largely Epicurus's ideas.[34] Lucretius was also admired by America's "Epicurean" third president, Thomas Jefferson.

More generally, in addition to providing an *ancient foundation for our contemporary atomic theory of matter, the views of the classical atomists were also key to the development of probably the most influential theoretical "materialist" of modern times, Karl Marx*, who, in 1841, wrote his doctoral dissertation on *The Difference Between the Democritean and Epicurean Philosophy of Nature* (German: *Differenz der demokritischen und epikureischen Naturphilosophie*), in which it is argued that theology must yield to the "superior wisdom of philosophy."[35]

The Sophists

The Greek word *sophistēs* formed from the noun *sophia*, "wisdom" or "learning," originally referred to "one who exercises wisdom or learning."

During the 5th century BCE, the term "sophist" was applied to professional educators and teachers who toured the Greek world offering instruction in a wide range of subjects, with particular emphasis on skill in public speaking and the successful conduct of life. The emergence of this new profession, which was an extension to new areas of the tradition of the itinerant rhapsodes (reciters of epic poems, especially of Homer), was a response to various social, economic, political, and cultural developments of the period. The increasing wealth and intellectual sophistication of the dominant elites in Greek city-states (*polises*), especially Athens, created a demand for higher education beyond the traditional pedagogical instruction in literacy, arithmetic, music, and physical training. *To some extent, this involved the popularization of Ionian philosophical speculation about the physical world, which was extended into such areas as history, geography, ethics, rhetoric, logic, and sociopolitical analysis.* The "Golden Age of Greece," centered in Athens during the 5th century BCE, also saw the flourishing of a skeptical, rationalistic climate of thought on questions including those of morality, religion, and political conduct, to which the sophists both responded and contributed.

Athenian political dynamics, which alternated between democratic and oligarchical governance, led to a demand for success in political and forensic oratory, and hence to the development of specialized techniques of persuasion and argument. For the sophist Gorgias (487–376 BCE), for example, the power of speech *(logos)* succeeded and incorporated the inspired effects of magic and poetry. The spell of words—their effects both on the speaker and on his or her audience—was akin to witchcraft. *To the sophists, the verbal sorcery of linguistic reason (logos) was thus best revealed in the show of eloquence.* But speech had more than a purely rational or persuasive function. It could arrest fear or inspire pity; induce sorrow or pleasure; call forth pride or contempt. *In short, rhetoric could evoke the world of pathos (emotion), a world hitherto ruled by tragedy, poetry, and magic.*[36]

Few writings by and about the first sophists survive. Most of the major sophists were not Athenians, but they made Athens the center of their activities, although travelling frequently. In addition to Gorgias, prominent sophists include Protagoras (490–420 BCE), Hippias (485–415 BCE), Prodicus (465–390 BCE), and Thrasymachus (459–400 BCE).

The early sophists charged money, sometimes a considerable amount, in exchange for tutoring, and so were usually employed by wealthy people. They were also politically active. Protagoras drew up the law-code for the foundation of an Athenian colony and Gorgias, Hippias, Prodicus, and possibly also Thrasymachus, acted as diplomatic representatives. But their wealth and celebrity status ignited the negative reaction they aroused in more traditional and aristocratic Greek circles.

Plato, for example, saw the sophists, possibly unfairly, as subversive of conventional morality and religion and a bad influence on the young.[37] According to Plato in the *Apology*, it was that sophistical climate of opinion that led eventually to the condemnation of Socrates on the grounds of irreligion and the corruption of the young. Consequently, Plato's defense of Socrates included the sharp contrast between the "genuine philosopher," Socrates, and the sophists, depicted mainly as charlatans. Plato's hostile portrait provided a foundation for the contemporary meaning of the sophist as a dishonest argumentative trickster, a characterization that remains the primary sense of the word, but that may be historically inaccurate. Due largely to the influence of Plato and Aristotle, philosophy eventually came to be regarded as distinct from sophistry, the latter considered untrustworthy and merely rhetorical.

The sophists were highly individualistic, belonged to no organization, shared no common body of beliefs, and founded no schools. Socrates was reputed to have started his philosophical theories and lifestyle as a sophist, only later to have rejected many ideas attributed to them, especially their epistemological skepticism and moral relativism. It is widely believed, however, that the sophists claimed that all thought rests solely on the apprehensions of the senses and on subjective impressions, and that therefore humans have no other standards of action than convention.[38] The search for an objective, stable order in nature had been shifted onto the human stage by intellectual novelty and socioeconomic necessities. But many Greek educators, particularly the sophists, remained skeptical about humanity's chances for uncovering a truth or law applicable to people in all societies.

While relativism, particularly in the area of morality, is popularly seen as characteristic of sophists, Protagoras is the only sophist to whom ancient sources ascribe relativistic views, and even in his case the evidence is ambiguous. A text of Protagoras called "Truth" commences with the legendary opening sentence: "Man is the measure of all things, of the things that are that they are and of the things that are not that they are not."[39] This statement is usually interpreted by Platonic scholars a claim for the relativity of the truth of all judgments—which are completely dependent on the experiences or beliefs of the individual making the judgment—or as epistemological subjectivism. From this point of view—*which has become perhaps the preeminent epistemological and moral attitude for perhaps the majority of people in the contemporary Occidental world—the way things seem, or feel, to an individual is the way they are, at least for that person.*

This relativistic (and hence false, in Plato's opinion) view of human perception and knowledge-acquisition is mentioned in the *Theaetetus* by Socrates, who quotes Protagoras's sentence as a claim concerning sensory appearances, e.g., that if the room *feels* cold to me and warm to you, then it *is* cold for me and warm for you. In this *Dialogue*, Socrates is portrayed by Plato as expanding this sophistic claim to all judgments, including itself, yielding the result that every belief is "true" for the person who holds it (and only for them), and hence that there is no objective truth whatsoever. This subjectivist interpretation of truth and knowledge was later illustrated by Aristotle's attribution to Protagoras of the view that "it is equally possible to affirm and to deny anything of anything."[40]

In the passage of the *Theaetetus* (167b–c), where, according to Socrates, Protagoras maintains the social relativity of moral judgments, a pragmatic justification of the role of the "expert" is also advocated, both in individual and political contexts. In the individual case, while no physical appearance is truer than any other, some appearances are better than others, and it is the role of the expert (for instance, the doctor or the educator) to produce better appearances instead of worse (as those appearances are then judged, even by a patient or a student); while in the case of city-states (*polises*, from which the English word "politics" is derived), some judgments of what is just etc. are better than others, and it is the role of the expert (in this case, the expert orator) to persuade the city to adopt the better judgment. *The role of expert judgment suggests that there* are *matters of fact* regarding what is better and worse *independent of the judgments of those whom the expert persuades*. Protagorean relativism, at least as represented in Plato's *Theaetetus*, is therefore ambiguous or even self-refuting, since Protagoras is represented as simultaneously maintaining: (a) universal subjectivism and (b) moral relativism, but also (c) objective realism on matters of perceived personal advantage. Even today, with subjective relativism being the default position of many people, it is important to note the self-refuting nature of statements such as "Everything is relative," which simultaneously claims both universality ("everything is") and relativity ("relative").

Protagoras's account of morality, according to which the universal acceptance of justice and self-restraint is necessary for the perpetuation of a stable polity, and thereby for the preservation of the human species, places him on one side—what might now be called the "conservative" side—of the debate about the relation between law and convention (*nomos*) on the one hand, and nature or reality (*phusis*) on the other, which was central to Greek thought in the 5th and 4th centuries BCE. The debate was fundamentally about the status of moral and other social norms; are such norms ever, in some sense, part of or grounded in the nature of things (*phusis*), and therefore universal—as claimed by most of the prominent Presocratic thinkers—or are they always mere products of human customs, conventions, or beliefs (*nomoi*)?

A stark expression of the opposition between *nomos* and *phusis* is expressed by Callicles (a pupil of Gorgias) in Plato's Dialogue *Gorgias*. According to Plato, Callicles holds that conventional morality is a contrivance devised by the weak and unintelligent to inhibit the strong and intelligent from doing what they are entitled by nature to do, viz. exploit their inferiors for their own advantage. From this point of view, *what it is really right to do is what it is conventionally wrong to do*—keeping in mind the "convention" in much of Socrates' lifetime meant democratic rule, which both Socrates and Plato found gravely defective. In contrast, universal norms are those that prevail in nature, as shown by the behavior of such non-human animals as beasts of prey, who act as nature has programmed them to behave, viz., by instinct. From a Platonic, or non-sophistical viewpoint, people who act in accord with universal, natural norms act in accord with the nature of justice and ... the law of nature, but perhaps not in accord "with this one which we lay down," which are mere conventions.[41]

Despite some superficial resemblances between them, there is no uniform sophistic position in on the nature of truth, nature/convention, knowledge, morality, or the world as a whole. Different sophists, or their associates, are found among the disputants on both sides of the debates, at least according to Plato. And even in Plato's own works, collectively entitled the *Dialogues*, the status and composition of the natural and political worlds, are, despite pronouncements in the *Republic* to the contrary, changeable.

Socrates' World

"Socrates was the first," according to Cicero "to call philosophy down from the sky, and to settle it in the city and even introduce it within the house, and compel it to inquire concerning life and death and things good and ill."[42] Socrates' mission is perhaps better indicated by Aristotle's statement that it was Socrates who invented definition. In Plato's *Dialogues*, Socrates is represented as striving to define courage, temperance, piety, beauty, justice, wisdom, the principles of right conduct, and the laws of leading a good life, inter alia. But in stark contrast with most of his Greek-speaking philosophical predecessors, *the Platonic Socrates focused almost exclusively on the human ethical and political "micro-cosmos," not on the physical universe as a whole. In this way, he shifted the conversation of the Presocratics from the heavens to the Earth and initiated the transition in Western philosophical discourse from the cosmos to the world.*

Socrates (469–399 BCE) was also socially distinctive in that he neither labored to earn a living, nor participated voluntarily in affairs of state, although he did serve honorably as a soldier for Athens during the Peloponnesian war against Sparta (431–404 BC). Rather, he embraced poverty and, although the "best and brightest" sons from Athenian social elites kept company with him and tried to live according to his principles, Socrates insisted he was *not a teacher* (Plato, *Apology*, 33a–b) and refused to take money for what he did. Because Socrates was not a transmitter of information that others were passively to receive, he resists the comparison to the commonplace image of professors as "sages on stages." Rather, he helped others to think for themselves and, with his prodding, to come up on their own with rational definitions of what is real, true, good, etc. (Plato, *Meno*, *Theaetetus*)—a new, and still suspect, approach to education. Socrates was known for confusing, stinging, and stunning his conversation partners into the unpleasant experience of realizing their own ignorance, a state sometimes superseded by genuine intellectual curiosity.[43]

Socrates was usually to be found in the marketplace and other public areas, conversing with a variety of different people—young and old, male and female, slave and free, rich and poor—with virtually anyone he could persuade to join in his question-and-answer mode of probing serious matters. Socrates's lifework consisted in the examination of people's lives, his own and others, because "the unexamined life is not worth living for a human being," as he is reported to have said at his trial (Plato, *Apology*, 38a).

Socrates questioned people about what he believed matters most in our world, e.g., courage, love, reverence, moderation, justice, and the state of their souls generally. He did this regardless of whether his respondents wanted to be questioned or resisted him. Many noble Athenian youths imitated Socrates's questioning style, much to the annoyance of some of their elders. He had a reputation for irony, illustrated by his declaration that he knew nothing of importance and wanted to listen to others, yet keeping the upper hand in every discussion. He also refused to align himself politically with oligarchs, tyrants, or democrats; rather, he had friends and enemies among, and variously supported or opposed, the views of all political castes.

The Socratic Problem: Who Was Socrates?

Various people wrote *about* Socrates, and their accounts often differ, leaving one to wonder which, if any, are accurate representations of the historical Socrates. Descriptions of Socrates appear in Xenophon's *Memorabilia* and *Symposium*, by Aristotle in his *Metaphysics*, *Politic*s, and *Ethics*, as well as a satirical caricature of Socrates by Aristophanes in his play "The Clouds." But, of course, it is the Platonic *Dialogues*, in particular with those from the "Socratic period" (roughly, the works composed before Plato's first journey to Sicily and return to Athens in 387 BCE), and, arguably, the *Republic* (*Politeia*) and *Phaedrus* (circa 367 BCE), that provide the basis for virtually all commentaries on Socrates.[44]

Even among those who knew Socrates, there was profound disagreement about what his actual views and methods were. Plato was about 25 when Socrates was tried and executed, and had probably known him most of his life. The *Ion*, *Lysis*, *Euthydemus*, *Meno*, *Menexenus*, *Theaetetus*, *Euthyphro*, the frames of the *Symposium* and *Parmenides*, *Apology*, *Crito*, and the *Phaedo* (although Plato says he was not himself present at Socrates's execution) are the *Dialogues* in which Plato had greatest access to the Athenians he depicted, of whom the most important for Plato's purposes, of course, was Socrates.

It is not clear, however, that Plato represented the views and methods of Socrates as he recalled them, much less as they were originally uttered, since Plato may have shaped the character Socrates (and all his characters) to serve his own purposes, whether philosophical, or literary, or both. Philosophers have often decided to bypass the historical problems altogether and to assume for the sake of argument that Plato's Socrates is *the* Socrates who is relevant for Western intellectual history.

But this raises the general question: Why is the history of philosophy, or of ideas more generally, valuable? A possible response is that our study of some of our intellectual predecessors is *intrinsically valuable*. When we contemplate the words of a dead philosopher, for example, we seek to understand not merely what the thinker said and assumed, but whether their claims are true. Truly great thinkers, including the Platonic Socrates, are still capable of becoming our dialectical partners in thoughtful conversations. Because Plato has Socrates intelligently addressing what philosophers, at least, have considered timeless, universal, and fundamental questions, our own understanding may be heightened. Many commentators would say it is not Plato's but Socrates's ideas and methods that mark the real beginning of philosophy in the West, and that what is Socratic in the *Dialogues* should be distinguished from what is Platonic. But *how* can one ever know the difference, and does it matter philosophically? That is the *Socratic problem*.

Socratic Definitions, Ethics, and Politics

Socrates seems to have been the first significant Occidental philosopher to treat ethics—as opposed to cosmology and natural philosophy (physics)—as a distinct area of inquiry. He largely rejected (or ignored) the Milesian tradition of a metaphysical philosophy of nature on the one hand and the Sophists' rhetorical skepticism on the other hand. For Socrates, deductive physical science had yielded no certain knowledge or infallible precepts to guide moral action. The Sophists, while having concerned themselves with language, virtue, and human nature, had nevertheless subordinated the philosophical search for "*areté*" (excellence, especially moral) and truth to the eristic game of verbal disputation. The Socratic dialectic—the proposal that there could be a supreme method for reaching ultimate truth about essences or forms—was an epistemological method operating solely by conversation in the form of question and answer; it incorporated and transcended the logical rules and strategies devised by the Eleatics and Sophists. In so doing, it established for philosophers and many educators to come a method of reasoning to be emulated and an ideal of knowledge to be pursued.

According to Aristotle, Socrates was the first philosopher to concentrate on the problem of definition. He asked the question, "What is x?" and attempted to discover by dialectical reasoning what a thing or virtue was in its most general form. Socrates also employed another kind of reasoning, inductive or epagogic reasoning, as part of the dialectical process. This was a process of "leading a mind to grasp general truths by pointing out particular causes or instances of it." By arguing from particular examples or names to general definitions, Socrates hoped to prove that the essence (Greek: *ousia*) of an object or a virtue lay in its invariant, universal character, its Form (Greek: *idea*). To demonstrate this, Socrates cross-examined a person regarding statements made about a particular notion (piety, courage, happiness, etc.) and then proceeded to deduce contradictory conclusions from the premise(s) of the statement. He repeated this procedure until either no satisfactory definition could be provided (as in the "inconclusive" early *Dialogues*) or until he had elaborated his own theory that derived the premise of the first statement from the theory of the Forms.[45]

As depicted by Plato, Socrates' search for such definitions led invariably to a concern with *knowledge of how best to live*, as not only one of the conventional virtues, such as piety or courage, but also as underpinning them all. That elevation of moral knowledge in turn led Socrates to contest the practices of rhetoric and decision-making of the political institutions of Athens—the law-courts, Assembly, and Council. As an alternative to these "democratic" forms of decision-making, which he deemed defective and unjust, Socrates posited the existence, or at least the possibility, of political expertise, claiming (in Plato's *Gorgias*) to be the one person in Athens who at least tried to pursue such a true *politikê technê*, or "political technique," a meaning of politics as a kind of professional expertise (*Gorgias*, 521d). The notion of political knowledge limited to one or a few experts, as opposed to the public opinions generated and practiced by the whole *dêmos* (the "people," or, more precisely, the voting electorate) of Athens in their judgments and political roles called into question the central premises of Athenian democracy and those of Greek politics more generally (in oligarchies, e.g., wealth rather than knowledge was the relevant criterion for rule; in tyrannies, sheer power). Thus, Socrates' concern with ethics led him directly into political philosophizing. The relation between politics and knowledge, the meaning of justice as a virtue, and the respective values of physical and moral courage were central topics of Socratic conversations in Plato's *Dialogues*.

Socrates' Trial, Death, and Political Philosophy

When Socrates was 70, he was arraigned, tried, and sentenced to death by an Athenian popular jury. In a prosecution brought by a group of his fellow citizens who claimed to be shouldering that burden for the sake of the city of Athens, the charges laid were against him were threefold: not acknowledging the city's gods; introducing new gods; and corrupting the young (*Apology*, 24b). Each of these had a political dimension. Western political philosophy may be said to have resulted from Plato's narrative of Socrates' trial and death.

Socrates' speeches in the court trial—literary versions of which were produced by Plato, Xenophon, and a number of other followers—forced him to confront directly the question of his role in an Athens as defined by its democratic institutions and norms. In Plato's account, after countering the religious accusations, Socrates acknowledged his abstention from public affairs but claimed to have had a more significant mission laid on him by the god Apollo, when that god's human "voice," the oracle at Delphi, declared that no man was wiser than Socrates and that Socrates' mission was to annoy the city like a gadfly (*Apology*, 30e). Socrates fulfilled this mission by discussing virtue and related matters (*Apology*, 38a), and also by ostensibly benefiting each person by "trying to persuade him" to care for virtue and their soul, rather than to pursue wealth for himself and for the city (*Apology*, 36c–d). Socrates argued that as a civic benefactor, he deserved not death but the lifelong publicly – provided meals commonly awarded to an Olympic champion (*Apology*, 36e–37a). Socrates is represented by Plato as a new kind of citizen, one who conceptualizes the public good in a novel way and serves the community best through unconventional and frequently unpopular actions, in sharp contrast with the conventionally defined paths to political and personal "success."

In Plato's *Apology*, Socrates' speeches of his own defense at his trial present him as a new kind of virtuous citizen. Socrates made several remarks indicating the limits he put on the civic obligation to obey the law. These statements have engendered a view of Socrates as endorsing civil disobedience in certain circumstances, thus initiating subsequent debates about the justifications for civil disobedience and the reasons for political obligation. But, in fact, Socrates did not disobey his own death sentence, and, when the time came, he drank the poisonous hemlock as prescribed by the jury. Furthermore, Plato (in his *Crito*) depicted Socrates as having been visited in prison by his friend Crito, who suggested that Socrates flee Athens and go into exile, a not uncommon practice at that time. But instead of saving his own skin, Socrates chose to obey a (unjust) jury verdict that commanded him to suffer what might be an injustice, and forfeit his own life, but not to commit a possibly greater injustice by betraying the city that had nurtured him for 70 years (*Crito*, 47a–50a). The enduring image of Socrates tried, (unjustly) convicted, and compelled to die (by his own hand) at the command of a democratic, but ignorant, citizenry has come to be a symbol of sometimes conflicted relationship between political philosophy and political authority, between individual and community interests, with, at least in this case, the latter being accorded priority.

Socratic "Ignorance" and Virtues

What Socrates sought to express in his behavior and speech (according to Plato) was the profound transformation of a man who had discovered his rationality not in the problems he had "solved" but in the standards he had established in the process of asking questions "wise" men could not satisfactorily answer. *Socrates exposed the ignorance of human empirical, common-sense beliefs about the world.* His dialectical method of questioning, refuting (Greek: *elenchos*), reformulating, and defining anew assumed that verbal reasoning was the only reliable way to ground and justify ethical and political concepts.[46]

Socrates readily confessed his own "ignorance" and agnosticism, but his rational skepticism towards prefabricated accounts of things (Greek: *logoi*) implied not that truth and knowledge were fictions, but that people had used the wrong tools to find them. Correct definitions were humanity's inadequate approximations of "Formal" objects (such as Justice, "*diké*") but they were not the transcendental Ideas, or Forms, themselves. The Socratic dialectic was therefore a negative dialectic; the question-and-answer language game was in fact the "midwife's art" (*maiutiké*) of bringing forth true statements from pregnant illusions.

By bringing philosophy down from the sacred canopy and thrusting it into the everyday world of the marketplace, Socrates substituted an ascetic standard of personal conduct for traditional Greek longing for physical pleasure. *In the human world, with all its imperfections and injustices, Socrates believed that the sole aim of the true philosopher was to seek wisdom and self-knowledge by means of struggling to know and to put into practice universal concepts or definitions (the Platonic Forms).* Socrates was convinced that the practical application of those Forms would inevitably improve both the self and the other. He hoped thereby to transform the beliefs and customs of his contemporaries by dialectically exposing the irrationality of hedonistic morality and democratic politics.

In the opinions of many of his commentators, Socrates' contributions to ethics were also contributions to human knowledge or to what would later be called science. Socrates insisted upon clear accounts (*logoi*) rational hypotheses (*Republic*, 335a), and the systematic and accurate classification of facts and data. But is the Socratic dialectical method for exposing illusions and errors and proposing a way to reach the truths of things that is best remembered.

The Post-Socratic World

In recent political life, both locally and globally, Socrates has been invoked for widely different purposes. In his 1963 "Letter from Birmingham Jail," Martin Luther King, Jr. wrote, "To a degree, academic freedom is a reality today because Socrates practiced civil disobedience."[47] The South African statesman, Nelson Mandela, 11 of whose 27 prison years were spent at hard labor in rock quarries, described the efforts of the prisoners to educate themselves by forming study groups in the quarries. "The style of teaching was Socratic in nature," he says (*Long Walk to Freedom*), with questions posed by leaders to their study groups.[48] Even more contemporary, but contemptuous of Socrates, is this excerpt from the introduction of *The Al Qaeda Training Manual*:

> The confrontation that we are calling for with the apostate regimes does not know Socratic debates …, Platonic ideals …, nor Aristotelian diplomacy. But it knows the dialogue of bullets, the ideals of assassination, bombing, and destruction, and the diplomacy of the cannon and machine-gun.[49]

While this martial declaration may appear to be jarringly dissimilar from the Socratic spirit of verbal dialogue instead of violent confrontation, it should be recalled that Socrates himself appears not to have shrunk from combat when his "homeland" (Athens) bid him to do so, and also that *Socrates' world, especially during the periods of war and plague, is, in its ethical and political dimensions, remarkably similar to the parlous and polarized world of today.*[50]

Plato: The Form(s) of a Better World

In perhaps the most famous tribute made by one philosopher to another, the 20th-century American thinker Alfred North Whitehead provided this assessment of Plato's influence:[51]

> The safest general characterization of the European philosophical tradition is that it consists of a series of footnotes to Plato. I do not mean the systematic scheme of thought which scholars have doubtfully extracted from his writings. I allude to the wealth of general ideas scattered through them. His personal endowments, his wide opportunities for experience at a great period of civilization, his inheritance of an intellectual tradition not yet stiffened by excessive systematization, have made his writing an inexhaustible mine of suggestion.

Plato's given name was apparently that of his grandfather Aristocles. "Plato" may have started as a nickname (for *platos*, or "broad"), perhaps first given to him by his wrestling teacher for his physique, or for his forehead. Although the name Aristocles was still given as Plato's name on one of the epitaphs on his tomb, history knows him as Plato.[52]

Although it is widely believed that Plato was born in 428 BCE and died about 348 BCE, these dates, however, like many chronologies in the ancient world, are not entirely certain. It is clear, though, that Plato came from one of the wealthiest and most politically active families in Athens. One of Plato's uncles (Charmides) was a member of the "Thirty Tyrants," who overthrew the Athenian democracy in 404 BCE. Charmides' own uncle, Critias, was the leader of the Thirty. Plato's relatives were not exclusively associated with the oligarchic faction in Athens, however. His stepfather was said to have been a close associate of Pericles, when he was the leader of the democratic faction. Plato's aristocratic social and heterogeneous political background is necessary but not sufficient for understanding his views of the world in general and of politics in particular.

Plato's surviving works, which are far more extensive than those any other preceding Western philosopher, are also distinctive in that, with the exception of the *Apology* (the "defense" Socrates gave during his trial), nearly everything he wrote takes the form of a dialogue. In addition, a collection of 13 letters has been included among his collected works, but their authenticity as compositions of Plato is not universally accepted among scholars.[53]

Platonic Discourse

Starting about the beginning of the 7th century BCE, two types of discourse emerged that were set in opposition to the epic poetry and mythological references of Homer and Hesiod and the more personal poetry of Sappho, namely: history (as shaped by, most notably, Thucydides, in his *History of the Peloponnesian War*) and philosophy (as developed by the naturalistic *Presocratic* tradition). Presocratic historical and natural philosophical discourses were more impersonal and proto-objective alternatives to the poetic accounts of the human and physical worlds.

Plato broke to some extent from these rival discursive traditions of Presocratic Greek thought in attempting to overcome the traditional opposition between *mythos* and *logos* by using both myths of his own invention as well as sporadic references to classical mythology. He also drew selectively on the writings of the Presocratics, especially such idealists as Pythagoras and Parmenides, as well as the teachings of the sophists and Socrates. These two ideal-typical forms of discourse, historical and mathematical objectivity on the one hand (*logos*), and poetic and mythological narratives on the other hand (*mythos*), were, figuratively speaking, "worlds apart" before Plato, and reappeared in a metaphysical guise in Plato's legendary doctrine of "The Two Worlds."

The Two Worlds in Plato's Thought

As previously noted, at least until his "Middle Dialogues"—notably the *Republic*—Plato's views are difficult to distinguish from those of his teacher, Socrates. With this in mind, it is nonetheless clear one of the most fundamental distinctions in Plato's philosophy is between the *many* observable objects that *appear* beautiful (good, just, true, etc.) and the one thing, or Form, (*idea* in Greek) that *really is* beautiful (good, just, true, etc.), and from which the many *apparently* beautiful (good, just, true, etc.) things receive their names and characteristics. Nearly every major work of Plato is, in some way, devoted to or dependent on this distinction. Many of them also explore the ethical and political consequences of conceiving of reality as a whole, and of our world in particular, in this bifurcated, dualistic way.

Plato's *Dialogues* may, collectively, *be interpreted as injunctions to transform our thoughts, perceptions, values, and behavior by recognizing the unreliability and untrustworthiness of the corporeal world and everything in it, and the superior reality of the timeless, incorporeal world of the Forms.* From this basic assumption follows Plato's claim that the soul (*psyché*) is *essentially* different from the body—so much so that it does not depend on the existence of the body for its functioning and existence, and *can more easily grasp the otherworldly nature of the Forms when it is not encumbered by its attachment to anything corporeal.* Some of Plato's works, especially the *Meno*, demonstrate that the soul always potentially retains the ability to recollect what it once grasped of the Forms, when it was disembodied prior to one's birth, and that the lives we lead are to some extent a punishment or reward for choices we made in a previous existence—eerily parallel to some key ideas in ancient Persian and Indian philosophical and spiritual traditions, with which Plato may have been indirectly familiar, due to his knowledge of Pythagoreanism.[54]

For Plato, real philosophers—those who are able to distinguish the *One* (the one enduring and ideal entity that goodness, or virtue, or justice *is*) from the many (the many temporary and illusory things that *are called* good or virtuous or just)—are, after a long education, capable not only of becoming knowledgeable and virtuous, but are also best qualified and entitled to rule the vast majority of human beings, who are unenlightened and whose behavior is principally motived by passions and desires. *Plato's dualistic understanding of the cosmos as a whole also entailed the division of the political world in which we live between the small minority—enlightened philosophers who should be kings—and the ignorant masses and corrupt politicians whose proper function is to serve the intellectual and ethical elite, and not to govern. It also underlies his depiction of the human psyche, as cleft between rational self-control, and excessive desires and impulses.*

For Plato, the world that appears to our senses is hence defective and filled with error, but there is a more real and perfect world, comprising the entities—the Forms or Ideas—that are eternal, changeless, and paradigmatic for the structure and character of the world presented to our senses. Among the most important of these abstractions—because they are not located in space or time—are goodness, beauty, equality, unity, being, sameness, virtue, and justice.

In the *Phaedo* and *Republic* (Greek: *Politeia*), Plato develops his two world theory, proceeding from this crucial distinction between the *sensible world* and the *intelligible world.* This binary opposition is foundational for much subsequent Western epistemology and metaphysics, particularly for what would later be called Cartesian idealism and Kantian transcendental idealism (to be discussed later in this book). Dualistic, categorical oppositions are also present in many modern Occidental political, metaphysical, psychological, anthropological, and social theories, including Hegel's master/slave and Marx's labor/capital antagonisms; the existential and phenomenological ontologies of such influential 20th-century thinkers as Edmund Husserl and Jean-Paul Sartre; the psychoanalytic theories of Sigmund Freud; Claude Lévi-Strauss's structuralist analyses, and the sociological framework underlying the notion of "the rationalization of the world" in the work of Max Weber.

The intelligible world, which is invisible, non-physical, and consisting of the Forms, can only be apprehended by the soul, because the soul, at least its rational part, is capable of illumination by the timeless Forms. *The sensible word, which is visible, physical, and constructed of "imitations" of the Forms, is perceived by our corporeal senses. The seeming, illusionary state of the sensible world—the world in which humankind presides—is a consequence of physical objects, especially our own bodies and perceptual organs, inadequately imitating the Forms.* A physical thing only exists to the extent that it "participates" in the Forms. This is so because Forms, by definition, exist by and for themselves. They are universal, eternal, and singular, existing in another, "better and purer," world. *What exists in the world as we perceive it is due to the Forms, for they are causes of the physical world. Thus, the physical world is an imperfect imitation of the world of the pure Forms. However, it is exceedingly difficult to ascertain if there are any intermediaries between the two worlds and, if so, how they may interact.*

In the *Phaedo*, Plato seems to indicate that absolute knowledge cannot be obtained in worldly life, but only after the death of the body and the subsequent "liberation" of the rational soul. This eternal afterlife is possible, however, only for those whose souls are pure and uncontaminated by the temptations and weaknesses of the body. For Plato, *the soul is pure because of its essential connection to the intelligible world, which is also pure; but when the soul enters the body, it becomes tainted. Only through purifying the soul, which is achieved by preparing the soul in this life for its separation from the body at the moment of death, will one be capable of apprehending the eternal Forms.*

Plato's theory of two worlds is further developed through his utilization of three key concepts: that the Form of the Good, the Divided Line, and the Allegory of the Cave. As the Forms generate and maintain the physical objects of the visible world, the Form of the Good is the cause and sustaining force of the other Forms. The Form of the Good also permits psychological illumination, which is our knowledge of the Forms. As Plato explains:

> The Form of the Good adds truth to what is known [explaining their essence and existence] and gives the knowing subject the power to know, but is itself more beautiful than truth or knowledge ... and must be honored more than them.
>
> (Republic, 509a)

With the introduction of the Divided Line, separating the sensory and intelligible worlds, as well as the higher and lower parts of the psyche (*Republic*, 509d–513a), Plato provides a diagram of the two worlds and of how the educated minority of humans can apprehend the intelligible realm.[55]

Finally, in the Allegory of the Cave (*Republic*, 514a–520a), Plato analogically compares the ignorance of human beings within the sensible world to shadows within the cave. Like the evanescent, insubstantial, and flickering shadows against the walls of the inner cave, our sensory perceptions are poor imitations of the intangible but more real world of Forms, as visually represented as the Sun, the source of life and the world, symbolizing, a "super-Form," the Form of the Good, the sustaining and creating force behind both worlds.[56] Forms in the intelligible world are maintained by the self-sufficient Form of the Good, which also potentially illuminates the objects of the visible world, and, ultimately, provides dialectically cultivated human beings with the opportunity of understanding this celestial realm. *Plato's divisions and dualisms extend beyond the division between the visible and invisible worlds to the universe (cosmos) as a whole.*

Plato's Cosmos

From a Platonic perspective, wherever it is possible to decipher or observe causal regularities in Nature (*phusis*), we are witnessing the work of *nous* (universal, or cosmic Mind).[57] If at the level of mere appearances, the world seems chaotic and events appear random, at the basic ontological level of the Forms, there exist cosmological law and an immanent rational design. Moreover, *nous* not only orders and interconnects temporal events, but also ascribes a value to them. Everything serves a purpose and fulfills a function—even the circular revolutions of heavenly bodies—and the purpose or function of something is "the Good" (Greek: *ton agathon*) of the thing. Even such "brute facts" as the motion of the Earth and the number of hours in a day are preordained conjunctions of matter and motion within a larger, teleological plan, an order whose specific instances are "persuaded by *nous*" to adhere to a cosmic program.

The cosmic events that humanity can observe, but cannot fathom, collectively constitute a *phusis, the natural order* as a whole, dependent for its perpetuation on the work of Mind (*nous*). The underlying structure of the physical universe is ultimately a logical or mathematical set of relationships that promote a moral as well as a natural law. The Good of the whole is furthered by the indwelling teleology of each part. Nature is an emanation of Mind. Its many facts and laws comprise a material order animated by one mental ruler. The ontological unity of Mind (*nous*), or Reason (*logos*), imparts a purpose or goal (*telos*) to an inherently aimless or irrational material necessity.[58]

For Plato, it is the attainment of veridical knowledge (Greek: *epistēmē*) that enables the philosopher—in sharp contrast with the masses who tend to rely on popular opinions and common sense to form their often erroneous beliefs (Greek: *doxa*)—eventually to intuit the divine order of the spheres and thereby to purify one's soul of its material contamination. Music, gymnastics, and geometry are three techniques recommended by Plato to discipline the body and to purify and enlighten the soul. Plato's vision of an underlying cosmic order is complemented by his utopian faith in an idealized human, the philosopher-king, who, ideally, should rule the Republic.

Plato's Politics of Human Nature

Plato's philosophy of politics is a reflection of his vision of human psychology. Furthermore, for Plato, the rational norms of the ethical life form the conceptual scaffolding of his ideal polis, the *Republic*. Plato's science of politics—his philosophy of the state—is the "applied ethics" of his moral science.

For Plato, humanity's present and potential political nature (*phusis*) is contingent upon the realization of the "higher" part of its essential nature, the moral excellence, or virtue (Greek: *aretḗ*), of one's soul. The permanent, rational principles that *should* guide individual moral conduct also are meant to construct the best possible government. Human nature is the "given" raw material of moral reeducation; its splits and tensions are to be coordinated by the formulation and implementation of moral laws based upon the rational understanding (Greek: *nóēsis*) of the nature of the cosmos as a whole and the composition of the human psyche in particular.

Just as the physical universe is split between the intelligible legislation of a universal rational Mind and the seeming randomness of blind necessity, so is human nature divided between the powers of reason and the appetites of desires and "high spirit" (Greek: *thumos*). Plato thought he had specified the origin of the irrational by pointing out the "tyranny" of the needs of the flesh. He furthermore argued (*Republic*, 431–6, 579–83, and 671–7) that people are "by nature" artisans, slaves, moneymakers, soldiers, or guardians.

There are three parts of the soul, but only the rational part ought to govern, in Plato's view. There are also three classes of humans—the lovers of wisdom, the pursuers of gain, and the seekers of victory—but only the wise are qualified to rule: those of the first rank are those "who are best born and best educated" (*Republic*, 431c). They make up a "natural aristocracy," who are able to curb their "simple and moderate appetites with the aid of reason and right opinion." It is essential to Plato's philosophy of politics that the "naturally best or wisest" class of citizens should perform their "natural or just" function: to rule over the sober (artisans) and the brave (soldiers).

The duality of an individual split between reason and unreason is technically a triple alliance among reason, appetite, and *thumos;* similarly, the structure of the well-ordered Republic is a top-down coordination of the rulers (the guardians from whom the philosopher-kings will be selected) and the ruled (the workers, soldiers, and such non-citizens as slaves, women, resident aliens in Greek: "metics", and children). Plato's portrait of a human nature divided among the planning faculty (Greek: *to logismos*, comprising both reason and will), the appetitive faculty (Greek: *epithumia*), and the aggressive faculty (Greek: *to thumoeides*) is used as an argument to justify the domination of the many by the few. Since it is just and wise for reason to rule the body, it is just and moral for the excellent (those possessing "*areté*") to rule and the passionate (Greek: *eros*) to obey. Presaging Freud's "structural theory of the mind," Plato claims: "But the point that we have to notice is this, that in fact there exists in every one of us, even in some reputed most respectable, a terrible, fierce, and lawless brood of desires, which it seems are revealed in our sleep" (*Republic* 572b).

Plato reduced the lawlessness he perceived in the Greek social world to the intemperance and licentiousness of the body and the *corpus democraticus* (democratic political body). Human nature was seen as a fixed substance and a civil war between the highest manifestation of Nature, the charioteer of Mind (*Republic*, 444), and the biological needs of matter (*Phaedrus*, 247–8). People are virtuous insofar as they fulfill their natural functions to think or to work, and the hierarchical and inflexible structure of the *Republic*, the *Politeia*, is justified by this "necessary" inheritance.

To Plato, history offers no help in attempting to explain the problems of human interaction and motivation. Quite the contrary. Our "modern" ideas of progress, evolution, and culture are alien to Plato's idea of human nature in particular and to classical Greek culture more generally. To Plato, the past, with all its wars and miseries, demonstrated the logical necessity to freeze the *Republic* outside human time and historical change. Human nature had hitherto been the chief obstacle to moral virtue. Its "lower parts"—such negative components as emotions, ungrounded beliefs, and irrational habits common to all humans—must be arrested and bound by rational prudence (Greek: *sophrosuné*). If humans were linked by the lower halves of their souls, they are differentiated according to the extent to which the rational part of the psyche comes to dominate and harmonize the senses and passions.

Plato thought that to be a good person, one had to become a rational individual; and to become rational, one had to learn to subordinate the ignorance of the multitude and the deceptions of sense perception to the universal and true insights of objective knowledge (*Republic*, 428–36). Human virtue, or personal merit, could thus be assessed according to a scale of epistemic values: the more advanced the dialectical ascent towards knowledge of the Form of the Good (Greek: *idea agathos*), the greater was the probability that an individual's actions were coordinated and guided by reason.[59]

According to Plato, while the democratic mass man and demagogue exalted their passions and high spirits by debasing their reason, the philosopher-king would enthrone reason by banishing, or at least taming, emotion. At the very least, he (or she) would prevent the *Republic* from sinking lower than it had under "democracy." Just as the philosopher-king's rational nature would come to resemble the eternal object of contemplation—the Form of the Good—so should the structure of the Republic come to imitate the just harmony of the philosopher's mental attunement. Unity and harmony, both in the soul of the individual and in the State of Humanity, therefore became the *telos* of Plato's desired philosophical rulers.

A just and rational political order led by philosopher-kings performs its proper function merely by perpetuating itself. Since reason and justice are stable and abiding, and since the *Republic* must be created to incorporate the principles reason and justice, the ideal state must be unchanging and inflexible. *The Republic abandons the existing world to immortalize a*

Utopia; Plato's last major work, the Laws, modifies the Republic better to accommodate an irredeemable world.

While the makeup of the ideal statesman or lawgiver remains constant in all of Plato's work—the statesman's art of political expertise (Greek: *politikê technê*) is always intended to make citizens good, wise, and happy—the *Laws* presents the case in a somewhat different fashion. Here, Plato proposes a kind of *Concordia Ordinum* among the strata of the polity.[60] These often clashing interest groups are legally coordinated in the same way as the arrangement of the parts of the best soul. The functional isomorphism between the structure of the healthy soul governed by reason, and the composition of the political community ruled by philosophers and enforced by the laws, is present throughout Plato's thought.

Plato's philosophy of politics is at bottom a metaphysical kingdom "not of this world" (*John*, 18, 36), whose monarch, the Form of the Good, bends the material universe and human nature to fulfill its hidden cosmic plan.[61]

Classical Greek thinkers introduced deductive mathematics, medical science, speculative philosophy, and civic democracy to the ancient Western world. Plato's cosmic vision is the culmination but not the end of this process. It continues the Ancients' quest for the "nature of things," but locates this final cause in an ideal *Cosmos*. It maintains the traditional Greek concern with the nature of human virtue, but abandons the Greek tradition's emphases on social excellence and the cultivation of the pleasures of this life and instead situates human goodness in the overcoming of social and corporeal motivations. Plato's *Dialogues* incorporate the religious, mystical, and other-worldly aspects of Greek philosophy and mythology and promote the ideal unity of speculative theory and political practice. Centuries later, Plato's mystical Idealism would be adopted by generations of Christian theologians and philosophers to attempt explain the nature of God, the perils of this world, and the path to redemption. Before then, however, Plato's onetime pupil Aristotle created an alternative vision of the world.

Aristotle's World

Aristotle (in Greek: *Aristotélēs*) (384–322 BCE) was one of the most influential intellectuals in Western history. He created a philosophical and scientific system that became the framework and vehicle for both Christian Scholasticism and medieval Islamic philosophy. Even after the intellectual revolutions of the Renaissance, the Reformation, and the Enlightenment, Aristotelian concepts, especially in Ethics, Logic, Epistemology, Aesthetics, and Political Philosophy, have remain embedded in Occidental thought and culture.

Aristotle's intellectual range was vast, covering most of the sciences of his day and many of the arts. He is considered a significant contributor to biology, ethics, history, logic, metaphysics, rhetoric, philosophy of mind, physics, poetics, political theory, psychology, and zoology. He was the founder of formal logic, and he pioneered the study of zoology, both observational and theoretical, in which some of his work remained unsurpassed until the 19th century. But he is, of course, most known as a philosopher. His works on ethics and political theory, as well as in metaphysics and what was called natural philosophy, continue to be studied. Aristotle left a great body of work, perhaps numbering as many as 200 treatises, from which approximately 31 survive.[62]

Born in 384 BCE in the Macedonian region of northeastern Greece in the small city of Stagira (which is why he was frequently called "the Stagirite"), Aristotle was sent to Athens at about the age of 17 to study in Plato's Academy. Once in Athens, Aristotle remained associated with the Academy for about 20 years, until Plato's death in 347, at which time he left for Asia Minor, on the northwest coast of present-day Turkey. There, he continued the philosophical activity he had begun in the Academy, but also began to expand his research into marine biology.[63] In 343, upon the request of Philip, the king of Macedon, Aristotle left the island of Lesbos, where he furthered his biological research, for Pella, the Macedonian capital, in order to tutor the king's 13-year-old son, Alexander—the boy who was eventually to become Alexander the Great. The extent of Aristotle's influence on Alexander is unclear.

After returning to Athens, in 335 BCE, Aristotle set up his own school in a public exercise area dedicated to the god Apollo, the *Lykeios*, hence the name, the *Lyceum* (many academic high schools in much of Europe are named after Aristotle's Lyceum). Those affiliated with Aristotle's school later came to be called *Peripatetics*, probably because of the existence of a walkway (Greek: *peripatos*) on the school's property. Based largely on Aristotle's own interests, members of the Lyceum conducted research into a wide range of subjects. In all these areas, the Lyceum collected manuscripts, thereby, according to some ancient accounts, assembling the first great library of antiquity. After 13 years in Athens, Aristotle left the city in 323 BCE, possibly because of anti-Macedonian sentiment. As a native-born Macedonian, Aristotle feared for his safety, allegedly remarking that he saw no reason to permit Athens to sin twice against philosophy (the first time was the execution of Socrates). Aristotle departed to Chalcis, on Euboea, an island off the mainland Greek coast, and died there of natural causes in 322 BCE.

Aristotle's Philosophy and Methodology

"Human beings began to do philosophy," said Aristotle, "even as they do now, because of wonder, at first because they wondered about the strange things right in front of them, and then later, advancing little by little, because they came to find greater things puzzling" (*Metaphysics*, 982b12).[64] The enigmas and mysteries we encounter in thinking about the universe and our place within it often challenge our current understanding of the nature of things and induce us to philosophize.

Aristotle seemed to assume that our perceptual and cognitive faculties are basically dependable, that they for the most part put us into direct contact with the features and divisions of our world. Accordingly, Aristotle frequently began his inquiries by considering *how the world appears*, reflecting on the puzzles those appearances throw up, and then reviewing what has been said about them to date. These methods comprise his twin appeals to *phainomena* and the *endoxic* method. In this way, we must assess the credible opinions (Greek: *endoxa*) about our perceptions and experiences—especially those that appear to be the most important.

Aristotle assumed that appearances tend to track the truth. We are outfitted with sense organs and mental capacities so structured as to put us into contact with the world, and thus to provide us with data regarding its basic constituents and divisions. While our cognitive and perceptual faculties are not infallible, neither are they systematically deceptive (in contrast with Plato, who distrusted sense perceptions). Since philosophy's aim is truth, and much of what appears to us proves upon analysis to be correct, appearances (Greek: *phainomena*) provide both an impetus to philosophize and a check on some of its more speculative impulses (presaging Kant's *Critique of Pure Reason*). Aristotle further claimed that science, or knowledge per se (Greek: *epistêmê*)—which extends to fields of inquiry like mathematics, politics, ethics, and metaphysics as well as the empirical sciences—not only reports the facts but also explains them (*Posterior Analytics*, 78a22–8). That is, science explains what is less well-known by what is better known and more fundamental.

Aristotle also claimed that there are scientific first principles that can be grasped through rigorous and systematic inquiry. He described the process by which knowers move from perception to memory, and from memory to experience (Greek: *empeiria*)—reflecting the point at which a single universal is apprehended by the mind—and finally from experience to a grasp of first principles (*Posterior Analytics*, ii 19). This final intellectual state Aristotle characterizes as a kind of unmediated mental apprehension (*nous*) of first principles (*Posterior Analytics*, 100a10–b6).

Aristotle's Natural Philosophy

Aristotle was a lifelong observer of nature. He investigated a variety of different topics, ranging from such general natural processes as motion, causation, place, and time, to the systematic exploration and explanation of natural phenomena across different kinds of physical entities. His natural philosophy also incorporated such specialized sciences as biology, botany, and astronomical and cosmological theories (but unlike many contemporary biological scientists, Aristotle apparently was not committed to a rigorous experimental method). Many contemporary commentators think that Aristotle treated psychology as a sub-branch of natural philosophy, because he regarded the soul (*psychê*) as the basic principle of life, including all animal and plant life.

Nature, according to Aristotle, is an inner principle of change and being at rest (*Physics*, 2.1, 192b 20–3). These principles of change and rest are contrasted with active powers or potentialities (Greek: *dynameis*), which are external principles of change and being at rest (*Metaphysics*, 9.8, 1049b 5–10), operative on the corresponding internal passive capacities or potentialities (*Metaphysics*, 9.1, 1046a 11–13).

To analyze nature as a whole and to describe the myriad physical objects and biological entities constituting the natural world, Aristotle integrated different models of scientific investigation into a single overarching enterprise. In his *Physics*, Aristotle laid out the conceptual apparatus for his analysis, offered definitions of his fundamental concepts, and argued for specific theses about motion, causation, place, and time, *including arguments for the existence of the "unmoved mover" of the universe, a supraphysical entity, without which the material world as a whole could not remain in existence.* (*Physics*, book 8).

The science of physics, Aristotle stressed, contains almost all there is to know about the world. For Aristotle, *this world is what he called:*

> *a shared place … which contains all bodies or to the particular place which immediately contains a body: For instance, you are now in the world, because you are in the air and the air is in the world; and you are in the world because you are on the earth*; and … You are on the earth because you are in this particular place, which contains nothing more than you.[65]

For Aristotle, *the world appears to be that in which we already find ourselves but which also remains separable from everything that we find. Aristotle describes this place as a "container."*[66] *Physics, therefore, according to Aristotle, studies the physical world in which we and all other beings and things are "contained."*

Furthermore, without such entities as the unmoved mover at the pinnacle of the cosmos (who later would be identified as a deity by Christianity and Islam)—which are without matter and are not part of the physical world—physics would be what Aristotle calls "Metaphysics," or "first philosophy" (*Metaphysics*, 6.1, 1026a 27–31). But since there are such separate immaterial entities, in Aristotle's view, physics is dependent on these nonphysical substances, and is only a "second philosophy," dependent on the "first philosophy," metaphysics (*Metaphysics*, 7.11, 1037a 14f).

The interaction between the two "philosophies" of physics and metaphysics is not exhausted by the causal influence exerted on the world by the supraphysical entity—the prime mover. The distinctive task of "first philosophy," or metaphysics, is an inquiry into first, imperceptible, entities; which also subsequently entails a metaphysical investigation of physical entities. Hence, there is an overlap between physics and metaphysics in Aristotle's natural philosophy, a position that would dominate Western thought until the scientific revolutions of the 16th and 17th centuries CE.

The Aristotelian Cosmos

Based on these general physical and metaphysical assumptions and procedures, Aristotle posited a *geocentric universe (cosmos) in which the fixed, spherical Earth is at the center, surrounded by concentric celestial spheres of planets and stars. Although he believed the universe to be finite in size, Aristotle claimed that it exists unchanged and static throughout eternity.*

Aristotle also argued that, in addition to the four elements of fire, air, earth, and water—identified by his Presocratic predecessors—they were acted on by two forces, gravity (the tendency of earth and water to sink) and levity (the tendency of air and fire to rise). He later added a fifth element, known as the "aether," a space-filling medium or field, to describe the void that fills the universe above the terrestrial sphere. The "aether" would not be disproved for millennia.

By combining rigorous logic with the systematic observation and analysis (breaking down and classification) of the components of the natural world, Aristotle believed we can make "true" statements about the cosmos and thereby understand:

1 The nature of essences (*what something is*);
2 The nature of causes (*why things occur*).

To understand how the natural world works, both in its physical and immaterial dimensions, Aristotle created what we would now call "science" and "scientific" (but not "experimental") "methods," as well as the "first philosophy" of metaphysics, which experimental scientists and empiricist philosophers would jettison two millennia after Aristotle's death.

The Eternity of the World

In his *Physics*, Aristotle argues that *the world must have existed from eternity and will eternally continue to exist.* He claims that everything that comes into existence does so from a "substratum," a base or foundation. Therefore, if the underlying matter of the universe came into existence, it would come into existence from a substratum. But the nature of matter is precisely *to be the substratum* from which other things arise. Consequently, the underlying matter of the universe could have come into existence only from an already existing matter exactly like itself; to assume that the underlying matter of the universe came into existence would require assuming that an underlying matter already existed. As this assumption is self-contradictory, Aristotle argued, matter must be eternal. Aristotle also argued that since motion is necessarily eternal, and since the universe is, by definition, in motion, the universe is eternal.

The question of the eternity of the world was a concern for both ancient philosophers as well as for many prominent medieval Christian theologians and philosophers, who debated the question whether the world has a beginning in time or if it has always existed. The problem became a focus of a dispute in the 13th century, when some of the works of Aristotle were rediscovered in the Latin-speaking West. Since this Aristotelian view conflicted with the prevalent view of the Catholic Church—that the world had a beginning in time, Aristotelian teaching on this subject was prohibited in the Condemnations of 1210–77.[67]

The Soul According to Aristotle

Aristotle defined the soul (*psychê*) as the form of a living compound, the principle or source (Greek: *archê*) of all life. For Aristotle, in fact, *all living things*, and not only human beings, have souls: "what is ensouled is distinguished from what is unensouled by living" (*De Anima*, 431a 20–2; *Metaphysics*, 1075a 16–25). The soul is the cause and source of the living body.[68]

From Aristotle's biological point of view, *the soul is not—as it was in some of Plato's Dialogues—an immaterial refugee from an ideal world entrapped in a corruptible body in this world.* The soul's essence is defined by its relationship to an organic structure. According to Aristotle, animals and plants also have souls—the "forms" (or what we might call "programs," though decidedly not the Platonic "Forms") of animal and vegetable life. A soul is "the actuality of a body that has life" (*De Anima*, 412a.20, b5–6), where life means the capacity for self-sufficiency, growth, and reproduction. If one regards a living substance as a composite of matter and form, then the soul is the form of a natural—or, organic—body. Organic bodies have organs—parts that have specific functions, such as the jaws of mammals and the roots of trees.[69]

The souls of living beings are ordered by Aristotle in a hierarchy. Plants have a vegetative or nutritive soul, which consists of the powers of growth, nutrition, and reproduction. Animals have, in addition to those vegetative powers, the powers of perception and locomotion—they possess a "sensitive" soul, and every animal has at least one sense-faculty, touch being the most universal. Whatever can feel, can experience pleasure and pain; animals, which have senses, also have desires. Humans, in addition to the previously mentioned powers, also have the powers of reason and thought (Greek: *logismos kai dianoia*), which Aristotle called the "rational soul."

Similarly, Aristotle claimed, that the soul is a cause, the source of motion, and that the soul and body are just special cases of form and matter: Furthermore, the soul, as the *telos*—the goal/end/purpose (the root of *teleology*, a key component of Aristotle's worldview)—of a compound organism is also the final cause of the body. Any given body is the body that it is because it is organized around a function (or *telos*), whose purpose is to unify the entire organism. The body is *organic* (Greek: *organikon*; *De Anima*, 412a 28) matter; it serves as a tool (Greek: *organon)* for implementing the specific life activities of the kind to which the organism belongs. Aristotle's depiction the soul, its faculties, and its powers influenced both philosophy and natural science for nearly two millennia.

Aristotelian Ethics and Political Theory

Aristotle's underlying teleological framework extends to his ethical and political theories, which he, like Socrates and Plato, regarded as complementary. He assumed that most people wish to lead "good lives"; but what a "good life" is for humans needs to be fleshed out.

In his *Nicomachean Ethics* (*EN*, 1094–7), Aristotle argued that the best life for a human being is not a matter of subjective preference, and that people can, and often do, choose to lead suboptimal lives.[70] Aristotle recommended that we reflect on the criteria any successful candidate for the best life must satisfy. If we do so, we might agree that only one kind of life meets those criteria and is, therefore, *the superior form of human life. This is a life lived in accordance with reason (nous), and other forms of leading a human life, including, for instance, a life led to maximize pleasure or to achieve fame or honor, are deemed inferior.*

According to Aristotle, that only thing all humans naturally seek, and that also meets his criteria for a fully human, or "good," life is what he calls *eudaimonia*, which has been variously translated into English as "happiness," "human flourishing," or "living well," but unlike the connotation of "happiness" as a transient, subjective mental, or emotional state, denotes an enduring stable and objectively virtuous psychophysical condition.[71]

Happiness (*eudaimonia)* is achieved, according to Aristotle, by fully realizing our natures, by actualizing to the highest degree our human capacities. Most people will agree with the proposition that "happiness" is a good everyone desires—even while differing about how they understand happiness. So, while seeming to agree, people in fact disagree about the highest human good. Consequently, it is necessary to reflect on the nature of "happiness." To do so, Aristotle says we need to identify the function (*ergon*) of a human being. Just as there seems to be a particular function for the eye, the hand, and for every part of a human being (and for any living being, for that matter), Aristotle argues that there is a higher function for the human being as a whole. *And that is an active life for a soul that has reason (nous*). (*EN*, 1097 b22–1098a4).

Consequently, Aristotle equates the human highest function with reason, which then enables him to describe the happy life as involving the exercise of reason, whether practical (i.e., involving action) or theoretical (involving knowledge). Happiness is, therefore, according to Aristotle, an activity of a rational soul in accord with virtue or excellence (Greek: *kat' aretên*), or is a rational activity executed excellently (*EN*, 1098a, 61–17). In the Greek tradition of moral philosophy, Aristotle's word for "virtue," *aretê*, is broader than the dominant sense of the English word "virtue," since it denotes many excellences, not just moral virtues.

Aristotle's claim that only *excellently* executed or *virtuously* performed rational activity constitutes human happiness is the cornerstone of what has come to be called *virtue ethics*. For Aristotle, the good life is *a life of activity*, not just a felicitous mental state, and humans are rightly praised or criticized only for things they *do*, not for what they think (*EN*, 1105b20–1106a13). To be truly happy from an Aristotelian perspective, we must not only act, but act excellently or virtuously. But what constitutes virtue or excellence with respect to such specific individual human virtues as honestly, courage, friendship, and prudence, for example, needs further explication. Aristotle therefore explores these virtues in detail, in both their practical and theoretical forms.

Aristotle concludes his discussion of human happiness in the *Nicomachean Ethics* by claiming that political theory is a continuation and completion of ethical theory. Ethical theory characterizes the best form of human life; political theory characterizes the forms of collective organization best suited to its realization (*EN*, 1181b12–23).

For Aristotle and the entire classical Greek tradition, the basic political unit is the *polis*, from, which is both a *city-state* in the sense of being an authority-wielding and powerful administrative unit, and a *civil society* in the sense of being an organized community whose constituents have varying degrees of converging interests.[72]

Aristotle's political theory is unlike later, "liberal" theories, in that he does not think that the *polis* requires justification when it might infringe on what most modern political theorists deem "universal human rights." Rather, as in his philosophy in general, in his *Politics*, Aristotle holds to a kind of *political naturalism*, in which human beings are regarded as *political animals by nature*, in the sense that it is only possible for human beings to flourish and to lead virtuous lives resulting in *eudaimonia*, *within the framework of an organized and beneficent polis*. The *polis* comes into being for the sake of our living per se, but it remains in existence for the sake of our living *well*. (*Pol.* 1252b29–30; cf. 1253a31–7).

The *polis* is thus to be evaluated according to how well it promotes human happiness. A superior form of political organization enhances human life; an inferior form hampers and hinders it. But which kind(s) of political arrangement best meet the goal of developing and augmenting human flourishing? Aristotle considers a number of differing forms of political organization, and sets most aside as inimical to the goal of facilitating genuine human happiness. For example, he rejecting contractarianism (such as those that would many centuries later be called "social contract" theories) on the grounds that it treats as merely instrumental those forms of political activity that are in fact partially constitutive of human flourishing (*Pol.*, iii 9).

In thinking about the possible kinds of political organization, Aristotle observes that rulers may be one, few, or many, and that their forms of rule may be legitimate or illegitimate, as measured against the goal of promoting human flourishing (*Pol.*, 1279a26–31). Taken together, these factors yield six possible forms of government, three beneficent and three malign:

	Beneficent	*Malign*
One ruler	Kingship	Tyranny
Few rulers	Aristocracy	Oligarchy
Many rulers	Polity	Democracy

The beneficent are differentiated from the malign by their relative abilities to realize the basic function of the *polis*: living well. Given that we prize human happiness, we should, Aristotle claims, prefer forms of political association best suited to this goal.

Necessary for human flourishing, argues Aristotle, is the maintenance of a suitable level of *distributive justice*. He claims that virtually everyone may agree to the proposition that we should prefer a just state to an unjust state, as well as to the proposal that a fair and equitable distribution of justice requires treating equal claims similarly and unequal claims dissimilarly. But like their notions of happiness, people will differ about what constitutes an equal or an unequal claim or, more generally, an equal or an unequal person. A democrat will assume that all citizens are politically and legally equal, whereas an aristocrat will maintain that the best citizens are, by definition superior to the inferior. Accordingly, the democrat will expect justice to provide equal distribution of goods and rights to all, whereas the aristocrat will assume that the best citizens are entitled to more than the worst members of the polity.

Aristotle proposed his own, meritocratic, account of distributive justice (*Nicomachean Ethics*, Vol. 3.) He accordingly disparaged oligarchs and plutocrats, who suppose that justice requires preferential claims for the rich, but he also criticized democrats, who contend that the state must support liberty for all citizens irrespective of merit. The best *polis* has neither function: its goal is to enhance human flourishing, an end to which liberty and justice are instrumental, and not ideals to be pursued for their own sake. The just city-state may also combine the best features of each form of beneficent political organization, resulting in a "mixed constitution" of laws and rights, rather than an unmixed, pure, but less beneficent polis.

Aristotle and Plato

Although Aristotle revered Plato, his philosophy eventually departed from that of his teacher in important respects. The most fundamental difference between Plato and Aristotle concerns their theories of forms.[73] For Plato, the Forms are perfect exemplars, or ideal types, of the properties and kinds of entities that are found in the world. Corresponding to every such property or kind is a Form that is its perfect exemplar or ideal type. Thus, the properties "beautiful" and "just" correspond to the Forms the Beautiful and Justice the kinds "dog" and "triangle" correspond to the Forms the Dog and the Triangle; and so on.

For Plato, Forms are abstract objects, existing completely outside space and time. Thus, they are knowable only through the mind, not through sense experience. Moreover, because they are changeless, the Forms possess a higher degree of reality than do things in the world, which are changeable and always coming into or going out of existence. The task of philosophy, for Plato, is to discover through reason (via "dialectical thinking") the nature of the Forms, the only true reality, and their connections, culminating in an understanding of the most fundamental Form, the Form of the Good, or the One. Aristotle rejected Plato's theory of Forms but not the notion of form itself. For Aristotle, forms do not exist independently of things—every form is the form of something.

Also, in contrast with Plato's abstract and often geometrical way of comprehending the cosmos, Aristotle's *teleology rendered the natural world susceptible to empirical, factual investigation at the material level, in addition to logical, formal analysis favored by Plato*. For Aristotle, both logical proof—the *reasons for* something—and systematic observation—*the facts* about something—were indispensable preconditions for scientific knowledge. History, as Aristotle argued in the *Poetics*, was not a science because it deals merely with facts, with non-repeatable, lawless events. Nature (and even poetry) on the other hand, evidenced both facts and reasons, an abiding lawfulness, or *telos*, which directed all its material forces towards an underlying formal end.

And, at the deepest metaphysical or cosmological level, Aristotle did not view the cosmos as riven between the "two worlds" of the ideal (the Forms) and the corruptible (the world we perceive), as did Plato. Instead, he saw the universe as a whole and the living world in particular as logically and organically linked. For Aristotle, there was one world, the one in which we live and die. But it is an eternal world, without beginning or end. And according to Sean Gaston, Aristotle's world:

> contains all bodies or [refers] to the particular place which immediately contains a body: For instance, you are now in the world, because you are in the air and the air is in the world; and you are in the world because you are on the earth; and ... You are on the earth because you are in this particular place, which contains nothing more than you ... the concept of world as a structure of containment ... the possibility of s subject-oriented presence, or a present moment, my moment, here and now, in the world ... World appears to be that in which we already find ourselves but which also remains separable from everything that we find ... A "container."[74]

This is far removed from Plato's "two worlds."

On the other hand, like Plato, Aristotle believed that the philosophical life, the life of reason and contemplation, was the highest mode of human existence and that rationality constituted the noblest virtue (*areté*) of the good soul. But Aristotle divided the Platonic idea of a unitary rational philosophy into two autonomous competencies, that of Theoretical Reason (Greek: *Nous Theoretikos*), the province of philosophy, physics, and mathematics, and that of Practical Reason (Greek: *Nous Praktikos*), whose domain included economics, politics, and ethics. Theoretical Reason took the form of "scientific" argumentation and was designed to afford the investigator with universal and necessary truths *about the natural world.* Practical Reason assumed the guise of moral arguments and was to provide rational agents with unchanging rules for the guidance of human conduct *in this political world.* Both Theoretical and Practical Reason were aspects of a cosmic Mind (*Nous* or *Logos*) and were concretized within the process of human rational thought.[75]

Also like Plato, Aristotle divided the human soul (*psuké*) into parts, a rational, temperate, and virtuous "upper" half and an inferior, irrational, and emotional lower sphere that is "moved" by insatiable desires, subhuman dispositions, and selfish wants. Both men regarded children, slaves, the handicapped etc. as *naturally* lacking in the virtues making up rationality (although Plato afforded some women the possibility for a philosophical education and possible political leadership, whereas Aristotle seems not to have shared this view). And Reason (*nous*) should be the only master of the soul, while the prime duty of the "base material" layers of the soul—and the lower classes of the polis—consists in obedience to the edicts of their rational sovereign.

Politically speaking, both Plato and Aristotle were in fundamental agreement concerning the nature and functions of the polis and the knowledge of statesmanship, political science. They both revered the (idealized) classical Greek federation of small, autarchic civic communities that could provide the material and intellectual conditions for civilized life. Although Aristotle witnessed the Athenian polis only during its decline, he nonetheless proclaimed the priority of the political community over the individual and argued that the most perfect, just, and best state was that which could give its citizens the training needed to achieve the good life of selfless rationality.

To Plato, politics was a branch of ethics and ethics was a category of the Good. To Aristotle, ethics was a branch of politics, and the true function (*telos*) of politics consists in the inculcation of temperance or rational self-control.[76] The just state, in Aristotle's view, like Plato's state of laws, was therefore to be seen as an organic, hierarchical association of interdependent, stratified classes (Aristotle mentions five classes in the *Politics*). But Aristotle's *Politeia* (*Republic*) was designed to *evolve*, both constitutionally and politically, as the times might warrant, and to foster the "common good" and happiness of *the entire citizenry*, and not just the welfare of the ruling class. Plato, on the other hand, believed that while the *Politeia* was the optimal political structure for all people, in practice it would have chiefly benefited the elites. Both Plato and Aristotle revered the traditional *aristocratic* ideals of temperance, deliberation, and rational debate, ideals embodied only by those "naturally fit" for the cultivation of dialectical reason.

Practical activity was valued highly by Aristotle, but it was not his highest value. Truth was rated higher than action, as was contemplation. The rational pursuit of truth and moral wisdom constituted the goal of human existence for Aristotle, Plato, and Socrates. But Aristotle's worldview was fundamentally biological, organic, and developmentally oriented, whereas Plato's was static and mathematically oriented. Ultimately, Aristotle rejected much of Plato's theory of knowledge and the Utopian otherworldliness of Plato's philosophy in *favor of a more realistic and naturalistic approach to understanding the cosmos as a whole and our world in particular.*

Aristotle's Influence

Following Aristotle's death, his name and works were spread through much of the ancient Mediterranean and Persian worlds by his student Alexander the Great. From the 6th through the 12th centuries, although the bulk of Aristotle's writings were lost to the West, they received extensive attention by prominent Byzantine and Islamic philosophers, for whom Aristotle was so prominent that be became known simply as "The First Teacher." In this tradition, the commentaries of such distinguished Islamic thinkers as Avicenna (Ibn-Sīnā, ca. 970–1037 CE) and Averroes (Ibn Rushd, 1126–1198 CE) interpreted and developed Aristotle's views in remarkable ways. These commentaries in turn proved extremely influential for the initial reception in the 12th century of the Aristotelian corpus in the Latin West. Among Aristotle's greatest exponents during this period of his reintroduction to the West, were the great theologians Albertus Magnus (Albert the Great), and, above all, Albert's student St. Thomas Aquinas, who sought to reconcile Aristotle's philosophy with Christian doctrine.

Since then, Aristotle's cosmology has largely been discarded by most physical scientists and many Roman Catholic theologians. However, beginning about the mid-20th century, Aristotle's virtue ethics—or eudaemonism—his theory of human well-being, has been revived in moral philosophy as a plausible alternative to the more mainstream ethical doctrines of deontology (following Kantian ethics) and consequentialism (in the utilitarian tradition).[77] A significant number of biological and social scientists (especially psychologists) also are calling upon some of his observations of the natural world and the human psyche to inform their theories and research.[78] And, even his political theory is being revived in some "Center-Left," progressive circles.[79]

Aristotle is clearly one of the greatest thinkers in Western history, and his stature is unlikely to diminish. In fact, there may be at present more references to and citations of Aristotle and Aristotelianism than to any other philosopher or philosophical movement. Only Plato comes close. For millennia, the Occidental intellectual and scientific world was changed by Aristotle.

The Stoics

Following Aristotle's death and the decline of the classical Greek polis, especially Athens, Stoicism became a major intellectual movements during the Hellenistic period between the death of Alexander the Great in 323 BCE and the emergence of the Roman Empire in 31 BCE, and the ensuing first two centuries of Roman imperial rule. The name "Stoicism" derives from the porch (Greek: *stoa poikilê*) in the agora (central marketplace) in Athens, which is where Stoic teachers and students congregated, and lectures were held.[80]

The sources of our knowledge about Stoicism are far from exhaustive, since there is no surviving single complete work by any of the first three heads of the Stoic school: the "founder," Zeno of Citium (344–262 BCE), Cleanthes (d. 232 BCE), and Chrysippus (died. c. 206 BCE). The only extant complete works by Stoic philosophers are those by writers from Roman imperial times, Seneca (4 BCE–65 CE), Epictetus (c. 55–135 CE), and the Emperor Marcus Aurelius (121–180 CE), and their books are principally focused on ethics. The Stoics' moral philosophy famously emphasized what they have come to be known as the four cardinal virtues, stemming from classical Greek ethics, namely, prudence, justice, fortitude, and temperance, or, alternatively called, wisdom, morality, courage, and moderation.

The Stoics—living up to the contemporary meaning of the English term "stoicism"—held that emotions like fear or envy (or impassioned sexual attachments and passionate love of anything whatsoever) either were, or arose from, false judgments, and that the sage—a person who had attained moral and intellectual illumination—would not be controlled by them. Stoic ethics and their psychological theories, which incorporated many of Aristotle's key ideas, have had a lasting impact on Occidental thought. But it is the Stoic view of the cosmos as a whole and of the world we inhabit that is of greatest relevance for my purposes.

The Stoic Universe

The Stoic philosophers believed in a kind of island universe in which a finite cosmos is surrounded by an infinite void (not dissimilar in principle from a galaxy). They held that the cosmos is in a constant state of flux, pulsates in size, and periodically passes through upheavals and conflagrations. In the Stoic view, *the universe is like a giant living body, with its leading elements being the stars and the Sun, but in which all parts are interconnected, so that what happens in one place affects what happens elsewhere. They also held a cyclical view of history, in which the world was once pure fire and would become fire again* (similar to the ideas of the Presocratic thinker Heraclitus).

To the Stoics, *the universe was divine material substance*, a *pneuma* (or "breath"), which is the basis of everything that exists. This *pneuma*, which is the active part or Reason (*logos*) of God, conveys form and motion to matter, and is the origin of the elements, life, our souls (*psyché* in Ancient

Greek and *anima* in Latin), and human rationality. From their cosmology, the Stoics explained the birth, development, and ultimately, the destruction of the universe as a never-ending cycle (Greek: *palingenesis*). *Therefore, the same events play out again and are repeated endlessly. Any subsequent world is bound to be identical to the previous one.*

The Stoic God

The Stoics often identified the cosmos with Reason (*Logos* in Ancient Greek, *Ratio* in Latin) and with God (*Zeus* in Greek, or *Jupiter* in Latin), the ruler and upholder, and at the same time the law, of the universe. The Stoic God is not a transcendent omniscient being standing outside nature, but rather is material and immersed in nature. Not only was the primary universal substance God, but divinity could be ascribed to its manifestations—to the heavenly bodies, to the forces of nature, even to deified persons; and thus the world was peopled with divine agencies. *The Stoic God orders the universe for the good, and every element of the world contains a portion of the divine element that accounts for its behavior.*[81]

However, the universe is good only as far as it is optimally rational and virtuous. Whatever exists in the world may be suitable for it, but is also what the Stoics call "morally indifferent," i.e., neither good nor bad. Thus, human and animal pain, suffering, and termination occur, even while the universe unfolds according to the rational good. *Events in the world unfold according to the best possible reason and are knowable by reason. The good for humans lies in their understanding and acceptance of this cosmic design and their virtuous conduct.*

Stoic Fate and Freedom

To the Stoics, nothing happens without a cause; there is a reason (*logos*) for everything in the universe. Because of the Stoics' belief in the unity and cohesion of the cosmos and its all-encompassing Reason, they embraced what might today be called a form of "soft determinism." However, instead of a single chain of causal events, there is a multidimensional network of events interacting within the overall framework of fate, of fortune (in Latin, *fatum, fortuna*, or *sors*, respectively, depending on the context). Humans appear to have free will because individual actions participate in the determined chain of events. Hence, for the Stoics, we humans responsible for our own actions, and are free to choose between virtue and vice, thus moderating the apparent contingency of fate/fortune (Latin: *fortuna*).[82]

The Stoic cosmos, especially as enunciated by Seneca, is ordered by two supreme powers—fortune and virtue (Latin: *virtus*).[83] Everything in the world is governed by the "law of nature" (Latin: *lex naturae*), Earlier Stoics considered natural law to be partly rooted in what they called a theory of appropriate action, and partly in a physical account of how divine Reason (alternately, *Zeus*, or *Logos*)—pervades the world. And our role in this cosmos is to make full use of the divine element in us, human reason, to do our best to navigate a path between these universal forces.

The Neoplatonists

The term "Neoplatonism" refers to a philosophical tradition that emerged and flourished in the Greco-Roman world of late antiquity, roughly from the middle of the 3rd century CE to the beginning of Islam in the middle of the 7th century CE. After the decline of Epicureanism and Stoicism, Neoplatonism became the dominant philosophical orientation of the later Roman and early Byzantine Empires, offering a comprehensive understanding of the universe and the individual human being's place in it. However, in contrast to labels such as "Stoic," or "Platonic," the designation "Neoplatonic" is of modern coinage and to some extent a misnomer. Late antique philosophers, now counted among "the Neoplatonists," did not think of themselves as engaged in an effort specifically to revive the reading of Plato's *Dialogues*.[84] They did call themselves "Platonists" and esteemed Plato's views more than those of any other thinker. And, they introduced the cosmological and ethical theories of Plato, Aristotle, and the Stoics into the theological doctrines and religious practices of early Christianity.

The Egyptian-born Plotinus (204/5–270 CE) is regarded as the founder of Neoplatonism. Plotinus may have been well aware of the rise of Christianity, but as someone Christians would deem a "pagan," he was not as alarmed by it as his pupil Porphyry. While contributing to the dissemination of Greco-Roman philosophical ideas, especially those of Plato, among the educated Roman elite, Plotinus, perhaps ironically, may have also contributed to the acceptance of a Neoplatonized version of Christianity by many literate "pagan" Roman citizens, including, most notably, the young Augustine of Hippo (354–430 CE). By the end of the 5th century CE, Neoplatonism, increasingly in early-Christian garb, was taught across the Latin-speaking cities of the Mediterranean basin and even in places we now call "The Middle East," which about 200 years later would be largely Islamicized,

The Neoplatonic Universe, Mind, and Soul

For the Neoplatonists, *the process of the emergence of the universe from the divine principle has gone on forever, just as it continues at this very moment, and will continue to do so, sustaining a world without end.* What they called "reality" emerged from "the First" in coherent stages, in such a way that one stage functions as creative principle of the next. *This kind of emergent cosmology rests on the belief that every activity in the world has both an inner and an outer aspect.* For example, the inner activity of the sun (nuclear fusion, in our terms) has the outer effects of heat and light, themselves activities. Or the inner activity of a plant or animal that is determined by the kind of plant or animal it is (its genetic code, as we would say). Following the classical Greek philosophical tradition, the Neoplatonists called this universal generative principle, *Logos*.

The Neoplatonic *Logos* facilitates the verbal and physical outward expression of our internal thoughts and feelings, which, over time, constitute both a person's biography and a society's history. Any inner activity prefigures the character and nature of its outer effect. The "first principle of reality" was conceived by the Neoplatonists as an entity that transcends all physical reality, an absolute Unity, about which little meaningful can be stated. *The entire universe is the effect of this One, the supreme and eternal agent and cause of all cosmic activity and energy.*

Also consistent with the Platonic cosmology in which a universal mental agency (ultimately, the Form of the Good) generates matter, the Neoplatonists asserted that all physical events and material effects in the universe must be due to *Nous*, which might in this context be understood as pure and absolute Mind. According to Neoplatonic theory, *Nous* is the first effect of the activity of the One, the supreme essence (*ousia*, or "Being" in ancient Greek) of reality as a whole.

Neoplatonists referred to *Nous* as the second "Hypostasis," an abstract noun derived from a verb meaning "to place oneself under or beneath," with the connotation of "standing one's ground." The word "hypostasis" therefore denotes a distinct essential being or underlying reality which, in the case of *Nous*, is the derivative outer activity of the generative first principle.[85] The Neoplatonic notion of "Hypostasis" is indebted to Stoicism and had a major influence on early Christian theologians.

The Neoplatonists also made a distinction between "Nature" (*phusis*) and "Soul" (*psyché* or *anima*), which, again following classical Greek philosophy, involves a separation between higher and lower mental functions. For the Neoplatonists, "Nature" denotes not only the essence or nature of each natural being or the entirety of the natural world (Nature as a whole), but also, a "lower" aspect of conscious life (what we would call the autonomic life activities not under an organism's conscious control). For them, *every aspect of the natural world, from inorganic matter to human souls, has an eternal and divine moment. In addition to the other-worldly dimensions of Neoplatonic philosophy, the material world was viewed as essentially good and beautiful, the product of cosmic providence and divine Nous, and hence deserving of our awe and respect.*

Neoplatonism's Influence

Neoplatonism has been quite influential in Occidental philosophy and theology. Through St. Augustine (354–430 CE) in the West and the 4th-century Cappadocian Fathers (St Basil the Great, St. Gregory of Nyssa, and St. Gregory of Nazianzus, who advanced the idea of the "Holy Trinity") in the Eastern Roman Empire, Neoplatonism profoundly influenced such mainstream and less orthodox late medieval Christian theologians as St.

Thomas Aquinas, John Duns Scotus, and Meister Eckhart. Neoplatonic thought also facilitated the integration of classical Greco-Roman philosophy and science into both Islam, especially through the works of such Muslim philosophers as Al-Kindi, Al-Farabi, and Avicenna (Ibn Sina), and into medieval Jewish thought, principally through the works of Moses ben Maimon (known as Maimonides).

During the Italian Renaissance, classical Greek philosophy and cosmology as a whole, and Neoplatonism in particular, were revived, notably by the Florentine thinker Marsilio Ficino (1433–99), whose late 15th-century translations and interpretations of Plato and Plotinus influenced not only the philosophy but also the art and literature of the period. Neoplatonic ideas have also made an impact on such "idealist" Western intellectuals as the 17th-century English thinkers Henry More and Ralph Cudworth (known as the "Cambridge Platonists"); German "idealist" philosophers Gottfried Wilhelm Leibniz, Georg Wilhelm Friedrich Hegel, Friedrich Wilhelm Joseph von Schelling, and Johann Gottlieb Fichte; the late 19th- and early 20th-century French philosopher Henri Bergson, and the mid-20th-century French Jesuit theologian and paleontologist Pierre Teilhard de Chardin, whose work is said to have influenced the two most recent popes, Benedict XVI and Francis.

Philosophical, Cosmological, and Theological Interlude

It has often been said that philosophy begins in wonder. This was clearly the case for those ancient and medieval philosophers and theologians who wondered what constituted the basic stuff of the world around them, how this basic stuff changed into the diverse forms they experienced, and how it came to be. It is still the case for many philosophers and cosmologists today. *Those origination questions are related to the puzzle of existence of the universe as a whole, and of the world's and human existence—or non-existence—in particular.*

These ontological questions include: why is there anything at all? Why is there something, no matter what it is, even if different or even radically different from what currently exists? Why is there something rather than nothing? Some doubt whether we can ask this latter question because there being nothing, or non-existence existing, is either paradoxical or logical nonsensical.

The related question, "why does the universe exist?" had been addressed by traditional religions as well as in contemporary cosmological theories. As long as humans have been trying to make sense of the universe, they have been proposing cosmological theories. Furthermore, the notion of a deity often plays a central role in these cosmological theories.

According to most monotheistic religions, God is the sole creator and sustainer of the universe. While some contemporary thinkers claim that scientific theories and evidence have superseded all theological answers, others claim that scientific discoveries reinforce the claim that God created the universe.

Christianity and other monotheistic religions (Islam and Judaism) assume a transcendent and sovereign God who created the universe and continually maintains its existence. *The world only exists because of an ultimate and supernatural cause*, which is, as the physicist Sir Isaac Newton said in 1692 to his friend Richard Bentley, “not blind and fortuitous, but very well skilled in Mechanicks and Geometry.”[86]

Whether in a philosophical sense or in a scientific sense, cosmology has usually been related to theism (religious, faith-based reasons for believing in a deity), but it is only relatively recently that cosmology based on physics and astronomy has entered the discussion concerning the existence and role of God. A limited application of physics to the study of the universe can be found in the second half of the 19th century, when the cosmological consequences of the law of entropy increase were eagerly discussed in relation to the Christian doctrines of a world with a beginning and end in time.[87]

The theological problems related to an infinitely large universe are not specifically related to modern physical cosmology but have been discussed since the early days of Christianity. The theological implications of an infinite universe were discussed by the church fathers and, in greater detail, by the Neoplatonic thinker Johannes Philoponus in the 6th century. Infinity was seen exclusively as a divine attribute; to claim that nature is infinite would be to endow it with divinity, a view characteristic of pantheism. While the generally accepted view among theists was, and to some extent still is, that an infinite universe is philosophically absurd and theologically heretical, there has been consensus on the issue. In fact, several prominent Christian thinkers, from Descartes in the 17th century through Kant in the 18th to Edward Milne in the 20th, have argued that an infinite universe is in better agreement with God’s will and omnipotence than a finite one.

Cosmological Arguments in Theology

The earliest formulation of a version of the cosmological argument is usually taken to be found in Plato’s *Laws* (893–6); the classical Greek elaboration of this argument is rooted in Aristotle’s *Physics* (VIII, 4–6), and *Metaphysics* (XII, 1–6). Islamic philosophy has also contributed to the debate, notably by Ibn Sina (Avicenna in Latin), who developed the argument from contingency, which was taken up by Thomas Aquinas in his *Summa Theologica* (Iq.2a.3) and his *Summa Contra Gentiles* (I, 13).

Influenced by John Philoponus, the *mutakallimūm*—Islamic theologians who used reason and argumentation to support their ostensibly revealed beliefs—developed the temporal version of the argument from the impossibility of an infinite regress. For example, the Sunni Islamic philosopher Al-Ghāzāli argued that everything that begins to exist requires a cause of its origin. The world is composed of temporal phenomena preceded by other temporally ordered phenomena. Since such a series of temporal phenomena cannot continue to infinity because an actual infinite is impossible, the world must have had a beginning and a cause of its existence, namely, God. This version of the argument entered the medieval Christian tradition through St. Bonaventure (1221–74) in his *Sentences*.[88]

Although the cosmological argument does not figure prominently in Asian philosophy, an abbreviated version of it can be found in the *Nyāyakusumāñjali* of the 10th century Hindu thinker Udayana.[89] In general, philosophers in the Nyāya (literally "rule or method of reasoning") Hindu spiritual tradition argued that since the universe has parts that come into existence at one time and not another, it must have a cause.[90] One could admit an infinite regress of causes if we had evidence, but lacking such evidence, it necessarily follows that God must exist as the non-dependent cause. Many of the objections to this argument claim that God is an inappropriate cause of the universe because of God's nature. For example, since God is immobile and has no body, God cannot properly be said to cause anything at all, much less all of creation. The Naiyāyikas reply that God could assume a body at certain times, and, in any case, God need not create in the same way humans do. It is, however, in medieval Christian and Islamic theology and philosophy the cosmological issues and arguments for God's and the universe's existence become most influential.

Logos in Early Christianity

The *logos* that originated in ancient Western mythology and philosophy, while also present, to varying degrees, in ancient Indian, Egyptian, and Persian thought, became particularly significant in Christian writings and doctrines, which attempt to describe or define the role of Jesus Christ as the agent in the divine trinity that is active in the creation and continuous structuring of the cosmos and in revealing the divine plan of salvation to humanity.

In the first chapter of the *Gospel of John*, Jesus Christ is identified as "the Word" (Greek: *logos*) incarnated, or made flesh. This identification of Jesus with the *logos* is based on such Old Testament concepts of revelation as implied by the phrase "the Word of the Lord"—which connoted ideas of God's activity and power—and the Jewish view that Wisdom is the divine agent that draws humanity to God and is identified with the Word of God.[91]

The author of the *Gospel of John* may have used *logos* to emphasize the redemptive character of the person of Christ, who is described as "the way, and the truth, and the life." Just as the Jews had viewed the Torah (the Law) as preexistent with God, so also may have the author of *John* viewed Jesus as the personified source of life, the incarnate *logos*, and the illumination of humankind. *Logos* is therefore inseparable from *the person* of Jesus and not just the revelation that Jesus proclaims. Such early Christian theologians as St. Jerome (c. 347–420 CE) and St. Augustine referred to *logos* as *the living word. Logos* was also later used in Sufism—a mystical and ascetic Muslim community arising during the days of early Islam—and, in the 20th century, in the analytical psychology of Carl Jung.

St. Augustine

Augustine (354–430 CE), Bishop of Hippo, also known as St. Augustine, was perhaps the greatest and most influential Christian philosopher of the ancient Western world. He was born in Roman North Africa (modern Souk Ahras in Algeria). His mother Monica (died 388), a devout Christian, seems to have exerted a deep influence on his religious development.

As a saint of the Catholic Church, Augustine's authority in theological matters was universally accepted in the Latin Middle Ages and remained virtually unchallenged in the Western Christian tradition until the 19th century. The impact of his views on sin, grace, freedom, sexuality, time, the soul, *and, especially, the two conflicting worlds—the "City of God" and the "Earthly City"*—on Western culture was considerable. But some of his views, especially regarding sin, sexuality, and free will were deeply at variance with ancient Greco-Roman philosophical and cultural traditions and have provoked fierce criticism and vigorous opposition from the late 18th-century Enlightenment on, especially by proponents of humanist, liberal, feminist, and secular positions.[92]

Augustine's most famous work, the *Confessions* (Latin: *Confessiones*), is unique in the premodern Occidental literary tradition and greatly influenced the modern tradition of autobiography. Augustine is often considered by historians as the first medieval philosopher. But even though he was born several decades after the emperor Constantine I had terminated the anti-Christian persecutions and, in his mature years, saw the anti-pagan and anti-heretic legislation, which virtually made Christianity—called "Roman Catholicism" after the Protestant Reformation of the 16th century—the official religion of the Roman Empire, Augustine did not live in a "medieval" Christian world. Pagan religious, cultural, and social traditions were much alive, as he often deplored in his sermons, and his own cultural outlook was, like that of most of his learned upper-class contemporaries, shaped by the classical Latin authors, poets, and philosophers he had studied long before he encountered the Bible and Christian writers. Augustine continuously engaged intellectually with pre- and non-Christian philosophy. Neo-Platonism in particular remained an important of his thought. Accordingly, Augustine was a Christian philosopher of late antiquity shaped by and in constant dialogue with the classical Greco-Roman intellectual tradition.

After moving to Milan, in 386 CE, Augustine converted to ascetic Christianity, for which he has become known as its principal defender. In 410, when the city of Rome had been sacked by Alaric and the Goths, he wrote The *City of God*, Augustine's great apology (defense, in Plato's sense of the term) of Christianity. Augustine's life ended when the Vandals besieged Hippo; he is said to have died with a word of Plotinus on his lips.

The *Confessions* shows how an individual life—Augustine's own—is illuminated by God's providence and grace, and the possibility of salvation. His apologetic treatise *The City of God* (Latin: *De civitate dei*, begun in 412, two years after the sack of Rome, and completed in 426) argues that happiness can be found neither in the Roman nor in the Greco-Roman philosophical tradition but only through membership in the city of God, whose founder is Christ. *The City of God* also has sections on the secular state and on a good Christian's life in a secular society.

Augustine inherited from classical Greek philosophy the notion that philosophy is "love of wisdom" (*Confessions*, 3.8), i.e., an attempt to pursue happiness—or, salvation—by seeking knowledge of the nature of things and living accordingly. He became convinced that the true philosopher is a lover of God because true wisdom is, in the last resort, identical with faith in God. This is why Augustine thought that Christianity is "the true philosophy." Whereas many contemporary thinkers tend to regard faith and reason as alternative or even mutually exclusive ways to (religious) truth, in Augustine's view, the two are inseparable.

Following the classical Greek philosophical tradition, Augustine thought that the human being is a compound of body and soul and that the soul (Latin: *anima*)—conceived as both the life-giving element and the center of consciousness, perception, and thought—is, or ought to be, the ruling part. The rational soul should control the sensual desires and passions; it can become wise if it turns to God, who is at the same time the Supreme Being and the Supreme Good. The human soul, which is mutable in time but immutable in space, occupies a middle position between God, who is unchangeable immaterial being, and our bodies, which are subject to temporal and spatial change. The soul is of divine origin and even god-like; it is not divine itself but is created by God.

Love is an indispensable notion in Augustine's ethics. It is closely related to virtue and often used synonymously with will or intention (*intentio*). Augustine's reference point is the biblical command to love God and one's neighbor (*Matthew*, 22.37; 39). And, although other Roman-era philosophers, especially the Stoic Seneca, had made use of the concept of will (Latin: *voluntas*) before him, Augustine comes closer than any earlier philosopher to positing will as a faculty of choice that is reducible neither to reason nor to non-rational desire. It has therefore been claimed that Augustine "discovered" the will.

From the Middle Ages onwards, Augustine's theology of grace has been regarded as the heart of his Christian teaching. His conviction that human beings in their present condition are unable to do or even to will the good by their own efforts is his most fundamental disagreement with ancient, especially Stoic, virtue ethics. After and because of the disobedience of Adam and Eve—the "original sin" signaling our "fall" from innocence to the road to perdition—we have lost our natural ability of self-determination, which can only be repaired and restored by the divine grace that has manifested itself in the incarnation and sacrifice of Christ and works to free our will from its enslavement to sin. Confession of sins and humility are, therefore, basic Christian virtues, in sharp contrast with the sinful pride that puts the ego-centered self in the place of God and was at the core of the evil angels' primal sin.

Augustine's *City of God* is an extended plea to persuade people "to enter the city of God or to persist in it." The criterion of membership in the city of God (a metaphor Augustine takes from the biblical Psalm 46) and its antagonist, the earthly city, is right or wrong love. A person belongs to the city of God if and only if he or she directs their love toward God, even at the expense of self-love, and belongs to the earthly city, or city of the devil, if and only if they subordinate love of God for self-love, proudly making themselves the greatest good. Augustine argues that real, enduring happiness, which is sought by every human being, cannot be found outside the city of God founded by Christ.

In complete contrast with Christian happiness, according to Augustine, are the conceptions of happiness in the Roman political tradition—which equates happiness with the prosperity of the Empire, thus falling prey to the "evil demons" who posed as the defenders of Rome but in fact ruined it morally and politically—and in Greek, especially Platonic, philosophy which, despite its insight into the nature of God, failed to accept the mediation of Christ incarnate and turned to false mediators, i.e., "deceptive demons," in Augustine's words.

Augustine's approach to world history is scriptural, creationist, and eschatological; it purports to cover all history, from its beginnings until the end of the world.[93] *The history of the two cities begins with the creation of the world, the defection of the devil and the sin of Adam and Eve*; it continues with the providentially governed vicissitudes of the People of Israel (the first earthly representative of the city of God) and, after the coming of Christ, of the Church (*City of God*, bks. 15–17); and it ends with the final destination (*finis*, to be understood both ethically as "ultimate goal" and eschatologically as "end of times") of the two cities in eternal damnation and eternal bliss (*City of God*, bks. 19–22).

To a great extent, Augustine's approach is exegetical.[94] For him, the history of the city of God is essentially sacred history as laid down in Scripture. *However, the heavenly and earthly cities must not be completely identified with the worldly institutions of the church and the state.*

As a principal founder of what has come to be called "Just War Theory," Augustine also argued that war results from sin and is the privileged means of satisfying our lust for power. Nevertheless, Augustine wrote a letter to refute the claim that Christianity advocated a politically impracticable pacifism (*Letter*, 138). His Christian reinterpretation of the traditional Roman Just War Theory should be read in the framework of his general theory of virtue and peace. To be truly just, according to the standards proposed by Augustine, a war would have to be waged for the benefit of the adversary and without any vindictiveness, in short, out of love of neighbor, which, in a fallen world, seems utopian (*Letter*, 138.14). Wars may, however, be relatively just if they are defensive and properly declared, or if commanded by a just authority, not necessarily the wars of the People of Israel, which, according to Augustine, were commanded by God himself.

Creation and Time

Just as the Neo Platonists developed their cosmological thinking by commenting on Plato's *Timaeus*, Augustine's natural philosophy, as depicted in his *Confessions*, is largely a theory of creation based on his exegesis of the opening chapters of the book of *Genesis,* the beginning of the *Old Testament;* God does not create in time but creates time together with changeable being while existing in timeless eternity himself. Creation occurs instantaneously; the seven days of creation are not to be taken literally but are a didactic means to make plain the intrinsic order of reality. Changeable being is not generated from God (which, according to the Nicene Creed of the early 4th century, is true only of God's son, Jesus Christ) but created out of nothing, which, in Augustine's view, partly accounts for its susceptibility to evil.

God "first" creates formless matter out of nothing and "then" forms it by conveying to it the rational principles (Latin: *rationes*) that eternally exist in his mind, or, as Augustine put it, in God's Word (*logos*), i.e., the Second Person of the Trinity, Jesus Christ. This formative process is Augustine's exegesis of the biblical "word" (*logos*) of God (Genesis 1:1 and John 1:1). Incorporeal and purely intellectual beings, i.e., the angels, are created from intelligible matter, which is created out of nothing. Corporeal being is created when the Forms, or rational principles contained in God and contemplated by the angels, are externalized so as to generate not only intelligible but also physical matter.

Augustine also posed the question, "What is time?" but did not provide a clear definition. Whereas his accounts of time in *The City of God* focused on cosmic or physical time, in his *Confessions,* Augustine described *how we experience time* from a first-person perspective and what it means for us and our relationship to ourselves and to God to exist temporally. Augustine argued that although the three familiar "parts" of time—past, present, and future—do not really exist (the past having ceased to exist, the future not existing yet, and the present being without extension), *time has subjective reality for us*. This is because time is present to us in the form of our present memory of the past, our present attention to the present, and our present expectation of the future. The stability of this precarious unity is however dependent on God—we cannot make sense of the memory of our life unless we perceive the ceaseless presence of God's providence and grace in it. Toward the end of his *Confessions*, Augustine exhorted his readers to turn from the distractions of temporal existence to the timeless eternity of God, which alone guarantees truth and stability.

Augustine's lifework is a grand attempt to reconcile classical Greek and Neoplatonic philosophies with early Christian theology. *Augustine translated and transmitted Plato's concept of Truth and its relationship to the natural world into Christian terms. For Augustine, however, this world, the City of Earth, is a shadow, a fallen version of the eternal City of God,* and the prideful pursuit of knowledge by Adam and Eve, whose "original sin" has set humanity on the road to perdition. *His dualistic view of the universe in general, divided between the "world of the spirit" ruled by God, and the "world of the flesh," governed by Satan, has remained foundational not only for Neoscholasticism and Neothomism—theoretical reactions by late 19th-century conservative Catholic theologians against the Enlightenment and European Idealism—but also, in the 20th century, for reactionary versions of Christian fundamentalism.*

On the other hand, Augustine's philosophy of time has inspired such distinguished secular philosophers as the phenomenologist Edmund Husserl (1859–1938), as well as Martin Heidegger (1889–1976) and Paul Ricœur (1913–2005). The noted political theorist Hannah Arendt (1906–75) wrote her doctoral dissertation on Augustine's philosophy of love. Ludwig Wittgenstein (1889–1951), widely considered by "professional philosophers" to have been the most influential philosopher of the 20th century, in his *Philosophical Investigations* critically assessed what he considered to have been Augustine's view of language and language acquisition. And such contemporary philosophers of religion as Jean-Luc Marion (1946–) and John Milbank (1952–) have favorably contrasted Augustine's notion of self, in which love is perceived as an integral part, against the purported egoism and isolation of the Cartesian subject, which is considered the hallmark and the "birth defect" of "Modernity." On the whole, however, outside conservative Christian circles, contemporary secular Western culture has little sympathy for Augustine's yearning for an inner divine light or for his less-than-optimistic views about human nature and the world as a whole.

Aquinas and Medieval Scholasticism

St. Thomas Aquinas (1225–74) made his intellectual lifework a grand attempt to reconcile Aristotelian philosophy with medieval Christian theology. He lived at a critical juncture of Western culture, when the arrival of the Aristotelian *corpus* in Latin translation reopened the question of the relation between faith and reason.

After Aquinas joined the Dominican Order while living in southern Italy, he went to Paris to study with the German philosophical theologian St. Albertus Magnus (c. 1200–85), among whose many works were commentaries on the Aristotelian *corpus*, and who argued that Aristotle's natural philosophy was compatible with Christian doctrine. Thomas defended the mendicant orders and also countered both the Averroistic (see below) interpretations of Aristotle and the Franciscan tendency to reject Greek philosophy. The result was a new *modus vivendi* between faith and philosophy, which survived until the rise of the new physics in the 16th century. The Catholic Church has over the centuries regularly and consistently reaffirmed the central importance of Thomas's work, both theological and philosophical, for understanding its teachings concerning Christian revelation, and his close textual commentaries on Aristotle remain virtually unrivaled in Western philosophy.[95]

When Aquinas returned to Italy from Paris, he completed his most well-known works, the *Summa contra gentiles*, and the *Summa theologiae*. In 1268, in Rome, he wrote *On the Soul*, and during the next five or six years commented on 11 more Aristotelian works. During this time, he was also writing such polemical works as *On the Eternity of the World* and *On There Being Only One Intellect*. To distinguish between philosophy and theology, Aquinas put the difference this way:

> the believer and the philosopher consider creatures differently. The philosopher considers what belongs to their proper natures, while the believer considers only what is true of creatures insofar as they are related to God, for example, that they are created by God and are subject to him, and the like.
>
> (Summa contra gentiles, Bk II, chap. 4)

Aquinas and Aristotle

As a philosopher, Thomas was clearly an Aristotelian, to such an extent that he famously referred to Aristotle as "The Philosopher." He adopted Aristotle's analysis of physical objects, his view of place, time, and motion, his proof of the prime mover, and his cosmology. He also held to Aristotle's account of sense perception and intellectual knowledge. His moral philosophy is closely based on what he learned from Aristotle, especially the latter's *Metaphysics*. Augustine and the Neoplatonists also greatly influenced Aquinas's attempt to make philosophy and Christian theology compatible—especially their conceptions of the mind/soul, and the universe as a whole.

Like Aristotle, Aquinas understood "philosophy" to be an umbrella term that covers an ordered set of sciences. Philosophical thinking is characterized by its argumentative structure. A practical use of the mind has as its object the guidance of some activity other than thinking—e.g., choosing in the case of moral action, or creating some product in the case of art. *Whereas practical mental activity aims to change the world, the theoretical use of the mind has truth as its object and seeks not to change but to understand it.* Also, like Aristotle, Aquinas declared that there is a plurality of both theoretical and practical sciences. Ethics, economics, and politics are practical sciences, while physics, mathematics, and metaphysics are theoretical sciences.

Aquinas, following both classical Greek idealism and Christian doctrine, argued that the soul is capable of existing apart from the living body after the body's death, because the soul is incorruptible. Its "incorruptibility" is due to the fact that mental understanding and willing are not the results of any bodily organ, but of the "rational form" of the human animal.

According to Aquinas, living things have various capacities to survive despite the ravages of the natural world around them; that is, in part, what it means to live is to sustain one's existence by one's own natural activities. And yet in nature, corruptible things inevitably die, and so do human beings. On the other hand, according to Aquinas, human souls are incorruptible, whereas the souls of other living beings are not.

The Material and Metaphysical Worlds, From Nature to Divinity in Aquinas

Like Aristotle, Aquinas asked: is there any science beyond natural science and mathematics? In doing natural philosophy, or what Aristotle called *Physics*, one gains the knowledge that not everything that exists is material. At the end of the *Physics*, Aristotle argued from the nature of moved movers that they require a first unmoved mover. Again, following Aristotle, Aquinas attempted to prove that there is a first mover of all moved movers, *and that the "unmoved mover" is not itself material, and does not exist within the material world, or what we call the physical universe.* For Aquinas, the activity of our intellects (or minds) provides sufficient grounds for demonstrating that, since the human soul is the substantial form of the body, it can exist apart from the body, that is, survive death, and is therefore an "immaterial existent." The Prime Mover and the immortal souls of human beings demonstrate that to exist, and to be material, are not identical. This is preeminently the case with the Prime Mover for Aristotle, and with God for Aquinas, and, therefore, for Roman Catholic theology as a whole.

According to Aquinas, God created an ordered natural world. God also created humanity's ability to use reason to understand the natural world, thereby celebrating the will of God as manifested in his creation, the universe as a whole. God could be an instantaneous and motionless creator, and could have created the world without preceding it in time. To Aquinas, *that the world began was an article of faith.*

The Enduring Influence of Aquinas's Worldview

In 1879, about 600 years after the death of Aquinas, Pope Leo XIII issued the encyclical *Aeterni Patris (Of the Eternal Father)*, and called for the revival of the study of Thomas Aquinas. This pope put forward Aquinas as the quintessence of philosophy, in strong contrast with what the traditional Roman Catholic theologians regarded as the vagaries of modern thought since Descartes. There ensued a sustained global revival of the works of Aquinas, especially by such prominent French Catholic thinkers as Jacques Maritain (1882–1973) and Etienne Gilson (1884–1978). The Second Vatican Council, or Vatican II (1962–5), convened by Pope John

XXIII, catalyzed sweeping reforms in the Catholic Church and also drew that Thomistic revival largely to a close.

In this postconciliar period following the end of the Second Vatican Council, many thinkers inside and outside the Roman Catholic Church have struggled to keep the faith at the same time that is widely believed that modernity in general, and the Enlightenment project to contribute to human progress through rationalism and science in particular, have failed. Subsequently, Catholics and non-Catholics alike have resumed invoking the thought of Aquinas in part to counter the perceived relativism and nihilism following the collapse of previously held "certainties." In 1998, for example, John Paul II issued an encyclical called *Fides et Ratio (Faith and Reason)*, which some have regarded as the charter of the Thomism for the third millennium after the death of Christ. How long a Thomistic revival of sorts endures is unclear. What is incontestable, however, is the enduring influence of Thomistic Aristotelianism, not just on Christianity, but on early Islamic and medieval Jewish cosmologies as well.

Medieval Islamic and Jewish Worldviews

The terms "Muslim world" and "Islamic world" commonly refer to the global *Islamic community* (*Ummah*), consisting of all those who adhere to the religion of Islam, or to societies where Islam is practiced.[96] In a contemporary geopolitical sense, these terms refer to countries where Islam is widespread.

The history of the Muslim world spans about 1400 years and includes a variety of sociopolitical developments, as well as advances in the arts, science, philosophy, and technology, particularly during the Islamic Golden Age (from about the 8th to the 14th centuries, roughly comparable to the timeframe of the Medieval World in the West).[97]

All practicing Muslims look for guidance to the *Quran* and believe in the prophetic mission of Muhammad, but disagreements on other matters have led to the appearance of different religious schools and branches within Islam. In the modern era, most of the Muslim world came under the influence of, or colonial domination by, European powers, which has profoundly affected their views of Occidental civilization, of which Muslims are often highly critical. On the other hand, some of the most prominent and influential Islamic theologians and philosophers—including Avicenna and Averroes—recognized and even venerated the contributions to their thought and traditions made by Western intellectuals, especially by "The Philosopher," Aristotle.

Avicenna

Abū-ʿAlī al-Ḥusayn ibn-ʿAbdallāh Ibn-Sīnā (Avicenna, c. 970–1037) is widely considered to be among the preeminent philosophers and physicians of the early Islamic world. In his work, he combined the disparate strands of philosophical/scientific thinking in late Greek antiquity and early Islam into a rigorous system that attempted to explain all reality, including the tenets of revealed religions. To some degree, this represents the culmination of the Hellenic tradition, defunct in Greek itself after the 6th century, but reborn in Arabic in the 9th century. Avicenna's work dominated intellectual life in the Islamic world for centuries to come. In Latin translation, beginning in the 12th century, Avicenna's philosophy also greatly influenced many late medieval and early Renaissance philosophers and scholars, just as the Latin translation of his medical texts formed the basis of medical instruction in European universities until the 17th century.[98]

The scientific and metaphysical view of the universe in Avicenna's day was largely framed by the Aristotelian view of the world combined with Ptolemaic cosmology and Neoplatonic cosmogonic emanationism.[99] According to Avicenna, all "intelligibles" (universal concepts, or "the forms of things as they are in themselves") were the eternal objects thought by "the First Principle," and then, in descending hierarchical order, by the intellects of the celestial spheres emanating from the "First Principle" and ending with the "active intellect" of the terrestrial realm. He attempted to demystify such concepts as inspiration, enthusiasm, mystical vision, and prophetic revelation, explaining them as natural functions of the rational soul. Systematic logical analysis was, according to Avicenna, accompanied by emotional states of joy and intellectual pleasure. The highest level of reflection was that of the Prophet Muhammad, who acquired the "intelligibles."

In the emanative language Avicenna inherited from the Neoplatonic tradition, and which he incorporated in his own understanding of the cosmology of the concentric spheres of the universe with their intercommunicating intellects and souls, he referred to the flow of knowledge from the supernal world to the human intellect as "divine effluence." This is possible because of the consubstantiality and congeneric nature of all intellects, human, and celestial alike.

According to Avicenna, this is also the case for other forms of communication from the supernal world. The prophet Muhammad acquired all the "intelligibles" comprising knowledge, because the intellective capacity of his rational soul was extraordinary and was coupled with a highly developed internal sense of imagination that translated this intellective knowledge into language and images (in the form of a revealed book) that the vast majority of humans can easily understand. The divine effluence from the intellects and the souls of the celestial spheres also includes information about events on earth, past, present, and future—which Avicenna calls "the unseen"—for which the intellects and souls of the celestial spheres are directly responsible.

Avicenna argued that the logistics of the reception of information from the supernal world varies, but the recipient has to be ready and predisposed to receive it. All humans have both the physical and mental apparatus to acquire intelligible and supernal knowledge and the means to do so, but they have to work for it, just as they have to prepare for their bliss in an afterlife in the eternal celestial spheres (Avicenna's "paradise"), while their immortal rational souls are still within the body.

Avicenna's analysis of the of the "supernal" (heavenly, divine, or celestial) and "sublunary" worlds, as well as his description of the rational soul, while grounded in Aristotelian theory, also go beyond it by supplementing Aristotelian metaphysics and cosmology with what he considered theological revelations. In addition, Avicenna also addressed the cognitive side of religion as well as the social component of practical reason, i.e., ethics. His vision of the universe in general and of our world in particular would endure for centuries of Islamic inquiries into the nature of the universe, the structure of our souls, and the practical arts of healing the body and caring for the soul.

Averroes

Ibn Rushd (Averroes, 1126—98) is still regarded by many Muslim and non-Muslim commentators and historians as the foremost Islamic philosopher. In the Islamic world, he played a decisive role in the defense of Greek philosophy against the views of Sunni religious philosopher al-Ghazali (c. 1056–1111), and in the rehabilitation of Aristotle, thereby contributing to the rediscovery of "the philosopher" in the Western world, after centuries of the nearly complete absence of Aristotle's thought from Occidental intellectual discourse.[100] That discovery was instrumental in launching late medieval Latin Scholasticism and, later, in Renaissance philosophy. Still, there has been comparatively little attention to Averroes' work in the English-speaking world, although greater interest has been shown in France.[101]

Averroes claimed that there is *no incompatibility* between religion and philosophy if both are properly understood. In his work *The Incoherence of the Incoherence* (Arabic: *Tahafut al-tahafut*), he defended Aristotelian philosophy against al-Ghazali's claims in *The Incoherence of the Philosophers* (Arabic: *Tahafut al-falasifa*).[102] Among his major works are *On the Soul, Physics, On the Heavens, On Generation and Corruption*, and the *Meteorology.* Averroes's *Physics*, his major treatise on natural philosophy, was his interpretation and revision of Aristotle's view of the cosmos and the origin of the world.

In *The Incoherence of the Incoherenc*e, Averroes he delineates his view of creation and, in his critique of al-Ghazzali's cosmology, and he also deals with the latter's argument against Aristotle's doctrine of the eternity of the physical universe. Ghazzali claimed that the Western philosophers had misunderstood the relationship between God and the world, especially since the *Qur'an* is clear on divine creation. Ghazzali questioned why God, being the ultimate agent, could not simply create the world *ex nihilo* and then destroy it at some future point in time. Averroes replied that the eternal realm works differently from the temporal sphere. As humans, we can decide to perform some action and then wait a period of time before completing it. For God, on the other hand, there can be no gap between decision and action; since for Him there is no differentiation between one time and another. Being omniscient and omnipotent, God also would have known from the eternal past what he had planned to create.

Averroes also argues that the universe, according to the human mind, works along certain causal principles, and the beings existing within the universe contain particular natures that define their existence. If these natures, principles, and characteristics of living beings were not definitive, this would lead to nihilism (i.e., what Averroes calls the "atheistic materialists" found in Greek and Arab philosophy and theology). *For Averroes, the world can neither be labeled preeternal nor originated, since the former would imply that the world is uncaused and the latter would imply that the world is perishable. Furthermore, according to Averroes, God's creation of the world is not simply the choice between two equal alternatives, but a choice between existence and non-existence.*

In the West, the "Latin Averroists," a group of philosophers writing in Paris in the middle of the 13th century supported Averroes's and Aristotle's doctrine of the eternity of the world against the position of the Scholastic philosopher Bonaventure and other theologians who argued the world can be logically proved to have begun in time.

Maimonides

Moses ben Maimon (1138–1204, known to English-speaking audiences as Maimonides) is the best-known medieval Jewish philosopher. His 14-volume compendium of Jewish law, the *Mishneh Torah*, established him as the leading rabbinic authority of his era and possibly of ours as well. His major philosophical work, the *Guide of the Perplexed*, is a sustained treatment of Jewish thought and practice, and is an effort to resolve the conflict between religious inspiration and secular knowledge. Although heavily influenced by the Neo-Platonized Aristotelianism that had taken root in Islamic circles, Maimonides departed from Aristotelian thought by emphasizing the limits of human knowledge and the questionable foundations of significant parts of Aristotle's astronomy and metaphysics. Like Avicenna, Maimonides was also a medical doctor who wrote treatises on a number of diseases and their cures.

Also like many major Islamic thinkers, Maimonides borrowed heavily from Aristotle, and he esteemed Aristotle's work as the pinnacle of unaided reason. Although Maimonides used many of Aristotle's concepts and categories, he challenged Aristotle's arguments in the *Physics* that "everything in existence comes from a substratum," and for the eternity of the universe, on the basis that Aristotle's reliance on induction and analogy is a fundamentally flawed means of explaining unobserved phenomena.

Maimonides analyzed what he took to *be the three main intellectual approaches to account for the world. They are: (1) the world is a free act of creation ex nihilo* (Latin: "out of nothing")*; (2) the world arises through the imposition of form on preexisting matter, and (3) the world is an eternal emanation, which is the Neoplatonic claim that the world did not come into being ex nihilo* or *de novo* (Latin: "anew"). Maimonides did not claim to have proof that God created the world *ex nihilo* and *de novo.* He also did not hold that Aristotle's claim that time is infinite was provable. Neither did he claim that he could refute the second and third approaches mentioned above. Instead, Maimonides held that we should accept the Biblical story of creation, interpreted in philosophical terms.

For Maimonides, creation is so important because the First Cause, *God (derived from Aristotle's "unmoved mover"), brought the world into existence through benevolence and wisdom, which are reflected in the created order*. By studying the created order, we can expand our understanding of God. Revelation is crucial because human beings receive help through divine graciousness.

Maimonides also believed that the Torah contains philosophical wisdom and that philosophy affords the best way to understand the Torah.[103] Thus, creation, revelation, and redemption are at the very core of Maimonides' understanding of reality as a whole. Through the Torah, human beings are provided with direction to perfection. This includes guidance regarding repentance and how to return to God when one sins. Redemption—understood by Maimonides as the culmination of providence—means that the created order is under divine governance. There is thus "ultimate" or "cosmic" justice. Human beings may not fully understand the wisdom and goodness of the created order (as in the biblical story of *Job*, for example), but they can be confident that it is governed by divine reason and justice.

Maimonides developed an original conception of how a tradition anchored in revelation—Judaism—can be understood in philosophical terms. Succeeding generations of philosophers and theologians wrote extensive commentaries on his works, which have influenced such major Western figures as Aquinas, Spinoza, Leibniz, and Newton.[104]

From the Cosmos to the World

In this chapter, I have attempted to provide an overview of the development of some major philosophical, scientific, and theological concepts of the world. Over the course of about 2500 years, what the ancient Greeks referred to as "the cosmos"—an orderly, imperceptible, unitary structure underlying the appearance of a pluralistic disorder of perceptible material things—was gradually transformed into "the world" to which we now refer.

That world has been conceived by theologians as God's creation, by scientists as the totality of entities on planet Earth, and by philosophers as that within which human beings come to be and pass away. The further development of the idea of the world, and of our place in the cosmos, would await the philosophical and scientific revolutions of the 17th and

18th centuries on the one hand, and the global expansion of and colonization by Occidental intellectual, economic, and political interests on the other hand.

Notes

1 See William A. Green, "Periodization in European and World History," *Journal of World History*, Vol. 3, No. 1 (Spring, 1992), pp. 13–53, University of Hawai'i Press on behalf of the World History Association Stable, www.jstor.org/stable/20078511. Also see: https://en.wikipedia.org/wiki/Early_modern_period; https://en.wikipedia.org/wiki/Modernity; and www.e-ir.info/pdf/67006.

2 Martin Heidegger, *Four Seminars* (Bloomington, IN: Indiana University Press, 2003), 38. One might regard Heidegger's "naturalistic," possibly Romanticized reading of early Greek philosophy as infused with his own views and not simply an exegesis of the relevant texts.

Also see Mark Payne, "The Natural World in Greek Literature and Philosophy," *Oxford Handbooks Online*, April, 2014, www.oxfordhandbooks.com/view/10.1093/oxfordhb/9780199935390.001.0001/oxfordhb-9780199935390-e-001.

3 These words from the *Critique of Practical Reason* are on Kant's tombstone near the cathedral of Kaliningrad, Russia (the former Prussian city of Königsberg).

4 The English noun cosmogony is derived from the ancient Greek *cosmogonia* (from the Greek noun *kosmos*, meaning "cosmos, the world") and the root of *gognomai/gegona* ("to come into a new state of being"), https://en.wikipedia.org/wiki/Cosmogony; and https://science.howstuffworks.com/dictionary/astronomy-terms/cosmogony-info.htm.

5 The word *ekpyrosis* (derived from the Ancient Greek, meaning "conflagration") refers to an ancient Stoic cosmological model in which the universe is caught in an eternal cycle of fiery birth, cooling, and rebirth. This theory addresses the fundamental question that remains unanswered by the Big Bang inflationary model: "What happened *before* the Big Bang?" According to the ekpyrotic theory, the Big Bang was actually a big *bounce*, a transition from a previous epoch of contraction to the present epoch of expansion. The key events that shaped our universe occurred before the bounce, and, in a cyclical version, the universe bounces at regular intervals. https://en.wikipedia.org/wiki/Ekpyrotic_universe.

6 See Edward J. Wollack, "Cosmology: The Study of the Universe," Universe 101: Big Bang Theory. *NASA*, December 10, 2010; and Sean Carroll, "A Universe from Nothing?" *Science for the Curious*, April 28, 2012. Also see: www.quantamagazine.org/physicists-debate-hawkings-idea-that-the-universe-had-no-beginning-20190606/, https://en.wikipedia.org/wiki/M-theory; www.space.com/17594-string-theory.html; and www.space.com/25075-cosmic-inflation-universe-expansion-big-bang-infographic.htmld.

7 Cassirer's major analyses of myth are in the second volume of his three volume work, *The Philosophy of Symbolic Forms*, titled *Mythical Thought,* trans. Ralph Manheim (New Haven, CT: Yale University Press, 1955); as well as in *The Myth of the State* (New Haven, CT: Yale University Press, 1946), esp. p. 280, where this quote appears. Also see Donald Verene "Cassirer's View of Myth and Symbol" *The Monist*, Vol. 50, No. 4, Symbol and Myth (October, 1966), 553–64. Oxford University Press, www.jstor.org/stable/27901663; and Jonathan Friedman, "Myth, History, and Political Identity," *Cultural Anthropology*, Vol. 7, No. 2 (May, 1992), 194–210. Wiley on behalf of the American Anthropological Association, www.jstor.org/stable/656282.

8 See https://en.wikipedia.org/wiki/Epic_of_Gilgamesh; and https://en.wikipedia.org/wiki/Pyramid_Texts.
9 See https://en.wikipedia.org/wiki/Enûma_Eliš.
10 See https://norse-mythology.org/cosmology/ginnungagap.
11 The *Book of Genesis*, the first book of the Hebrew *Bible* and the *Old Testament,* The name "Genesis" is from the Latin Vulgate, the commonly used late 4th-century Christian translation of the Hebrew, which was borrowed or transliterated from Greek "Genesis," meaning "Origin"; Hebrew "*Bərēšīṯ*," meaning "In [the] beginning," https://en.wikipedia.org/wiki/Book_of_Genesis; and www.vatican.va/archive/bible/genesis/documents/bible_genesis_en.html.
12 See https://en.wikipedia.org/wiki/Hindu_views_on_evolution; and https://en.wikipedia.org/wiki/Vedic_period.
13 See https://en.wikipedia.org/wiki/Theogony. The traditional *mythos* of Ancient Greece, was primarily a part of its oral tradition. The Greeks of this era eventually became a literate culture for the small educated minority of the populations of its city-states (*polises*) but produced no sacred texts. There were no definitive or authoritative versions of myths recorded in texts and preserved in an unchanging form. Instead, as in many ancient cultures, multiple variants of myths were in circulation. These variants were adapted into songs, dances, poetry, and visual art. Performers of myths could freely reshape their source material for a new work, adapting it to the needs of a new audience or in response to a new situation. Children in Ancient Greece were familiar with traditional myths from an early age. According to Plato, mothers and nursemaids narrated myths and stories to the children in their charge.
14 See www.britannica.com/topic/Greek-mythology/Types-of-myths-in-Greek-culture; and Joshua J. Mark, "Mythology." *Ancient History Encyclopedia*, 2018. Also see www.ancient.eu/mythology/.
15 The Greek noun *Logos* comes from the verb, *légō* meaning "I say" or "I speak." *Noos* (in Homeric Greek) and *nous* (in classical Attic Greek) probably derive from the verb *néō*, ("I spin"), here meaning "to spin the thread of the mind." See https://en.wiktionary.org/wiki/nous, and https://en.wiktionary.org/wiki/νόος.
16 See Charles Webel, *The Politics of Rationality Reason through Occidental History* (New York and London: Routledge, 2015), 19–20 and *passim*.
17 Hermann Diels and Walther. Kranz, *Die Fragmente der Vorsokratiker*, 6th ed. (Berlin: Weidmann, 1951–2), which will be abbreviated as DK: The Orpheus legend: DK B12; Demokritos: DK, A74, and Anaxagoras: DK, VII, 5–29.
18 See Dirk Baltzly, "Stoicism," *The Stanford Encyclopedia of Philosophy* (Spring 2019 Edition), Edward N. Zalta, ed., https://plato.stanford.edu/archives/spr2019/entries/stoicism/.
19 See Richard Bett, "Pyrrho," *The Stanford Encyclopedia of Philosophy* (Winter 2018 Edition), Edward N. Zalta, ed., https://plato.stanford.edu/archives/win2018/entries/pyrrho/.
20 For doxography, see Jaap Mansfeld, "Doxography of Ancient Philosophy," *The Stanford Encyclopedia of Philosophy* (Winter 2020 Edition), Edward N. Zalta, ed., https://plato.stanford.edu/archives/win2020/entries/doxography-ancient/.
21 See: DK, *op. cit*., Vol, I, 113–39; and G.S. Kirk, J.E. Raven, and M. Schofield, *The Presocratic Philosophers: A Critical History with a Selection of Texts*, 2nd Edition (Cambridge and New York: Cambridge University Press, 1983), 163–80; James Lesher, "Xenophanes," *The Stanford Encyclopedia of Philosophy*

(Summer 2019 Edition), Edward N. Zalta, ed., https://plato.stanford.edu/archives/sum2019/entries/xenophanes/, and https://history.hanover.edu/texts/presoc/Xenophan.html.

22 The Many-Worlds Interpretation of quantum mechanics holds that there are many worlds which exist in parallel at the same space and time as our own.

See Lev Vaidman, "Many-Worlds Interpretation of Quantum Mechanics," *The Stanford Encyclopedia of Philosophy* (Fall 2018 Edition), Edward N. Zalta, ed., https://plato.stanford.edu/archives/fall2018/entries/qm-manyworlds/; www.space.com/31465-is-our-universe-just-one-of-many-in-a-multiverse.html; and www.nature.com/articles/d41586-019-02602-8.

23 For Parmenides poem "On Nature," see: A.H. Coxon, *The Fragments of Parmenides: A Critical Text with Introduction, Translation, the Ancient Testimonia and a Commentary*. *Phronesis* Supplementary Volume iii (Assen/Maastricht: Van Gorcum, 1986). Other sources include: W.K.C. Guthrie, 1 *A History of Greek Philosophy*, I and II: *The Earlier Presocratics and the Pythagorean; and The Presocratic Tradition from Parmenides to Democritus* (Cambridge: Cambridge University Press, 1962 and 1965, respectively). Also see John Palmer, "Parmenides," *The Stanford Encyclopedia of Philosophy* (Winter 2020 Edition), Edward N. Zalta, ed., https://plato.stanford.edu/archives/win2020/entries/parmenides/ and www.iep.utm.edu/parmenides/.

24 See Aristotle's *Metaphysics*. 1.5.986^{b}14–18, and *Physics*. 1.2.184^{a}25-^{b}12, in J. Barnes, ed. *The Complete Works of Aristotle*, Volumes I and II (Princeton, NJ: Princeton University Press, 1984). A reliable translation of selections of Aristotle's works is T. Irwin, and G. Fine., eds., *Aristotle: Selections, Translated with Introduction, Notes, and Glossary* (Indianapolis, IN: Hackett, 1995). Aristotle's works are often referred by abbreviations.

25 Seehttps://archive.org/stream/in.ernet.dli.2015.46406/2015.46406.Nature-And-Human-Nature_djvu.txt; and https://archive.org/stream/monist01instgoog/monist01instgoog_djvu.txt.

26 See Carl Huffman, "Pythagoras," *The Stanford Encyclopedia of Philosophy* (Winter 2018 Edition), Edward N. Zalta, ed., https://plato.stanford.edu/archives/win2018/entries/pythagoras/.

27 See DK, *op. cit.*, Volume 1, Chapter 14, 96–105 https://en.wikipedia.org/wiki/Pythagoras; and Leonid. Zhmud, *Pythagoras and the Early Pythagoreans*, trans. Kevin Windle and Rosh Ireland (Oxford: Oxford University Press, 2012); and Robert Hahn, *The Metaphysics of the Pythagorean Theorem* (Albany: State University of New York Press, 2017).

The Pythagorean theorem (in China, it is called the "Gougu theorem") is a fundamental relation in Euclidean geometry among the three sides of a right triangle. It states that the area of the square whose side is the hypotenuse (the side opposite the right angle) is equal to the sum of the areas of the squares on the other two sides. This theorem can be written as an equation relating the lengths of the sides a, b, and c, often called the "Pythagorean equation:" $a^2 + b^2 = c^2$, where c represents the length of the hypotenuse and a and b the lengths of the triangle's other two sides. There are indications that this may have been known in other ancient cultures, including Egypt and Mesopotamia, in addition to ancient China, and that it was not "discovered" by Pythagoras. See https://en.wikipedia.org/wiki/Pythagorean_theorem.

Now known as the "Platonic solids," the five geometric solids whose faces are all identical, regular polygons meeting at the same three-dimensional angles. Also known as the five regular polyhedra, they consist of the tetrahedron (or pyramid), cube, octahedron, dodecahedron, and icosahedron. Pythagoras may have known about the tetrahedron, cube, and dodecahedron. See www.britannica.com/science/Platonic-solid.

Demonstrations of the Earth's sphericity, and its globular shape, were proposed to varying degrees by Eratosthenes (c. 276–194 BCE), who measured the Earth's circumference, Plato, Aristotle, and other Greek and Roman thinkers, but probably not by Pythagoras himself.

The theory of proportions is now attributed not to Pythagoras but to Eudoxus (c.350 BCE) and is preserved in Book V of Euclid's *Elements*. It established an exact relationship between rational magnitudes and arbitrary magnitudes by defining two magnitudes to be equal if the rational magnitudes less than them were the same. In other words, two magnitudes were different only if there was a rational magnitude strictly between them. This definition served mathematicians for two millennia and paved the way for the arithmetization of analysis in the 19th century. See www.britannica.com/science/analysis-mathematics/History-of-analysis#ref848197.

The discovery that the planet Venus is both the "evening star" and the "morning star," and not two separate orbs, has been attributed to Parmenides and other ancient thinkers, as well as to Pythagoras. See https://en.wikipedia.org/wiki/Venus.The "music of the spheres," or *musica universalis*, regards proportions in the movements of such celestial bodies as the Earth, Sun, and Moon as an inaudible harmonic, spiritual, or mathematical form of "music." According to some legends, Pythagoras discovered that consonant musical intervals can be expressed in simple intervals of small numbers. But it was Plato, and much later the early-modern German astronomer Johannes Kepler (1571–1630), who extended this model to the structure of the cosmos. See www.keplerstern.com/signature-of-the-celestial-spheres/music-of-the-spheres.

28 See K. Scarlett Kingsley and Richard Parry, "Empedocles." *The Stanford Encyclopedia of Philosophy* (Summer 2020 Edition), Edward N. Zalta, ed., https://plato.stanford.edu/archives/sum2020/entries/empedocles/.

29 See Patricia. Curd and Daniel H. Graham (eds.), *The Oxford Handbook of Presocratic Philosophy* (New York: Oxford University Press, 2009).

30 Fragments and doxographical reports about Leucippus are in DK. See also C.W. Taylor, *The Atomists: Leucippus and Democritus. Fragments, A Text and Translation with Commentary* (Toronto: University of Toronto Press, 1999). Also see: https://plato.stanford.edu/entries/atomism-ancient/; https://en.wikipedia.org/wiki/Leucippus, and the report on Democritus in: Diogenes Laertius, *Lives of Eminent Philosophers* (Loeb Classical Library), R.D. Hicks, trans. (Cambridge, MA: Harvard University Press, 1925), book 9, 34–49.

31 Sylvia Berryman, "Democritus," *The Stanford Encyclopedia of Philosophy* (Winter 2016 Edition), Edward N. Zalta, ed., https://plato.stanford.edu/archives/win2016/entries/democritus/.

32 The major source for Epicurean doctrine is Diogenes Laertius, *op. cit.* In his last book, devoted to Epicureanism, Diogenes preserved three of Epicurus' letters to his disciples. *Letter to Herodotus* summarizes Epicurus' physical theory, the *Letter to Menoeceus* offers a précis of Epicurean ethics, and the *Letter to Pythocles* treats astronomical and meteorological matters.

33 M.F. Smith, *Lucretius, 'De Rerum Natura,'* trans. W.H.D. Rouse, revised with new text, introduction, notes, and index (Loeb Classical Library) (London

and Cambridge, MA: 1975). Also see David Sedley, "Lucretius," *The Stanford Encyclopedia of Philosophy* (Winter 2018 Edition), Edward N. Zalta, ed., https://plato.stanford.edu/archives/win2018/entries/lucretius/.

34 See Salu Fisher, "Pierre Gassendi," *The Stanford Encyclopedia of Philosophy* (Spring 2014 Edition), Edward N. Zalta, ed., https://plato.stanford.edu/archives/spr2014/entries/gassendi/.

35 See www.jstor.org/stable/pdf/26213840.pdf, and www.marxists.org/archive/marx/works/1841/dr-theses/index.html.

36 See Charles Webel, *The Politics of Rationality*, *op. cit.*, 29.

37 Plato, *Meno* 89e–94e, in Plato's *Dialogues* and *Letters*, John M. Cooper, ed., *Plato: Complete Works* (Indianapolis, IN: Hackett, 1997). All textual references to Plato's works are from this edition.

38 See C.C.W. Taylor, and Mi-Kyoung Lee, "The Sophists," *The Stanford Encyclopedia of Philosophy* (Fall 2020 Edition), Edward N. Zalta, ed., https:plato.stanford.edu/archives/fall2020/entries/sophists/, and Webel, *op. cit.*, 28–30.

39 Plato, *Theaetetus* 151e and DK 80 B1.

40 Aristotle, *Metaphysics* 1007b20–22.

41 Plato, *Gorgias*, 483e.

42 Cicero, as quoted in Melissa Lane, "Ancient Political Philosophy," *The Stanford Encyclopedia of Philosophy* (Winter 2018 Edition), Edward N. Zalta, ed., https://plato.stanford.edu/archives/win2018/entries/ancient-political/.

43 See Debra Nails, "Socrates," *The Stanford Encyclopedia of Philosophy* (Spring 2020 Edition), Edward N. Zalta, ed., https://plato.stanford.edu/archives/spr2020/entries/socrates.

44 See Webel, *The Politics of Rationality*, *op. cit.*, 32, and Frederick Copleston, *A History of Philosophy*, Vol. 1 (Garden City, NY: Doubleday, 1968), 129–30.

45 See Webel, *The Politics of Rationality*, *op. cit.*, 30–7.

46 *Ibid.*

47 Martin Luther King, Jr. "Letter from a Birmingham Jail," cited in: https://lifeexaminations.wordpress.com/2010/09/19/socrates-according-to-martin-luther-king-jr/.

48 Nelson Mandala, *The Long Road to Freedom*, cited in: https://prismdecision.com/nelson-mandelas-inspiring-limestone-quarry-classroom/.

49 Translation of the *Al Qaeda Training Manual* is available at https://fas.org/irp/world/para/manualpart1.html.

50 The "Great Plague of Athens" of 430 BCE, during the Peloponnesian War, was one of many epidemics that plagued the ancient world and is strikingly similar in its political and psychological effects to contemporary pandemics: https://greektraveltellers.com/blog/the-plague-of-athens.

51 Alfred North Whitehead, *Process and Reality* (New York: Free Press, 1979), 39.

52 See www.iep.utm.edu/plato/.

53 Noted secondary sources and commentaries on Socrates and Plato include W.K.C. Guthrie, *A History of Greek Philosophy*, Volumes 4 and 5 (Cambridge: Cambridge University Press, 1975 and 1978); Gregory Vlastos, *Studies in Greek Philosophy* (Volume 2: *Socrates, Plato, and Their Tradition*), Daniel W. Graham (ed.) (Princeton, NJ: Princeton University Press, 1995.) Also see: Richard Kraut, "Plato," *The Stanford Encyclopedia of Philosophy* (Fall 2017 Edition), Edward N. Zalta, ed., https://plato.stanford.edu/archives/fall2017/entries/plato/.

54 The possible parallels between, or influences on, Plato's metaphysics in particular, and classical Greek thought in general, on the one hand, and Indian and Persian spiritual and intellectual traditions on the other hand, are still debated. See, for example, George P. Conger, "Did India Influence Early Greek Philosophies?" *Philosophy East and West*, Vol. 2, No. 2 (Jul., 1952), 102–28. University of Hawai'i Press, www.jstor.org/stable/1397302; and www.sabhlokcity.com/2012/09/the-direct-influence-of-hinduism-on-socrates-and-on-platos-republic/.

55 In the *Republic* (509D–513E), Plato described the visible world of perceived physical objects and the images we make of them (in our minds and in our drawings, for example). The sun not only provides the visibility of the objects, but also generates them and is the source of their growth and nurture. Many premodern religions and mythological traditions also identify the sun with God, for instance in ancient Egypt. Beyond this visible world, which later philosophers (especially Immanuel Kant) would call the *phenomenal* world, there is an intelligible world (that Kant would call the *noumenal world*), according to Plato. The intelligible world is (metaphorically) illuminated by "the Good" (Greek: *ton agasthon*) just as the visible world is illuminated by the sun. The division of Plato's Line between the Visible and the Intelligible is also a divide between the Material and the Ideal, the foundation of most Dualisms.

Plato used the word "idea" somewhat interchangeably with the Greek word for shape or form *eidos*. The Greek word *idea* derives from the Greek word for "to have seen." Plato's Line is also a division between Body and Mind.

Plato described two classes of things, those that can be seen but not thought, and those that can be thought but not seen. He also made the analogy between the role of the sun, whose light gives us our vision to see and the ability of visible things to be seen, and the role of the (Form of the) Good. The sun illuminates our vision and the things we see. The (Form of the) Good animates our (hypothetical) knowledge and the (real) objects of our knowledge (all Forms, or Ideas): "… the good is in the intelligible region to reason and the objects of reason, so is this (the sun) in the visible world to vision and the objects of vision." (*Republic*, 508B–C). Plato then described the line as divided into two sections not of the same size (*Republic*, 509D–510A). *For Plato, the Intelligible World (a unified Being) is to the Visible World (of concrete particulars) as the One is to the Many.* See: www.informationphilosopher.com/knowledge/divided_line.html; and https://en.wikipedia.org/wiki/Analogy_of_the_divided_line.

56 See: https://web.stanford.edu/class/ihum40/cave.pdf.

57 fuller treatment of Plato's worldview is in Webel, *The Politics of Rationality*, *op. cit.*, 34–53.

58 Plato's mathematical ideal of a universal rationality seems indebted to Orphic-Pythagorean beliefs. Among other Pythagorean ideas Plato may have adopted are the mythological or religious notion of stellar divinities rotating in perfect eternal circles, the systematic application of geometrical theorems of ratios, means, and proportions to astronomy and cosmogony, and the cosmic opposition between the ideas of limit (Greek: *peras*) and harmony on the one hand, and unlimited chaos on the other hand. Most importantly, Plato incorporated into his cosmology, ethics, and epistemology an Orphic insistence upon the immortality of an immaterial soul and a Pythagorean dichotomy between the

divine or rational form of that soul and the mortal and irrational materiality of the body. The cosmic oppositions between reason, Being, and law at the upper level, and necessity, Becoming, and chance at the "inferior" level, filter down into the human tensions between reason and appetite, between mind and body. Like his Pythagorean and Eleatic predecessors, Plato opted for the aristocratic sovereignty of "formal" Reason (*logos*). The strife (Greek: *eris*) between ontological and psychological polarities is harmonized by an order-conferring, mathematical *logos*; the conflicts between social classes and warring passions are resolved by their subordination to the sway of philosopher-kings and the rational part of the psyche, respectively.

59 Webel, *ibid.*

60 *Corcordia Ordinum* is a Latin expression used by the orator and statesman Cicero to denote the "harmony of order" needed to maintain a balance between the financial and oligarchical interests in ancient Rome.

61 The complete reference is from the *New Testament*, in the *Gospel of John* (18:36–7), in which Jesus is reported to have stated (in reply to Pilate): "My kingdom is not of this world; if it were, My servants would fight to prevent My arrest by the Jews. But now, My kingdom is not of this realm." "Then You are a king!" Pilate said. "You say that I am a king," Jesus answered. "For this reason I was born and have come into the world, to testify to the truth. Everyone who belongs to the truth listens to My voice."

62 Most of Aristotle's own writings disappeared over the centuries, and those that survived consisted largely of lecture notes taken by his students. They were also "lost" to the West for centuries, but were found and preserved by Islamic scholars, who "recovered" and reintroduced them to Occidental scholars and theologians. The "Recovery of Aristotle" (or "Rediscovery") refers to the copying or retranslating of most of Aristotle's works from Greek or Arabic texts into Latin. This spanned about 100 years, from the middle 12th century CE into the 13th century, when over 40 of Aristotle's surviving books were copied or translated, including Arabic texts from Muslim authors. See www.britannica.com/EBchecked/topic/34560/Aristotle; and https://en.wikipedia.org/wiki/Recovery_of_Aristotle.

63 See Christopher Shields, "Aristotle," *The Stanford Encyclopedia of Philosophy* (Fall 2020 Edition), Edward N. Zalta, ed., https://plato.stanford.edu/archives/fall2020/entries/aristotle/.

64 The standard English translation of Aristotle's complete works into English is: J. Barnes, ed. *The Complete Works of Aristotle*, Volumes I and II (Princeton, NJ: Princeton University Press, 1984). A reliable translation of selections of Aristotle's works is T. Irwin, and G. Fine., eds., *Aristotle: Selections, Translated with Introduction, Notes, and Glossary* (Indianapolis, IN: Hackett, 1995). Aristotle's works are often referred by abbreviations. For a complete list, see: https://faculty.washington.edu/smcohen/433/ariworks.htm.

65 Aristotle, *Physics*, 209a and 209b.

66 Aristotle, *Physics*, 209b, 22–8.

67 See https://en.wikipedia.org/wiki/Condemnations_of_1210–1277.

68 In addition to Aristotle's book *De Anima (On the Soul)*, see Hendrik Lorenz, "Ancient Theories of Soul," *The Stanford Encyclopedia of Philosophy* (Summer 2009 Edition), Edward N. Zalta, ed., https://plato.stanford.edu/archives/sum2009/entries/ancient-soul/; and https://plato.stanford.edu/entries/aristotle-psychology/active-mind.html.

69 See Anthony Kenny, *A New History of Western Philosophy* (Oxford: Clarendon Press, 2010), 192; Lorenz, *op. cit.*; and www.britannica.com/biography/Aristotle/Philosophy-of-mind.

70 Aristotle, *Nicomachean Ethics*, trans. W.D. Ross. Available at http://classics.mit.edu/Aristotle/nicomachaen.1.i.html.

71 For *Eudaimonia*, see www.britannica.com/topic/eudaimonia.

72 See https://en.wikipedia.org/wiki/Polis; and Aristotle, *Politics*, trans. Benjamin Jowett, (Kitchener, Ont., Batoche Books, 1999), available at https://socialsciences.mcmaster.ca/econ/ugcm/3ll3/aristotle/Politics.pdf. Also see Marc Overtrup, "Aristotle's Philosophy of Equality, Peace, & Democracy," *Philosophy Now*, Oct./Nov. 2016, https://philosophynow.org/issues/116/Aristotles_Philosophy_of_Equality_Peace_and_Democracy.

73 When used to refer to forms as Plato conceived them, the term "Form" is conventionally capitalized, as are the names of individual Platonic Forms. The term is lowercased when used to refer to forms as Aristotle conceived them.

74 Sean Gaston, *The Concept of World from Kant to Derrida*, *op. cit.*, referring to Aristotle's *Physics* (209a–b).

75 See Charles Webel, *The Politics of Rationality*, *op. cit.*, 66–8.

76 In addition to Aristotle's *Politics* and *Nicomachean Ethics*, see: A.W.H. Adkins, "The Connection between Aristotle's Ethics and Politics," *Political Theory,* Vol. 12, No. 1 (Feb., 1984), 29–49.

77 See, for example, Alasdair MacIntyre, *After Virtue A Study in Moral Theory,* Third Edition (Notre Dame: University of Notre Dame Press, 2007), and MacIntyre's *A Short History of Ethics*, Second Edition (Notre Dame: University of Notre Dame Press, 1998), xviii; and several books by Martha C. Nussbaum, especially *The Fragility of Goodness Luck and Ethics in Greek Tragedy and Philosophy* (Cambridge: Cambridge University Press, 2001).

78 Aristotle is often regarded as the father of psychology, and he wrote the first comprehensive psychological book, *De Anima* (*On the Soul*). Many historians of that field believe he contributed more to pre-empirical psychology than any other person. See www.intelltheory.org/aristotle.shtml. For Aristotle's influence on biology, see Armand Marie Leroi, *The Lagoon How Aristotle Invented Science* (London: Penguin, 2015).

79 See, Matt Qvortrup, "Aristotle's Philosophy of Equality, Peace, & Democracy," *Philosophy Now*, Issue 116.

80 See Dirk Batzly, "Stoicism," *The Stanford Encyclopedia of Philosophy* (Spring 2019 Edition), Edward N. Zalta, ed., https://plato.stanford.edu/archives/spr2019/entries/stoicism/; and https://medium.com/stoicism-philosophy-as-a-way-of-life/accepting-ones-fate-9a64da5776f.

81 The English word "God" is derived from the German *Gott*, stemming from old Germanic languages equivalents of the Latin word for God, *Deus*. The Christian *Deus* of the Roman Empire is etymologically related to Zeus, the chief Greek divinity, as well as *to Deus or Theos,* in various Ancient Greek dialects and traditions. See https://en.wikipedia.org/wiki/Deus; www.etymonline.com/word/god; and Katja Vogt, "Seneca," *The Stanford Encyclopedia of Philosophy* (Spring 2020 Edition), Edward N. Zalta, ed., https://plato.stanford.edu/archives/spr2020/entries/seneca/.

82 See https://medium.com/stoicism-philosophy-as-a-way-of-life/accepting-ones-fate-9a64da5776f; https://en.wiktionary.org/wiki/fate; www.quora.com/What-is-the-translation-of-the-word-destiny-in-Latin; and www.traditional stoicism.com/the-winds-of-fortuna/.

83 Machiavelli's cosmos is also ruled by the powers of *Fortuna* ("chance/fortune"), *Necessità* ("necessity"), and *Virtù* ("virtue"). See Charles Webel, *The Politics of Rationality op. cit.*, esp. 73–4, and 85–6; and Neal Wood, "Some Common Aspects of the Thought of Seneca and Machiavelli," *Renaissance Quarterly*, Vol. 21, No. 1 (Spring, 1968), 11–23, www.jstor.org/stable/2858873?seq=1#metadata_info_tab_contents.

84 See Christian Wildberg, "Neoplatonism," *The Stanford Encyclopedia of Philosophy* (Summer 2019 Edition), Edward N. Zalta, ed., https://plato.stanford.edu/archives/sum2019/entries/neoplatonism/.

85 See www.encyclopedia.com/medicine/diseases-and-conditions/pathology/hypostasis.

86 Sir Isaac Newton, as quoted in I.B. Cohen, *Isaac Newton's Papers and Letters on Natural Philosophy and Related Document* (Cambridge, MA: Harvard University Press), 1978, 282.

87 See Hans Halvorson and Helge Kragh, "Cosmology and Theology," *The Stanford Encyclopedia of Philosophy* (Spring 2019 Edition), Edward N. Zalta, ed., https://plato.stanford.edu/archives/spr2019/entries/cosmology-theology/.

88 See Bruce Reichenbach, "Cosmological Argument," *The Stanford Encyclopedia of Philosophy* (Spring 2021 Edition), Edward N. Zalta, ed., forthcoming, https://plato.stanford.edu/archives/spr2021/entries/cosmological-argument/.

89 See https://en.wikipedia.org/wiki/Udayana.

90 See www.iep.utm.edu/nyaya/.

91 See www.britannica.com/topic/Gospel-According-to-John.

92 See Christian Tornau, "Saint Augustine," *The Stanford Encyclopedia of Philosophy* (Summer 2020 Edition), Edward N. Zalta, ed., https://plato.stanford.edu/archives/sum2020/entries/augustine/.

93 Eschatology is a part of theology concerned with the final events of history, or the ultimate destiny of humanity, especially the "end of the world" or "end times." The word is derived from the Greek *eschatos* meaning "last," and *-logy*, meaning "the study of," and it first appeared in English around 1844. The *Oxford English Dictionary* defines eschatology as "the part of theology concerned with death, judgment, and the final destiny of the soul and of humankind." In the context of mysticism, the term refers metaphorically to the end of everyday reality and to reunion with the Divine. Many religions treat eschatology as a future event prophesied in sacred texts or in folklore. Most modern religious and secular eschatological and apocalyptic texts predict the violent disruption or destruction of the world; Christian and Jewish eschatologies view the end times as the consummation or perfection of God's creation of the world. See https://en.wikipedia.org/wiki/Eschatology.

94 Exegesis comes from the Greek exegesis from (from the Greek verb for "to lead out") is a critical explanation or interpretation of a text, particularly a religious text. In the past, the term was used primarily for interpretations of the Bible; however, more recently *biblical exegesis* is used to distinguish it from any other broader critical, or exegetical, textual explanation. See https://en.wikipedia.org/wiki/Exegesis.

95 See Ralph McInerny and John O'Callaghan, "Saint Thomas Aquinas," *The Stanford Encyclopedia of Philosophy* (Summer 2018 Edition), Edward N. Zalta, https://plato.stanford.edu/archives/sum2018/entries/aquinas/.

96 "World" in Arabic is *alealam*. However, the word *dunia* expresses the meaning of "world" within Muslim culture in particular. "Existence" in Arabic *is*

alwujud and also implies the creator of existence, God (*Allah*), of whom we are all creatures. In Islam, the purpose of our lives is to worship God (*Allah*). The word for "Fate" is Arabic is *qadar.* In Arab culture, fate denotes divine destiny, and this also implies predestination. In Islam, as in Christianity, the relationship between fate, or predestination, and free will, or individual choice, is complex and often ambiguous. "Meaning" in Arabic is denoted by the word *maenaa.* While Muslims create their own meanings, both for themselves and for life in general, the general tendency in Islam is for believers submit to and accept their "Fate," which is Allah's will. My primary source for this section is a former student from Bahrain, Salman Rashed Al-Abbasi. Also see https://en.wikipedia.org/wiki/Muslim_world.

97 See https://en.wikipedia.org/wiki/Islamic_Golden_Age.

98 See Dimitri Gutas, "Ibn Sina [Avicenna]," *The Stanford Encyclopedia of Philosophy* (Fall 2016 Edition), Edward N. Zalta, ed., https://plato.stanford.edu/archives/fall2016/entries/ibn-sina/.

99 For "emanationism," see www.newadvent.org/cathen/05397b.htm.

100 For al-Ghazali, see Frank Griffel, "al-Ghazali," *The Stanford Encyclopedia of Philosophy* (Summer 2020 Edition), Edward N. Zalta, ed., https://plato.stanford.edu/archives/sum2020/entries/al-ghazali/.

101 See www.muslimphilosophy.com/ir/.

102 See www.iep.utm.edu/ibnrushd/.

103 *Torah* (Hebrew for "Instruction," "Teaching," or "Law") can denote the first five books (*Pentateuch*, or five books of Moses) of the 24 books of the Hebrew *Bible*, commonly known as the *Written Torah*. It can also mean all 24 books, from the *Book of Genesis* to the end of the *Tanakh* (*Chronicles*), or even the totality of Jewish teaching, culture, and practices, whether derived from biblical texts or later rabbinic writings, often known as the *Oral Torah*. More generally, *Torah* refers to the origin of the Jewish people, their call into being by God, their trials and tribulations, and their covenant with their God, which involves following a way of life embodied in a set of moral and religious obligations and civil laws (*halakha*). See https://en.wikipedia.org/wiki/Torah.

104 See Jacob Rader Marcus, *The Jew in the Medieval World: A Sourcebook, 315-179* (New York: Hebrew Union College Press), 1999; Kenneth Seeskin, "Maimonides," *The Stanford Encyclopedia of Philosophy* (Spring 2021 Edition), Edward N. Zalta, ed. Forthcoming, https://plato.stanford.edu/archives/spr2021/entries/maimonides/; www.iep.utm.edu/maimonid/; and Howard Kreisel, *Judaism as Philosophy: Studies in Maimonides and the Medieval Jewish Philosophers of Provence* (Boston, MA: Academic Studies Press, 2015).

2 From the Existence of the World to Our Existence in This World

The Creation of the Modern Universe

For most "premodern" Occidental thinkers and scientists, the universe was usually viewed through the lenses of Aristotelian natural philosophy and Ptolemaic geocentric cosmology. For "medieval" Christians, God had created and set in motion what came to be known as "the world system." For both "believers" and "non-believers," the Earth was the fixed point around which all other worlds revolved. The manner in which we now tend to conceive the "modern universe" as a whole, and our terrestrial world and the solar system in which it moves in particular, took millennia to develop.

As the universe has expanded, both literally and metaphorically, the geocentrism of ancient and medieval philosophy and cosmology gradually receded, and the cosmic significance of our terrestrial world has declined dramatically. The recent discoveries of "other Earth-like" worlds, on which extraterrestrial life may exist, accentuate this diminishing importance of anthropocentric views of our place in the cosmos.

From the Heliocentric Universe to the Modern Universe

During the 3rd century BCE, the Greek astronomer and mathematician Aristarchus of Samos presented an explicit argument for a heliocentric model of the Solar System, placing the Sun, not the Earth, at the center of the cosmos.[1] He envisioned the Earth as rotating daily on its axis and revolving annually about the Sun in a circular orbit, along with a sphere of fixed stars. His heliocentric theory was later rejected by most ancient and medieval natural philosophers and theologians in favor of the geocentric theories of Aristotle and Ptolemy, until it was revived and modified nearly 1800 years later by Copernicus. However, there were exceptions along the way. Among those who supported Aristarchus's heliocentric cosmology were Seleucus of Seleucia,[2] who lived about a century after Aristarchus and referred to the motions of the tides and the influence of the Moon to explain heliocentricity; the Indian astronomer and mathematician Aryabhata, who at the end of the 5th century CE described elliptical orbits around the Sun;[3] and the 9th-century Persian astronomer Ja'far ibn Muhammad Abu Ma'shar al-Balkhi.[4]

DOI: 10.4324/9781315795171-3

The 2nd century Roman-Egyptian mathematician and astronomer Ptolemy (Claudius Ptolemaeus) created a geocentric model of the cosmos, largely based on Aristotelian ideas, in which the planets and the rest of the universe orbit about a stationary Earth in circular epicycles.[5] In terms of longevity, it was perhaps the most successful cosmological model of all time. Modifications to the basic Ptolemaic system were suggested by the Islamic Maragha School in the 13th, 14th, and 15th centuries, including the first accurate lunar model, by Ibn al-Shatir, and the rejection of a stationary Earth in favor of a rotating Earth, by Ali Qushji.[6]

Several medieval Christian, Muslim, and Jewish scholars put forward the idea of a universe that is finite in time. In the 6th century CE, the Christian philosopher John Philoponus of Alexandria argued against the Aristotelian notion of an infinite past, and was perhaps the first Occidental thinker to argue that the universe is finite in time and therefore had a beginning. Early Muslim theologians such as Al-Kindi (9th century) and Al-Ghazali (11th century) offered logical arguments supporting a finite universe, as did the 10th-century Jewish philosopher Saadya Gaon.[7] During the late 15th and early 16th centuries, Somayaji Nilakantha of the Kerala school of astronomy and mathematics in southern India, developed a computational system for a partially heliocentric planetary model, in which Mercury, Venus, Mars, Jupiter, and Saturn orbit the Sun, which in turn orbits the Earth. This was similar to the Tychonic system proposed later in the 16th century by the Danish astronomer Tycho Brahe as a kind of hybrid of the Ptolemaic and Copernican models.[8]

The Copernican Universe

In 1543, the Polish astronomer and polymath Nicolaus Copernicus adapted the geocentric model of Ibn al-Shatir to meet the requirements of the ancient heliocentric universe of Aristarchus.[9] His publication of a scientific theory of heliocentrism, in *De Revolutionibus (On the Revolutions)*, demonstrated that the motions of celestial objects can be explained without putting the Earth at rest in the center of the universe. This stimulated further scientific investigations, especially by Galileo, Kepler, and Newton, and became a landmark in the history of modern science, sometimes known as "The Copernican Revolution."[10]

The Copernican Revolution (that the Earth is not in a central, specially favored cosmic position) and its implication that celestial bodies obey physical laws identical to those on Earth, established cosmology as a science rather than a branch of metaphysics or natural philosophy. In 1576, the English astronomer Thomas Digges popularized Copernicus' ideas and also extended them by positing the existence of a multitude of stars extending to infinity, rather than just Copernicus' narrow band of fixed stars.[11]

The late Renaissance Italian philosopher Giordano Bruno took the Copernican principle a step further by suggesting that even our Solar System is not the center of the universe, but rather a relatively insignificant star system among an infinite multitude of other solar systems. He further proposed that the stars were distant suns surrounded by their own planets, and he raised the possibility that these planets might foster life of their own, a philosophical position known as cosmic pluralism. Bruno also insisted that the universe is infinite and could have no "center." For his efforts, Bruno was burned at the stake in 1600 in Rome.[12]

In 1605, the German astronomer/astrologist and mathematician Johannes Kepler (1571–1630) made further refinements to the Copernican cosmology by abandoning the classical assumption of circular orbits in favor of elliptical orbits, which could explain the strange apparent movements of the planets.[13] As a Copernican, Kepler defended this "new world system" on different fronts: against the old astronomers who still sustained the system of Ptolemy, against the Aristotelian natural philosophers, against the followers of the new "mixed system" of Tycho Brahe, and even against the standard Copernican position, according to which the new model of the solar system was considered merely as a computational device and not necessarily a physical reality. Kepler's cosmic view incorporated both Platonic and Aristotelian influences, the former by giving priority to the role of geometry in modeling the structure of the world system (more accurately, our Solar System), and the latter in accentuating the role of experience and causality in knowing the cosmos as a whole.

Importantly, Kepler's mathematical approach to the natural world would become a hallmark of what would later come to be known as "the new science." This "new science," according to Norman Melchert and David Morrow, "bequeaths to philosophers four deep and perplexing problems":

1. What is the place of mind in this world of matter?
2. What is the place of value in this world of fact?
3. What is the place of freedom in this world of mechanism?
4. Is there any room left for God at all?[14]

Although philosophers, theologians, and physical and human scientists since the early 17th century have put forward "answers" to these enduring questions, there is still no consensus, and, accordingly, these questions are still with us.

Later in the 17th century, Galileo Galilei (1564–1642), in the face of significant hostility by the Catholic Church, defended the Copernican system. Galileo used a telescope to see the moons of Jupiter and the mountains of the Moon. He is also renowned for inventing the microscope, dropping stones from towers and masts, playing with pendula and clocks, being the first "real" experimental scientist, advocating the relativity of motion, and creating mathematical physics. Galileo experimentally determined the parabolic path of projectiles and calculated the law of free fall of physical bodes. For Galileo, as for Copernicus and Kepler, *the great book of nature is written in mathematical language, and we, by using that language, can understand it*.[15]

Apart from his significant scientific contributions, Galileo is probably best remembered for his trial by the Roman Catholic Inquisition in 1633 for "heresy." This event (notwithstanding Galileo's "recantation" of his "heretical" views) has led to Galileo's purported role as the paradigmatic "heroic, rational, modern man" in the subsequent history of the conflict between modern experimental science and traditional religion.[16]

The Mechanical Universe of Sir Isaac Newton

In 1687, Sir Isaac Newton (1647–1727) published *The Mathematical Principles of Natural Philosophy* (aka the *Principia*), probably the single most important contribution to the transformation of early-modern natural philosophy into what is now considered modern physical science. In that work, he postulated his theory of universal gravity (Newton's "Three Laws"[17]), and described, among other things, a static, steady state, infinite universe, which Albert Einstein, in the early 20th century, initially took as a given cosmological model (at least until Einstein's special and general theories of relativity demonstrated the restricted validity of the static Newtonian framework).[18]

In Newton's mechanical universe, which, in the third part of his *Principia* he called "The System of the World," matter on the large scale is uniformly distributed, and the universe is gravitationally balanced but essentially unstable.[19] Interestingly, *when Newton referred to the "world" in that part of the Principia, he did so apparently to denote the cosmos as a whole as it was known in the early 18th century, and in the context of the contributions and limitations of ancient Greek and Egyptian philosophy, "philology," religion, and science in understanding the "system of the world."* Newton is also known for having invented the calculus in the mid-to-late 1660s, just before the German polymath Gottfried Wilhelm (von) Leibniz did so independently. He also made major discoveries in optics, and, curiously, was deeply interested in alchemy.

Newton argued that the "empirical world" must serve not only as the ultimate arbiter, but also as the sole basis for adopting any possible scientific theory. He was quite distrustful of what was then known as the "method of hypotheses," widely attributed to Descartes and the rationalist/deductive philosophical and scientific tradition as a whole (which will be discussed later in this chapter). This is the epistemological practice of putting forward hypotheses that reach beyond all known phenomena, and then testing them by deducing observable conclusions from them. Newton insisted instead on having specific phenomena test each element of a scientific theory, with the goal of limiting the provisional aspect of theory as much as possible to the step of inductively generalizing from the specific phenomena. This stance is perhaps best summarized by Newton's four "Rules of Reasoning," added in 1726 to the third edition of the *Principia*:

> Rule 1 We are to admit no more causes of natural things than such as are both true and sufficient to explain their appearances.

Rule 2 Therefore to the same natural effects we must, as far as possible, assign the same causes.

Rule 3 The qualities of bodies, which admit neither intensification nor remission of degrees, and which are found to belong to all bodies within the reach of our experiments, are to be esteemed the universal qualities of all bodies whatsoever.

Rule 4 In experimental philosophy, we are to look upon propositions inferred by general induction from phenomena as accurately or very nearly true, notwithstanding any contrary hypothesis that may be imagined, till such time as other phenomena occur, by which they may either be made more accurate, or liable to exceptions.[20]

To conduct "experimental philosophy" in this rigorous, scientific manner, Newton famously declared in an essay "General Scholium," appended to the second (1713) edition of the *Principia* that: "Hypotheses non fingo" (Latin for "I feign no hypotheses," "I frame no hypotheses," or "I contrive no hypotheses"):

> I have not as yet been able to discover the reason for these properties of gravity from phenomena, and I do not feign hypotheses. For whatever is not deduced from the phenomena must be called a hypothesis; and hypotheses, whether metaphysical or physical, or based on occult qualities, or mechanical, have no place in experimental philosophy. In this philosophy particular propositions are inferred from the phenomena, and afterwards rendered general by induction. [21]

Interestingly, Newton spent at least as much time thinking and writing about Biblical and other religious topics as he did about science. And Newton viewed God as the "cosmic mechanic," the ultimate cause setting in motion "*the system of the world.*"[22]

Newton's Legacy

Later in the 18th century, although generally based on the model of a Newtonian static universe, the depiction of matter in a hierarchical universe as clustered on even larger scales of hierarchy, and as endlessly being recycled, was first proposed in 1734 by the Swedish scientist and philosopher Emanuel Swedenborg, and developed further (independently) by Thomas Wright (1750), Immanuel Kant (1755), and Johann Heinrich Lambert (1761), and a similar model was proposed in 1796 by the French physicist–mathematician Pierre-Simon Laplace.[23] What has come to be known as the Kant-Laplace "nebular hypothesis" regarding the origins of the Solar System persisted until the 20th century, when Kant's contribution to cosmology was ultimately refuted, while Laplace's has been refined.[24]

Newton's "System of the World," as governed by gravitational laws, is still the basic framework for understanding large-scale cosmological phenomena, as modified by contributions made by Einstein and quantum electrodynamics in the 20h century. Specifically, Newton's views on the absolute, invariant status space, time, and motion dominated physics from the 17th Century until the advent of Einstein's theories of relativity in the early 20th century. Einstein proved that all motion is relative motion, and that space and time are not independent, absolute, and invariant, as Newton believed, but instead are interconnected dimensions of a relativistic and shifting universe. Nonetheless, Newton's mechanistic model of the "world system," and his inductive, experimental, observational approach to understanding the world writ large, still underlie the dominant "modern" scientific worldview.

The World and the Mind According to René Descartes

René Descartes (1596–1650) was an original mathematician, scientific thinker, and philosopher. In mathematics, he developed algebraic (or "analytic") geometry and the sine law of refraction. In the natural sciences, Descartes developed an empirical account of the rainbow and proposed a naturalistic account of the formation of the Earth and planets (a precursor to the nebular hypothesis of Laplace and Kant).[25] And, of course, Descartes is considered by many to be the "father of modern" (Western) philosophy because of his pathbreaking contributions to epistemology and metaphysics, mainly as presented in his *Discourse on Method* (1637, in French) and *Meditations on First Philosophy* (i.e., on metaphysics, in Latin, 1642). My emphasis, however, will be less on those works and more on his much less-cited book, *The World or Treatise on Light*, containing the core of his natural philosophy (in French, 1664).

In his "natural philosophy," Descartes envisioned *the natural world as a material world* possessing a few fundamental properties and operating according to a finite number universal laws. As Norman Melchert and David Morrow claim:

> Since the natural world can be geometrically represented in terms of the size, figure, volume, and spatial relations of natural things, analytic geometry promises an algebraic treatment of all of nature. Descartes realizes he has found a new way to read the "*great book of the world.*"[26]

This "book of the world" also can be "read" somehow to include an immaterial mind that, in human beings, was curiously related to the body.[27] *Descartes thereby posed the modern version of what has come to be called the mind–body problem.*

As a "substance dualist," Descartes argued that our minds, as non-physical finite substances, or "thinking things" (Latin: *res cogitans*), whose essence is spatially unextended, somehow interact with and direct our bodies, which are finite material substances, and whose essence is spatial extension (Latin: *res extensa*). In his strained effort to clarify the mind–body connection, Descartes provided arguments for the existence of God, as a kind of infinite, third substance, who enables our actions in and knowledge of ourselves and of the world.

Descartes also deployed a distinctive method, which was variously exhibited in mathematics, natural philosophy, and metaphysics, and which, in the latter part of his life, included, or was supplemented by, a method of radical skeptical doubt. Descartes eventually "resolved" his doubts about the existence of the "external world" in particular and the acquisition of knowledge more generally, and thereby laid the basis for modern epistemology.

The Cartesian Cosmos

The World, also called *Treatise on the Light and Other Principal Objects of Sense* (French title: *Le Monde, ou Le Traité de la Lumière et Des Autres Principaux objets des Sens*), contains a nearly complete version of Descartes' philosophy, including his method, metaphysics, theology, and contributions to physics, astronomy, and biology, as well as the other sciences of his day.[28]

The World assumes the Copernican heliocentric view. Descartes delayed the book's release upon news of the Inquisition's conviction of Galileo for "suspicion of heresy" and sentencing to house arrest. *The World*'s many chapter headings include: "Our Senses Are Not Aware of Certain Bodies"; "On the Number of Elements and on Their Qualities"; "On the Laws of Nature of this New World"; "On the Formation of the Sun and the Stars of the New World"; "On the Origin and the Course of the Planets and Comets in General; and of Comets in Particular"; "On the Planets in General, and in Particular on the Earth and Moon"; "On the Ebb and Flow of the Sea"; "On the Properties of Light"; and "That the Face of the Heaven of That New World Must Appear to Its Inhabitants Completely like That of Our World."

Descartes espoused a *mechanical* philosophy. He thought everything physical in the universe to be made of tiny "corpuscles" of matter, a view that is closely related to what we now call the atomic theory of matter.

According to Descartes' *corpuscularian cosmology, the universe as a whole arose from utter chaos and, according to its basic natural laws, had its particles (corpuscles) arranged so as to resemble the universe we observe today.* Once the particles in the chaotic universe began to move, the overall motion was circular because there is no void in nature, so whenever a single particle moves, another particle must also move to occupy the space where the previous particle once was. This type of circular motion, or vortex, created the orbits of the planets orbiting our Sun, with the heavier objects spinning out toward the outside of the vortex and the lighter objects remaining closer to the center.

Descartes thereby outlined a model of the universe with many of the characteristics of what Newton would later model as a static, infinite universe.[29] But, in contrast with Newton, according to Descartes, the vacuum of space is not empty at all, but is filled with matter that swirls around in large and small vortices.

Cartesian Actual and Possible, Old and New, Worlds

In Chapter 6 of *The World*, called "*Description of a new world, and the qualities of the matter of which it is composed,*" Descartes constructs a *hypothetical world* on the basis of the theory of matter set out in the first five chapters of that book. According to Stephen Gaukroger:

> The ultimate aim is to show that a world constructed in this manner, one without forms or qualities, and that is indistinguishable from the actual world. The traditional forms and qualities are excluded because they could not form part of a properly mechanist explanation.

If we strip the world of the traditional forms and qualities, what we would be left with would, in Descartes' view, would be its genuine properties. *His new world is to be conceived "a real, perfectly solid body which uniformly fills the entire length, breadth, and depth of the great space at the center of which we have halted our thought."*[30] In Descartes' own words:

> For a while, then, allow your thought to wander beyond this world to view another, wholly new, world, which I call forth in imaginary spaces before it. The Philosophers tell us that these spaces are infinite, and they should certainly be believed, since it is they themselves who invented them. But in order to keep this infinity from impeding and hampering us, let us not try to go all the way, but rather enter it only far enough to lose sight of all the creatures that God made five or six thousand years ago.[31]

Descartes also considers the theological issue dating back to medieval times of the *"plurality of worlds"—that is, whether it is possible for there to be a world completely outside this one, namely outside our cosmos, which exists in an "imaginary" space. And he dates the creation of the world on the basis of biblical chronology as it was commonly accepted in Descartes' time.*[32]

In addition, Descartes' *epistemological and metaphysical dualism—most often expressed in the radical cleft between the mental and material worlds—is also exhibited in his distinction between a "new, most perfect world" illuminated by God, and "the old, actual world" of physical appearances:*

> For God has established these laws in such a marvelous way that even if we suppose that He creates nothing more than what I have said … the laws of nature are sufficient to cause the parts of this chaos to disentangle themselves and arrange themselves in such a good order that they will have the form of a most perfect world, a world in which one

will be able to see not only light, but all the other things as well, both general and particular, that appear in the actual world …

Were I to put in this new world the least thing that is obscure, this obscurity might well conceal some hidden contradiction I had not perceived, and thus without thinking I might suppose something impossible.

Instead, since everything I propose here can be imagined distinctly, it is certain that even if there were nothing of this sort in the old world, God can nevertheless create it in a new one; for it is certain that He can create everything we imagine.[33]

Descartes and Cartesianism

Descartes preserved and extended the tradition dating back to Parmenides, Pythagoras, and Plato of *contrasting the ideal, rational world of mathematical logic, mental immortality, and imaginary speculation with the actual, material world in which our bodies, like all other physical "machines," come into being and pass away. This distinction between "actual" and "possible" worlds would be elaborated and debated in the centuries to come by thinkers as distinct as Leibniz, Kant, and, much more recently, David Lewis.*

What is Descartes' legacy now, and why has what has come to be called "Cartesianism" become so controversial, at least in philosophical circles?[34] Descartes' influence during the 17th century and 18th centuries was significant, and includes his specific contributions to mathematics and optics, his vision for a mechanistic physiology, and the model he offered to Newton of a unified celestial and terrestrial physics that assigns a few basic properties to cosmic matter, the motions of which are governed by a few simple natural laws.

But during the 20th century, thinkers from very different intellectual orientations—ranging from empiricist behaviorists and analytic philosophers of language on the one hand to Marxists, feminists, and poststructuralists on the other hand—have variously celebrated or criticized his famous "cogito" starting point, and have also called into question his radical epistemological skepticism. Many have also rejected the rationalism, dualisms, and theology underlying his metaphysics, while other, more conventional, historians of ideas, have looked to him as the very model of "the modern philosopher."

Descartes' legacy partly consists of problems he raised, or highlighted, but did not solve, the mind–body problem being a key example. Descartes himself argued from his ability clearly and distinctly to conceive mind and body as distinct entities to the conclusion that they really are separate substances. Since the time of Kant, few philosophers have believed that the clear and distinct thoughts of the human mind are a guide to the absolute reality of things. Moreover, few philosophers and even fewer psychologists and behavioral scientists today are substance dualists, tending instead to be physicalists, who often argue that "the mind is what the brain does."[35]

Descartes has been seen, at various times, as a hero and as a villain; as a "universal genius" who is the progenitor of modern thought and science, or as the harbinger of a disembodied, rationalistic, and mechanistic conception of human beings. "Cartesianism," or the rationalistic worldview posthumously attributed to Descartes, is, like the human world itself, ineluctably a mixed bag.

God's Design for "The Best of all Possible Worlds"

Gottfried Wilhelm Leibniz and Baruch Spinoza are usually grouped by historians of ideas with Descartes as the most influential members of the "Rationalists" (in contrast with the "Empiricists"—Hobbes, Locke, and Hume.)[36] Since Leibniz is the main "rationalist" contributor to Occidental theories of the world in general, and to "possible worlds," in particular, his relevant works on this theme will be highlighted, after a brief consideration of Spinoza.

The God-Permeated Universe of Baruch Spinoza

Bento (in Hebrew, Baruch; in Latin, Benedictus) Spinoza is one of the most important philosophers of what if often called "the early modern period" (in Western Europe from the end of the Renaissance until the Enlightenment, or, roughly, from 1550 to 1700). His thought combines a number of Cartesian metaphysical, theological, and epistemological principles with elements from ancient Stoicism, Hobbes, and medieval Jewish rationalism. His naturalistic views on God, the universe, "human nature," and the possibility of knowledge, anchor a moral philosophy centered on the very contemporary-sounding argument that control of the passions leads to virtue and happiness. Spinoza also provided logical, theological, and political arguments for democracy and state sovereignty.[37]

Spinoza is probably best known for his *Ethics*, an encyclopedic book that presents a moral–theological vision founded on a monistic metaphysics in which God and Nature are identified and the universe (the world as a whole) is infused with a divine spark. For Spinoza, *God is immanent within Nature itself, which is understood as an infinite, necessary, and fully deterministic system of which humans are a part.*[38]

Spinoza's *Ethics* is structured as a logically deductive series of interrelated claims and proofs. It proceeds from the assumptions, or "propositions," that:

> God … a being absolutely infinite—is, a substance consisting in infinite attributes, of which each expresses eternal and infinite essentiality …; there cannot exist in the universe two or more substances having the same nature or attribute …; the human mind is part of the infinite intellect of God …;

and, also perhaps surprisingly, given Spinoza's otherwise theocentric framework, in a stunning anticipation of Nietzsche's critique of Western ethics, metaphysics, and theology and Freud's metapsychology, Spinoza claims that "*desire is the actual essence of man*" ...; *and* "*the terms good and bad ... indicate no positive quality in things regarded in themselves, but are merely modes of thinking, or notions which we form from the comparison of things one with another.*"[39]

Spinoza concludes this magisterial, but for contemporary readers sometimes opaque and rigidly constructed work, with the following propositions:

> No one wishes to preserve his being for the sake of anything else ...; the mind's highest good is the knowledge of God, and the mind's highest virtue is to know God ...; men can differ in nature, in so far as they are assailed by those emotions, which are passions, or passive states ...; in the state of nature ..., sin is inconceivable; it can only exist in a state, where good and evil are pronounced on by common consent, and where everyone is bound to obey the State authority ...; in the state of nature, no one is by common consent master of anything, nor is there anything in nature, which can be said to belong to one man rather than another: all things are common to all ...; the man, who is guided by reason, is more free in a State, where he lives under a general system of law, than in solitude, where he is independent ... the human mind cannot be absolutely destroyed with the body, but there remains of it something which is eternal.[40]

Spinoza's relatively liberal views on "the state of nature" are both a response to the materialistic and conservative construal of this concept by Thomas Hobbes and an anticipation of the later positions of John Locke and Jean-Jacques Rousseau. And his perspective on the passions, emotions, desires in particular, and on human rationality and irrationality more generally, have provided a partial catalyst (along with Descartes) for the reflections of the contemporary neurobiologist Antonio Damasio on our conduct and nature.[41]

On account of the many provocative positions he advocated, Spinoza became a quite controversial figure. For many, he is a harbinger of enlightened modernity who calls us to live by the guidance of reason. For others, he is the enemy of the conventional religious, ethical, and political traditions that have provided a stable Judeo-Christian cultural framework.

While not offering us an explicit and comprehensive theory of the world akin to those of Descartes before him and Leibniz after him, Spinoza nonetheless formulated a theocentric vision of creation in which, despite the presence of sin and evil, we may indeed live in "the best of all possible worlds."

Leibniz and the "Best of All Possible Worlds"

Gottfried Wilhelm Leibniz (1646–1716) was one of the greatest philosopher–scientists of the "early-modern era" and is known as a "universal genius."

He made important contributions to metaphysics, epistemology, logic, and the philosophy of religion, as well as mathematics (the calculus and probability theory), physics, geology, jurisprudence, and history. Even the 18th-century French atheist and materialist Denis Diderot, whose views were frequently at odds with those of Leibniz, wrote in his entry on Leibniz in the *Encyclopedia*:

> Perhaps never has a man read as much, studied as much, meditated more, and written more than Leibniz ... What he has composed on the world, God, nature, and the soul is of the most sublime eloquence. If his ideas had been expressed with the flair of Plato, the philosopher of Leipzig would cede nothing to the philosopher of Athens.

The "Best of All Possible Worlds"

Leibniz famously (and for the great late-18th century French writer and satirist, Voltaire, infamously ...) deployed the phrase "the best of all possible worlds" (French: *le meilleur des mondes possibles*; German: *Die beste aller möglichen Welten*) in his 1710 work *Essais de Théodicée sur la bonté de Dieu, la liberté de l'homme et l'origine du mal* (*Essays of Theodicy on the Goodness of God, the Freedom of Man and the Origin of Evil*). Arguments in Christian theology/theodicy regarding this claim date back at least to Aquinas.[42] But Leibniz's discussion of this issue has endured and spawned an entire area of contemporary logic and metaphysics, namely "modal logic" and "possible worlds theories," whose most notable contributors may have been the American philosophers Saul Kripke and David Lewis.[43]

The claim that the actual world is the best of all possible worlds is the central argument in Leibniz's theodicy, which is his attempt to solve the problem of evil by arguing for the beneficence of an almighty God the creator. Leibniz's theodicy addresses the question that bedeviled theology and philosophy for millennia, especially during the "medieval" period: if God is omnibenevolent, omnipotent, and omniscient, how do we account for the suffering and injustice that exists in the world? Historically, attempts to answer the question have been made using various arguments, for example, by explaining away evil or reconciling evil with good.[44]

Leibniz presented his main argument for "the best of all possible worlds" in five theses:

1. God has the idea of infinitely many universes, or possible worlds.
2. Only one of these universes, or possible worlds, can actually exist.
3. God's choices are subject to the principle of sufficient reason, that is, God has a sufficient reason to choose one thing or another.
4. And God is good, all-knowing, all-powerful, and beneficent.
5. Therefore, the universe, or world, that God chose to exist is the best of all possible worlds.[45]

Although Leibniz conceded that God has created a world with evil in it and could have created another world without it, he argued that this is still the best world God could have made by claiming the existence of evil does not necessarily mean a worse world. In fact, he went as far as to claim *that evil's presence creates a better world*, as "it may happen that the evil is accompanied by a greater good." According to Leibniz, without the existence of evil, everything would be good; goodness would no longer appear good, it would simply be normal and expected, no longer to be praised. As put by Leibniz: "an imperfection in the part may be required for a perfection in the whole." God allowed evil in the world for us to understand goodness, which is achieved through contrasting it with evil. Once we understood evil and good, it gives us the ability to produce the "greatest possible good." Possibly paradoxically, Leibniz argues, evil fuels goodness, which leads to a "perfect whole."

According to Leibniz, there are three types of evil in this world: moral, physical, and metaphysical. Humans are able to tell the difference between right and wrong. God is not the creator of evil; evil is simply the lack of human-induced good. Often, humans make the claim that evil exists because God allowed it into the world; however, "God is not the author of sin." Sin is, paradoxically, necessary in creating the best of all possible worlds and is a result of our free will. There has to be a balance between good and evil in order to maintain the gap between humans and God. If humans were perfect, it would put them on the same level as God, which would destroy the need for grace. Instead, we are weighed down by our own free will, in contrast with God, who provides us with divine grace and endless mercy to address the sometimes unfortunate consequences of individuals' exercise of their free will.

Leibniz also claimed that from all the creative possibilities, God created the very best world not only because God is all-powerful but also because God is also all-good. If one supposes that this world is not the best possible world, then it follows that the creator of the universe is not sufficiently knowledgeable or powerful, or inherently good, for an inherently good God would have created the best world. *But since God's nature is inherently good, the universe that God has chosen to create must be the best of all possible worlds.*

Leibniz's cosmic speculation about the plurality of possible worlds is in some ways eerily similar to some current multiverse theories.[46] But Leibniz's multiverse is not identical to those postulated by some contemporary astrophysicists. Unlike modern multiverse theories or even unlike related ideas posed by Ancient Greek atomists, Leibniz did not believe that all the possible universes actually exist. For him, human decisions do not create branching universes depending on one's decision, and there is no other, less perfect universe, existing alongside our own. God's creation is only the perfect universe in which we can live. This is due to the nature of God, who has chosen to create this universe. *According to Leibniz, God envisioned all the possible universes and chose to create only the one God wants—the perfect one, the best of all possible worlds.*

The Fate of an Idea

In his *Theodicy*, Leibniz also raises the issue of fate (*fatum*) and states:

> To call that *fatum* ... is not contrary to freedom: *fatum* comes from *fari*, to speak, to pronounce; it signifies a judgement, a decree of God, the award of his wisdom. To say that one cannot do a thing, simply because one does not will it, is to misuse terms. The wise mind wills only the good ... Slavery comes from without, it leads to that which offends, and especially to that which offends with reason: the force of others and our own passions enslave us. God is never moved by anything outside himself, nor is he subject to inward passions, and he is never led to that which can cause him offence. [47]

So what has been the "fate" of Leibniz's theory of the "best of all possible worlds?" Such critics of Leibniz as Voltaire argue that the world contains an amount of suffering too great to justify optimism. The statement that "we live in the best of all possible worlds" was scorned by Voltaire, who ridiculed it in his comic novella *Candide* by having the character Dr. Pangloss (a parody of Leibniz and Maupertuis, an 18th-century French philosopher) repeat it like a mantra. Today, the adjective "Panglossian" describes a person who believes that the world in which we live, suffer, and die is the best possible.

While Leibniz argued that suffering is good because it catalyzes the human will, his critics argue that the degree of human (and animal) suffering is too severe to justify belief that God has created the "best of all possible worlds." Leibniz addresses this concern by considering what God desires to occur (God's "antecedent will") and what God allows to occur (his "consequent will"). Other commentators, including the Christian philosopher Alvin Plantinga, criticized Leibniz's theodicy by arguing that there probably is not such a thing as *the best* of all possible worlds, since one can always conceive a better world, such as a world with more morally righteous persons.[48]

The *Theodicy* was deemed illogical by the philosopher Bertrand Russell. Russell argued that moral and physical evil must result from metaphysical evil (imperfection). But imperfection is merely finitude or limitation; if existence is good by definition, as Leibniz maintains, then the mere existence of evil requires that evil also be good, a self-contradiction.[49]

Another philosopher who was significantly influenced by, if highly critical of, Leibniz's philosophy is Kant. Although Kant mentioned his debt to Leibniz (among other philosophers he respectfully criticizes, most notably David Hume), Kant found Leibniz's philosophy "misleading," due to the "one-sidedness" of his theory.

In sum, *Leibniz's vision of this world is far from the best of all feasible worldviews. Leibniz claims that the world we live in now is the best world that could be created. He says in* The Monadology *that God has the idea of infinitely many worlds. If that is true, why would God make this one? What makes this world so special? If there are infinite worlds, then why wouldn't there be at least one that is slightly, or even much, better than this one? And one better than that one? To account for this possibility, Leibniz concluded that "there can be no infinite continuum of worlds." But how does he know that, and why isn't this world, our world, actually the worst of all possible worlds, since, as far as we know, there are no other actual worlds with creatures sufficiently rational and free to choose to do evil? Only humans have, to date, actualized this abysmal possibility.*

Perhaps ironically, it has been the "fate" of Leibniz's idea of the "best of all possible worlds" to have become one of the most lampooned doctrines in the history of Western ideas. Leibniz's overall contributions to human knowledge in general, however, deserve … a much better fate.

Kant's Conceptual Worlds

Immanuel Kant (1724–1804) is widely considered the most influential Western philosopher of the past few hundred years. Only Descartes, Nietzsche, Marx, and Wittgenstein come close. And while Kant, Descartes, Nietzsche, and Wittgenstein differ in many ways, both personally and philosophically—*Kant, for instance, had little experience of the world outside his hometown, while Descartes and Wittgenstein had much more direct experience of the world outside their natal lands—one of the common threads winding its way through their works is … the concept of the world.*

By then end of the 17th century, two spheres of existence that had previously overlapped were distinguished: the "inner world" of religious and moral values and the "outer world" of physical bodies. The Swiss Protestant theologian John Calvin (1509–64),[50] as well as Descartes, Galileo, and Newton, separated ostensibly descriptive statements of fact, supposedly characteristic of genuine science, from prescriptive and theological principles of value—a polarization that not only abetted the liberation of capitalist business ethics and production methods from "other-worldly" concerns and fetters but also helped to shove non-factual statements into the zone of the "irrational." While critical of conventional religion as well as of both philosophical rationalism and empiricism, *Kant would retain the idea of separate but mysteriously interlinked "worlds" but would also transform the way in which they should be conceived.*

During the last half of the 18th century, a confluence of historical events created "modern" Western thought and civilization. These include the French and American Revolutions; the industrial "revolution" in the capitalist mode of production; and the spread of political liberalism and republicanism, especially during what has come to be known as the "Enlightenment," most prominently in the French-speaking world by thinkers classified as the Encyclopedists[51] and the Swiss philosopher Jean-Jacques Rousseau (1712–78). Prominent Enlightenment thinkers and movements also include, in the English-speaking world John Locke

(1632–1704) and such members of the "Scottish Enlightenment," as the philosopher David Hume (1711–76) and the economist Adam Smith (1723–90; in the German-speaking world, the rationalism of Leibniz and Christian Wolff (1679–1754) and the development of Romanticism;[52] and throughout Europe the extension of the Newtonian and other scientific revolutions into many areas of daily life, notably via their technological applications.

In addition, *what we now call the "Western world," initially centered in western Europe, came into being as a result of these trends as well as because of the colonialist and imperialist ventures by major Western powers, including the widespread deployment of slaves from Africa to the "New World," from the late 15th through the 19th centuries—notably by Portugal, Spain, France, and the United Kingdom—into previously unexplored and "uncivilized" parts of the geographical "world" as we know it today.*[53] *This led to the creation by these "Old-World" European empires of the "New World" of the Americas, as well the colonization of Australia, New Zealand, and large parts of Asia and Africa.*[54]

The critical philosophy of Immanuel Kant was the culmination of this era of scientific progress, philosophical rationalism, republican liberalism, and worldly materialism. *Conceptualizing the world (*German: *die Welt) became a central project in Kant's entire oeuvre, perhaps more so than any preceding "modern" philosopher.* Since Kant's death in the early 19th century, many other prominent thinkers have explicitly or tacitly responded to Kant's conceptual "worlds," notably Schopenhauer, Hegel, Nietzsche, Heidegger, Sartre, Foucault, Derrida, Nancy, and Habermas, from what has come to be called the "continental" (mostly German- and French-speaking) tradition in Occidental philosophy, as well as some philosophers in the "analytic," largely Anglophonic, tradition.[55]

Kant's Metaphysical and Epistemological Worlds

Kant philosophy is an immense effort to address questions he himself had raised; and it takes the form of a conceptual system designed to resolve problems partially of Kant's own making and in part adapted by Kant from his predecessors. In a letter written in 1793, Kant described the plan of his philosophical system as centered on four questions: "What can I know? What ought I to do? What may I hope for? and What is Man (*der Mensch*: "Humanity")?" Kant's metaphysics, epistemology, and philosophy of science ("natural philosophy") constitute his answer to the first question.[56] His ethics deals with the second; his philosophy of religion with the third; and his political, anthropological, and historical writings tackle the final question. While I cannot hope to cover in sufficient depth Kant's encyclopedic system, and will not discuss his theology except for issues related to Kant's metaphysics and ethics, I will summarize some of his most significant contributions to our understanding of the seemingly disparate "worlds" he conceptualized in his epistemology/metaphysics, moral philosophy, and political theory.[57]

It has been customary to divide Kant's work into periods. The scholarly consensus seems to be that there were two major periods in Kant's philosophy, the great divide being the year 1770, when the *Inaugural Dissertation* appeared. Before 1770, one has the option of further subdividing Kant's work into the stages of "dogmatic rationalism" (1755–60), when Kant was still under the sway of the German philosopher Christian Wolff (1679–1754), and "skeptical empiricism."

During his "precritical" period, Kant spent several productive years dealing with theoretical matters in the cosmology and astrophysics of his day.[58] In 1755, he published his first purely philosophical work, the *New Elucidation of the First Principles of Metaphysical Cognition*. There Kant argued that, according to the "Principle of Coexistence," *multiple substances can be said to coexist within the same world only if the unity of that world is grounded in the intellect of God. Although Kant would later claim that we can never have any metaphysical knowledge of the relation between God and the world* (not least of all because we can't even prove that God exists), *he would nonetheless continue to be occupied with the question of how multiple distinct substances can constitute a single, unified world.*

The final publication of Kant's precritical period was *On the Form and Principles of the Sensible and the Intelligible World* (1770), also referred to as the *Inaugural Dissertation*, since it marked Kant's appointment as the University of Königsberg's Professor of Logic and Metaphysics. Although Kant had not yet formulated the arguments that would lead to the development of his "transcendental idealism," many of the important elements of his mature metaphysics are prefigured in that work.

First, in a break from his philosophical predecessors, Kant distinguished between two mental "faculties": the sensibility, and the understanding. Kant argued that the sensibility represents the sensory world of "phenomena" while the understanding represents an intelligible world of "noumena" (derived from the Greek noun *noumenon*) or "things-in-themselves." Also, in describing the "form" of the sensible world, Kant argued that space and time are "not something objective and real," but are rather "subjective and ideal." *The claim that space and time refer to only the appearances of things in the physical world and not as they are "in themselves," would be one of the central theses of Kant's critical philosophy*. With the publication of the *Dissertation* and its critique of the "pan-mathematicism" represented by the Leibniz-Wolffian identification of rational knowledge and physical reality with deductive mathematical laws, Kant's "Socratic turn" is apparent. Kant moved from the study of the laws of nature to the analysis of human cognition and action.

From 1770 until 1781, Kant labored over the problems whose solutions were to find their published form in the first edition of the *Critique of Pure Reason*; this was a gestation period sparked by Kant's awakening from his "dogmatic slumber" in 1772, when he became acquainted with some of David Hume's work. In the *Prolegomena to Any Future Metaphysics* (1783), Kant says that his faith in his rationalist assumptions about knowledge of the world and the human soul was shaken by Hume, especially by Hume's skepticism regarding the possibility of any certain knowledge of necessary causal connections.

Hume argued that we can never have knowledge of necessary connections between causes and effects, because such knowledge can neither be given through the senses nor derived from *a priori* (prior to and independent of sense experience) reasoning. Kant realized that Hume's challenge to rationalist metaphysics was significant because Hume's skepticism about knowledge of the necessity of the connection between cause and effect *could be generalized to all metaphysical and scientific knowledge pertaining to necessity and universality*, not just to causation. For Kant, this crucially involved the questions why mathematical truths necessarily hold true in the natural world, and also whether we can know that a supreme being (God) necessarily exists.

Kant's proposed solution to Hume's skepticism, which would form a pillar of his "critical philosophy," was twofold. Kant initially agreed with Hume that metaphysical knowledge (such as knowledge of causation) is neither given through the senses nor provided *a priori* through pure conceptual analysis. Kant argued, however, that there is another kind of knowledge that is *a priori*—which is logically necessary and universally binding, but prior to and independent of any possible experience, as opposed to *a posteriori* knowledge, which is based on sense experience—yet that is not known simply by analyzing concepts. He referred to this as "synthetic *a priori*" knowledge, judgments, or propositions. Whereas "analytic" judgments, or propositions, are true by definition and are justified by the semantic relations between the important words in the specific proposition (e.g., "all circles are round"), synthetic judgments are verified by their correspondence to the particular object they describe (e.g., "the table over there is white"). Kant's challenge to Humean skepticism in particular and to empiricism more generally is illustrated by his claim that there is *synthetic a priori knowledge*, thereby implying that an object is not simply "given" to the human senses, but is in some sense *constructed* by what Kant would call the innate mental "categories" (such as time, space, relation, and causation) of the "faculty" Kant called the understanding. This argument underlies Kant's "transcendental idealism," in contrast with the varying idealisms of Plato, Descartes, and Leibniz, as well the "immaterialism," or "subjective idealism" of the Irish philosopher (Bishop) George Berkeley (1685–1753).[59]

The other component of Kant's response to Humean skepticism is his explanation of how synthetic *a priori* knowledge is possible. He described the key to this solution as a "Copernican Revolution," or shift, in his thinking about the relation between the mind and the world.[60] Prior to Kant, most philosophers and scientists, especially such empiricists as Locke and Hume, had claimed that the world fills the "blank slate" (*tabula rasa*) of the mind with sensations, intuitions, and impressions, which are then somehow organized into perceptual experiences. *Kant reversed this and argued that the human mind is not a blank slate but instead, because of its underlying conceptual structure, in some non-trivial way constructs the world.*

To construct, or mentally to have a world, according to Kant, requires both "concepts," or thoughts, and sense impressions, or the content of "intuitions." As Kant puts it:

> Thoughts without content are empty, intuitions without concepts are blind. It is, therefore, just as necessary to make our concepts sensible, that is, to add the object to them in intuition, as to make our intuitions sensible, that is to bring them under concepts. These two powers or capacities cannot exchange their functions.[61]

This shift in epistemology from, in Kant's words, the sensible to the intelligible world, is comparable in significance to the Copernican Revolution in physics (natural philosophy).

Copernicus had demonstrated that it only *appears* as though the Sun and stars revolve around the Earth, and that we might have knowledge of the way the solar system *really* is, if we acknowledge that the sky looks the way it does because we perceivers are moving as the Earth rotates on its axis and moves around the Sun. Analogously, Kant argued *against* the belief that the way things appear reflects how things are in themselves. He further claimed that we must investigate the most basic underlying but unexamined structures of knowledge and experience, that is, the cognitive preconditions for the way things appear to us and how we can know anything whatsoever. If it is only possible to have experience of an object if the object conforms to the cognitive preconditions of experience in general, then, Kant argued, knowing the preconditions of experience will enable us to have knowledge—in the form of universal and necessary synthetic *a priori* judgments—of every possible object of experience. Kant thought he had refuted Hume's skepticism by showing that we can have synthetic *a priori* knowledge of objects in general when we take as the object of our investigation the very forms of a possible object of experience. The *Critique of Pure Reason* is Kant's magisterial attempt to work through the important details of this basic philosophical strategy.

In his *Critique of Pure Reason* (considered as perhaps the single most influential book in the history of Western epistemology), to respond to Hume's critique of metaphysics, Kant thought it necessary initially to provide an extended critique of unaided reason's pretensions to "know" "ultimate realities." This is the "negative" or "critical" component of Kant's epistemological project. But while highly critical of what he considered ungrounded speculations about entities transcending human experience, Kant also provided a "positive" defense of "scientific metaphysics." By demonstrating the strict universality and logical necessity of synthetic *a priori* propositions in mathematics, physics, and metaphysics—like "every event has a cause" and "7+5=12"—*Kant believed that the claim to objectivity made by science in particular and human knowledge more generally as a defense against skeptical and empiricist disclaimers might be preserved. This is a unique contribution to Western thought in general and to "idealistic" worldviews in particular.*

The metaphysical system underlying this worldview is what Kant called "transcendental idealism." It is "transcendental" because, unlike such empiricists as Hume, Kant argued that there are "transcendental" (not "transcendent") preconditions for sensory experience and knowledge of the physical world, notably the "categories of pure understanding."[62] And like such previous "idealists" as Plato[63] and such "rationalists" as Descartes and Leibniz, Kant was a philosophical dualist.

In addition to defining the physical world as "the sum total" or "totality of all appearances," Kant also divided reality into two separate "worlds," the "phenomenal," "natural," "observable" and "sensible," world (mundus sensibilis) accessible to human sensory experience on the one hand, and the "noumenal," or "intelligible" world (mundus intelligibilis), which cannot be known in any sensory way but nonetheless exists, on the other hand.[64] *This "intelligible" world includes such "supersensible" objects, or "ideals," as God, freedom, and the human soul.*

But Kant also argued that we can only have knowledge of things we can experience. Accordingly, in answer to the question, "What can I know?" Kant replied that *we can only know the appearances of things in the natural, observable world.* However, and crucially, *we cannot know what underlies this physical world*, namely "things in themselves," or be able to prove the existence, or demonstrate the empirical "reality," of God, freedom, and the immortality of the human soul.

To address these and such other ontological questions as "why is there anything at all?" and "must there be some being, a divine entity humans call God that has created the world/universe and everything in it and endowed them with a purpose?," Kant proceeds from a number of assumptions. First, the phenomenal world of appearances is all we can ever know. Second, every effect in this world is the result of a cause, arising either from nature or from human freedom. Third, just as everything that happens within this world has a cause, so must the world itself—the totality of appearances—have a cause, of which it is the effect. Next, the cause of that world must either lie within the world of appearances or outside it. But since, this world-foundational cause cannot lie within the world itself (if it did, it would be an effect of another, more foundational cause), it must lie in "another" world, the noumenal world of things-in-themselves, of pure ideas.[65]

God is such a pure idea as well as an ultimate foundational cause and does not, by definition, lie within the natural world of appearances. So, God must be an idea we have by nature of our reason, not the effect of some experience we have in the sensible world. But our reason cannot prove the existence of God (or of such other ideas as freedom and the immortality of the soul), as demonstrated in Kant's devastating critique of ontological proofs for God's existence, in which Kant argued that existence is not a predicate of a logical/grammatical subject, or proper name, even of the substantive "God." And, God is also not within the world of human experience or the object of any possible knowledge. Accordingly, Kant concludes, in one of the most striking proclamations in the history of Western thought: *"I have therefore found it necessary to deny knowledge to make room for faith."*[66] For Kant, then, rather than being either a possible object of knowledge by human understanding or an entity whose existence is provable by logical demonstration, God is a matter of faith, and "... the ideal of the supreme being is nothing but a *regulative principle of reason*, which directs us to look upon all connection to the world *as if* it originated from an all-sufficient necessary cause."[67]

Furthermore, in addition to the phenomenal world of sense perceptions and cognitions, and the noumenal "world beyond this world" of such ideals as God, freedom, and immortality, Kant also calls:

> the world a moral world, in so far as it may be in accordance with all moral laws; and this is what by means of the freedom of the rational being it can be, and what according to the necessary laws of morality it ought to be ... This world is so far thought as an intelligible world only, ought to have an influence upon the sensible world, to bring that world, so far as it may be possible, into conformity with the idea. The idea of a moral world has, therefore, objective reality.[68]

Like all of creation, whatever happens in this "moral world" is the effect of a cause. But in this case, the cause is within us and due to our conduct, not as an effect of natural causes or of divine purposes. And, furthermore, the moral world is neither a past nor the present world, but rather what Kant calls "a future world." As Kant puts it:

> Now since we are necessarily constrained by reason to represent ourselves as belonging to such a world [a *moral* world], while the senses present to us nothing but a world of appearances, we must assume that moral world to be a consequence of our conduct in the world of sense ... and therefore to be for us a future world.[69]

Perhaps ironically, Kant spent hundreds of pages in his epistemological and metaphysical *chef d'oeuvre*, the *Critique of Pure Reason*, debunking the pretensions of unaided human reason to know anything beyond its ken. And, yet his conclusion is that we must "leave room for faith" and assume that "the idea of a moral world has ... objective reality!" It is to Kant's moral world(s) of the future that I now turn.

Kant's Moral Worlds

Just as Kant reframed how we view the world as a metaphysical and epistemological object of inquiry, so, too, did Kant revolutionize moral philosophy. His two major works on ethics are the *Groundwork of the Metaphysics of Morals* (1785) and the *Critique of Practical Reason* (1788), both major components of Kant's "critical philosophy."[70] They address in detail the second of Kant's foundational questions, "What ought I to do?"

Kant is justly renowned both for the depth and complexity of his thinking as well as for some remarkably pithy and striking declarations. This passage from the Conclusion of the *Critique of Practical Reason* epitomizes both of these stunning stylistic traits:

> *Two things overwhelm the mind with constantly new and increasing admiration and awe the more frequently and intently they are reflected upon: the starry heavens above me and the moral law within me* ... I see them before me and connect them immediately with the consciousness of my existence. The first begins from the place which I take up in the external world of our sense and expands the connection in which I stand into unfathomable magnitude with worlds beyond worlds and systems of systems, and even beyond that into the limitless times of their periodical movement and their beginning and continuation. The second begins with my invisible self, my personality, and displays me in a world which has true infinity, but which is sensible only to the understanding, and with which (but in that way also simultaneously with all those visible worlds) I recognize myself not in merely contingent, but rather in universal and necessary, connection. The first glance at innumerable masses of worlds destroys, as it were, my importance as an animal creature who must give back again to the planet (a mere point in the universe) the material from which it was made and after having been furnished (who knows how?) with vigorous life for a short time. The second, on the other hand, infinitely elevates my value as an intelligence through my personality, in which the moral law reveals to me a life independent of animalism and even from the entire world of our senses, at least as far as can be discerned from the purposeful determination of my existence through this law, which determination is not limited to conditions and boundaries of this life, but rather goes out to infinity.[71]

Once again, Kant has a theory of multiple worlds, external and internal, the former determined by the laws of nature, the latter governed by what Kant would call the "laws of freedom." Both go "out to infinity." But the infinite cosmos dwarfs in scale the terrestrial world in which we exist. And humanity, in Kant's view, is distinguished from both the "animalistic" world and the "entire world of the senses" by "the moral law." Kant, therefore, has an anthropocentric view of the human species' stature on the planet on which we, as well as all other life forms, exist.

This anthropocentrism extends so far in Kant's critical philosophy as to imply not only that humans can only see and interpret the world through their own cognitive and moral lenses, which is plausible, but also that humanity is the "center" and "end" of all creation. As Kant puts it in his often-neglected *Critique of Judgment*:

> Now of man (and so of every rational creature in the world) as a moral being it can no longer be asked: why he exists? His existence involves the highest purpose to which ..., he can subject the whole of nature; contrary to which at least he cannot regard himself as subject to any influence of nature.—If now things of the world, as beings dependent in their existence, need a supreme cause acting according to purposes, man is the final purpose of creation; since without him the chain of mutually subordinated purposes would not be complete as regards its ground. Only in man, and only in him as subject of morality, do we meet with unconditioned legislation in respect of purposes, which therefore alone renders him capable of being a final purpose, to which the whole of nature is teleologically subordinated ... The commonest Understanding, if it thinks over the presence of things in *the world, and the existence of the world itself,* cannot forbear from the judgement that *all the various creatures,* no matter how great the art displayed in their arrangement, and how various their purposive mutual connection—even the complex of their numerous systems (which we incorrectly call worlds)—*would be for nothing, if there were not also men (rational beings in general). Without men the whole creation would be a mere waste, in vain, and without final purpose.*[72]

Kant's moral philosophy is based on the premise that a "rational being" (*animal rationale*) is the "end" (*Zweck*), or final purpose, of creation, of God's design for the world as a whole. In this world, although it might appear that these rational creatures, or "men," are equated with the human race as a whole, Kant's disparaging remarks regarding "irrational" "rabble" (*Pöbel*), and others (children, "savages," and, probably, most women as well) who cannot gain "mastery over their inclinations," and therefore remain stuck in the "state of nature," imply that is primarily if not exclusively adult (northwestern European) males who are capable of fully exercising their reason in both "pure" (i.e., "metaphysical and epistemological") and "practical" (i.e., ethical) ways.[73] To exercise one's practical reason, for Kant, means to follow the "moral law." Kant's ethical philosophy is in large part a "practical" exploration of how we might become sufficiently rational to become moral law followers and thereby to justify "men's" unique teleological place in creation.

According to Kant, humans belong to both the physical world of sense experience and the intelligible world of the understanding. As members of the world of sense, our choices and actions, like everything else in the natural world, fall under the laws of nature, which causally determine physical events in that world of appearances. But as members of the world of the understanding, we are free, and so our wills are capable of being governed by the moral law. Since the world of the understanding contains the ground of the world of sense and its laws, it is in our capacity as members of the world of the understanding, to give laws to ourselves as members of the world of sense. And this is what gives us moral and other obligations. The conception of ourselves as members of the world of the understanding is a conception of ourselves as self-governing, and so as autonomous and morally rational, actors.[74]

As Kant says:

> As a rational being, and thus as a being belonging to the intelligible world, the human being can never think of the causality of his own will otherwise than under the idea of freedom; for, independence from the determining causes of the world of sense (which reason must always ascribe to itself) is freedom. With the idea of freedom the concept of *autonomy* is now inseparably combined, and with the concept of autonomy the universal principle of morality, which in idea is the ground of all actions of *rational beings,* just as the law of nature is the ground of all appearances.[75]

Kant argues in the *Groundwork of the Metaphysics of Morals* that, to the degree that we ourselves are the *authors* of our own thoughts and choices, we are promoting our own freedom and autonomy, our practical reason, or moral rationality. And insofar as we deem ourselves the ultimate sources of these cognitive and ethical "inner appearances," we think of ourselves as members of the noumenal world, "the world of the understanding." By freely choosing to follow or disobey the moral law, we demonstrate our autonomy from both the natural world of appearances and the wills of other actors.[76]

Kant attempted to establish and justify what he called "the supreme principle of morality." That supreme principle, or moral law, which Kant named the *categorical imperative*, commands unequivocally that our actions derive from universal principles. Whenever we act, we ought to ask ourselves whether the reasons for which we propose to act, or subjective "maxims," could be made universal, embodied in a binding, or "categorical," principle, allowing no exceptions, and constraining our behavior.

In his *Critique of Practical Reason*, Kant calls this categorical imperative, or moral law, the "Fundamental Principle of Pure Practical Reason," which commands us to "Act in such a way that the maxim of your will can at the same time always hold as a principle of a universal legislation … Pure reason is practical of itself alone and gives (humans) a universal law called the moral law."[77] Furthermore, according to Kant:

> because it is to be valid for everyone who can reason and will, the moral law is necessarily thought as objective. The maxim of self-love (prudence) merely counsels; the law of morality commands. *But there is a great difference between what we are advised to do and what we are obligated to do.*[78]

Indeed …

But *why* is that? Why don't (most?) human beings follow the moral law, obey the categorical imperative, not do what others "advise" us to do or what we feel like doing, and instead follow our obligation to, in Kant's words, "do our duty?"[79] Aristotle explained this as due to "weakness of the will" (*akrasia* in Greek), "bad luck," and as well to other *external* circumstances seemingly out of the control of individual actors, who, for a combination of reasons, may not act "rationally," or "prudently."[80] Kant, while also advocating the cultivation of practical rationality, in contrast with Aristotle, focused on the *internal*, mental tug of war between our subjective "inclinations," or what we feel like doing or not doing, and our "obligations" to do what is "right" and never, for any reason, to deviate from doing our duty to follow the categorical imperative—never, for example, for any reason to lie or to treat other human means instrumentally, as "means rather than as ends."

Kant's "objective" moral law and "knowledge" of ethical duties and obligations are frequently referred to as "deontology."[81] *If we can manage to do what is right, that is to do our duty by following the categorical imperative, consistently, we can tame the irrational, "animalistic" components of our nature and construct a moral world, a "kingdom of ends," in which we, as rational beings, live peaceably together obeying the "laws" of our own "freedom."*[82] But, as Gaston argues, for Kant:

> the moral world cannot be "grounded merely in nature" but in "the ideal of the highest good …" The moral world then gives us a "special" vantage point on the world as a whole … it is a regulative world … it is a categorical world.[83]

In other words, the moral world of rational beings is an "intelligible" and not a "natural" world, an "ideal" and not a "real" world. It is the "real" world of human anthropology, history, and politics, as analyzed by Kant, to which I now turn.

Knowing and Having the World: Kant's Anthropocentric Anthropology

As previously mentioned, Kant's view of the cosmos in general, and of this earthly world in particular, is anthropocentric. In Kant's vast opus, there is very little said about life on Earth apart from humanity. And we, humans, are endowed by God with reason and freedom. If we exercise our freedom and reason in accord with the moral law, we are rightly to be considered the lords of nature, and of the living world it contains, and, hence, as the "end of creation."

However, insofar as we are a part of nature as a whole, we also have a "nature," a *human* nature, which, in Kant's view, is prone to irrationality and rule-breaking rather than to rule-following. In his late work *Anthropology from a Pragmatic Point of View* (1798), Kant sees "anthropology" as the study of human nature, and "pragmatic anthropology" as useful, practical knowledge needed to successfully navigate the geographical and moral worlds. Kantian anthropology is based *not "on things in the world" but on "knowledge of the human being as a citizen of the world."* In that work, Kant discusses "knowledge of the world" as distinguished from "having a world":

> All cultural progress, by means of which the human being advances his education, has the goal of applying this acquired knowledge and skill for the world's use. *But the most important object in the world to which he can apply them is the human being: because the human being is his own final end.—Therefore to know the human being according to his species as an earthly being endowed with reason especially deserves to be called knowledge of the world, even though he constitutes only one part of the creatures on earth* ... Such an anthropology, considered as knowledge of the world, which must come after our schooling, is actually not yet called pragmatic when it contains an extensive knowledge of things in the world, for example, animals, plants, and minerals from various lands and climates, *but only when it contains knowledge of the human being as a citizen of the world*. Therefore, even knowledge of the races of human beings as products belonging to the play of nature is not yet counted as pragmatic knowledge of the world, *but only as theoretical knowledge of the world. In addition, the expressions "to know the world" and "to have the world" are rather far from each other in their meaning,* since one only understands the play that one has watched, while the other has participated in it.[84]

While a great deal of Kant's work focuses on what it means to "*know* a world," very little explicitly unpacks the phrase "to *have* the world," to participate in its grand "play." Such 19th- and 20th-century "continental" philosophers as Martin Heidegger, Jean-Paul Sartre, and Maurice Merleau-Ponty took great care in exploring what it means for human beings to "have a world" in which we exist as perceiving, interacting, and emotional beings, as well as knowing subjects.

In another work written toward the end of his career, *Religion within the Limits of Bare Reason* (1793), as well as in his scattered essays on politics, peace, and history, Kant addressed in detail humanity's strengths and weaknesses, the "good" and "evil" within our nature.[85]

Human Nature in the Historical and Political Worlds

Human nature, like scientific knowledge and virtuous action, is not "given" for Kant but is constructed. For such idealists and rationalists as Plato, Leibniz, and Kant, the "natural" part of human nature is the "base" element in mankind, the bane of human existence. Insofar, as humanity is a cog in the machine of nature, a mere creature of the senses (German: *Sinnenwesen*), it is insignificant and bestial.

As Kant saw it, in the "state of nature," humanity's inclinations are perverse and unregulated. And we have an unlimited propensity to overindulge our cravings for honor, pleasure, and power; and our childlike narcissism and "immaturity" veil our lack of virtue. Left without a trainer, humanity, as Kant thought, has all the passions, impulses, and instincts of animals and all the vices and violent predilections of "savages."

However, humanity, is also a unique kind of terrestrial animal, one endowed with the capacity of reason (*animal rationabilis*). Accordingly, unlike other animals, we can make of ourselves rational animals (*animal rationale*). In doing so, we first seek to preserve ourselves and the species as a whole. Then, we train and educate ourselves for domestic society. And finally, humanity may come to regard itself as a systematic whole (i.e., a unified species ordered by principles of reason), and traverses the long road to Enlightenment (German: *Aufklärung*).

But we are held back from civilizing ourselves by our "subhuman irrationalism." Kant defines irrationalism (German: *Unvernunft*) as an "absence of rules" or a lack of discipline; and humankind "needs to act according to rules imposed by reason" if it is to introduce law and justice into the state of nature. Among the list of bestial and irrational elements in the state of human nature are bad habits, unrestrained sense appetites, superstitious beliefs, wars, ignorance, misology, pathological inclinations, envy, malice, "horrible lust," masturbation, mental illusions and deficiencies, fanaticism, the "rabble" (German: *Pöbel*), and the two deepest impulses of human nature—the "love of life and the sexual drive" (the latter impulses remarkably presaging Nietzsche's and Freud's related ideas).

But what is even more reprehensible than "our dormant vices" and "vicious inclinations" is what Kant labels "immoral"—for here one *chooses* to do evil, as in the cases of lying and suicide; one chooses to act, not out of ignorance of the moral law, but contrary to it. And after all, it is the moral law that transforms an *animal rationabile*, a being capable of rationality, into the *animal rationale*, a "rational being." To choose freely to violate that moral law, which, according to Kant, is based on human narcissism, is universal, and reflects the "evil propensity in human nature."

Kant's conception of "the radical evil" (*Böse* in German) in human nature rests on three assumptions: (1) evil constitutes the underlying disposition of the human will (and hence is "radical"); (2) evil consists in the motivational primacy of the principle of self-love; and (3) there is a universal propensity to evil in all human beings, even the best.[86] In Kant's words from *Religion within the Limits of Bare Reason*:

> What I call "human nature" is the subjective basis of the exercise (under objective moral laws) of man's freedom … which … precedes every action that is apparent to the senses. But this subjective basis must also be an expression of freedom, because otherwise the resultant action … couldn't be morally good or bad. *So the basis of evil can't lie in anything that determines the will through inclination, or in any natural impulse; it can lie only in a rule that the will makes for itself, as something on which to exercise its freedom—i.e. a maxim* … If it were not ultimately a maxim but a mere natural impulse, the man's "free" action could be tracked back to determination by natural causes, which contradicts the very notion of freedom. *So when we say "He is by nature good or … bad," this means only: There is in him a rock-bottom basis (inscrutable to us) for the adoption of good maxims or of bad ones (i.e. maxims contrary to law); and he has this just because he is a man, so his having it expresses the character of his species*. So we shall characterize as innate the good or bad character that distinguishes man from other possible beings that have reason … you have to grasp how innate is being used here. *The rock-bottom basis for the adoption of our maxims must itself lie in free choice*, so it can't be something we meet with in experience; therefore, the good or evil in man … is termed "innate" only in the sense of being posited as the basis for—and thus being earlier than—every use of freedom in experience … so it is conceived of as present in man at the time of birth—though birth needn't be its cause …[87]

The way to overcome this propensity to do ill is to cultivate the development of people of "goodwill," that is, those who freely choose to follow the moral law and live together in a republican commonwealth, the "kingdom of ends" under the supreme moral legislator, God the creator.[88]

Kant also poses the timeless question, is "the world" getting better, getting worse, alternating between improvement or decline, or in endless repeating cycles? (as many ancient Greeks and Romans believed). In *Religion within the Limits of Bare Reason*, Kant struggles with that conundrum, as well as the "innate" and/or "freely chosen" roots of "good and evil":[89]

> The complaint that "the world lies in evil" is older than history … as old as … the religion of the priests. All religions agree that the world began in a good state … But they soon let this happiness vanish … and give place to a fall into evil (moral evil, always going hand in hand with physical evil), speeding mankind from bad to worse … *so that now we live in the final age, with the Last Day and the destruction of the world knocking at the door … More recent … is the opposite optimistic*

> *belief that the world is steadily … moving from bad to better,* or … that the predisposition to move in that way can be found in human nature … If this is a thesis about movement … from moral badness to moral goodness … it certainly hasn't been derived from experience—the history of all times speaks too loudly against it!

Kant then continues and poses the following logical possibilities:

> *Isn't it at least possible that … man as a species is neither good nor bad, or—partly good, partly bad. We call a man bad … not because his actions are bad (contrary to law) but because his actions show that he has bad maxims in him …* Man's predisposition to animality can be … "physical and purely mechanical self-love," for which reason isn't needed. It is threefold: for self-preservation; for the propagation of the species through sexual intercourse and the care of offspring … and for community … i.e. the social impulse. On these stems all kinds of vices can be grafted (but they don't spring from) (1) this predisposition itself as a root. They can be called vices of the coarseness of nature and their extreme cases are called the "bestial vices" of gluttony and drunkenness, lasciviousness and wild lawlessness … (2) The predisposition to humanity can be … "self-love" … we judge ourselves happy or unhappy only by comparing ourselves with others. This self-love creates the inclination to become worthy in the opinion of others.[90]

Kant further develops this idea and rephrases the three possibilities for the course of world history as follows:

> There are three possible forms which our prophecy might take. The human race is either continually regressing and deteriorating, continually progressing and improving, or at a permanent standstill, in relation to other created beings, at its present level of moral attainment … The first statement might be designated moral terrorism, the second eudaimonism …, while the third could be called abderitism [a Kantian neologism equivalent to "foolishness"]. For in the latter case, since a genuine standstill is impossible in moral affairs, rises and falls of equal magnitude constantly alternate, in endless fluctuation, and produce no more effect than if the subject of them had remained stationary in one place … *The terroristic conception of human history … A process of deterioration in the human race cannot go on indefinitely, for mankind would wear itself out after a certain point had been reached. Consequently, when enormities go on piling up and up and the evils they produce continue to increase, we say: "It can't get much worse now." It seems that the day of judgement is at hand, and the pious zealot already dreams of the rebirth of everything and of a world created anew after the present world has been destroyed by fire.* [91]

"The terroristic conception of history," according to Kant, is largely rooted in the "propensity to evil in human nature," with that inclination being either innate or acquired, and consisting in "the subjective basis for the possibility of its maxims (subjective preferences, or intentions) deviating from the moral law." If this propensity is characteristic of humankind as a whole, Kant deems it a "natural propensity" to "evil" and thus indicative of the "frailty of human nature." Thus, for Kant, the proposition "man is bad by nature" means that:

> he is conscious of the moral law but has nevertheless allowed occasional departures from it into his maxim. "He is bad by nature" means that badness can be predicated of man as a species ... [and is] rooted in humanity itself ... and ... because of freedom ... we can call it a radical innate evil in human nature, though one we have brought upon ourselves.[92]

On the other hand, Kant also thought it is the "aim of nature" to elevate the human race from its moral depravity and lawless perversity to freedom, dignity, and rationality. There is a purposiveness (German: *Zweckmässigkeit*) in nature as a whole, a "secret plan," that terminates in a moral *Rechtsstaat* (state of laws) and a perfectly stable and pacified society. The purpose of nature is to induce humanity to become fully human by outgrowing its nature, by becoming rational and good. History and civil society are the human vehicles that carry out nature's plan.

Kant believed it is by means of "our physiological deficiencies and our unsocial sociability," that nature has nudged us, generation by generation, to develop our capacity for reason and slowly to ascend from the fog of prehistory to the present age of Enlightenment. This development is not yet complete. In his essay: "An Answer to the Question: What is Enlightenment?" (1784),[93] Kant argued that to be enlightened is to determine one's beliefs and actions in accordance with the free use of one's reason. The process of enlightenment is humanity's "emergence from its self-incurred immaturity," that is, our emergence from an uncritical reliance on the authority of others (e.g., parents, monarchs, priests, and other authorities). This is a slow, ongoing historical process. Kant thought that his own age was a time (late 18th-century Western Europe) of enlightenment, but not yet a fully enlightened age.

The goal of humanity is, according to Kant, to become fully enlightened, an historical endpoint when all interpersonal interactions are conducted in accord with reason, and hence in harmony with the moral law (this is the idea of a kingdom of ends). Kant believed that there are two significant conditions that must be in place before such an enlightened age can come to be. *First, humans must live in a perfectly just society under a republican constitution. Second, the nations of the world must coexist as an international federation in a state of "perpetual peace."*

Following Rousseau, Kant argued that republicanism is the best form of government.[94] In a republic, voters elect representatives and these representatives decide on specific laws on behalf of the people. However, perhaps reflecting the cultural prejudices of his own age and society, Kant also claimed that neither women nor the poor should be full citizens with voting rights. Among the freedoms that ought to be respected in a just society (republican or otherwise) are the freedom to pursue happiness in any way one chooses (as long as this pursuit does not infringe the rights of others), freedom of religion, and freedom of speech. The last two freedoms were quite important to Kant and he associated them with the ongoing enlightenment of humanity. In "What is Enlightenment?" Kant argued that it "would be a crime against human nature" to legislate religious doctrine, because doing so would be to deny to humans the freedom of the "public use of reason" that makes them human. Similarly, restrictions on the "public use of one's reason" are contrary to the most basic teleology of the human species, namely, the development of reason.

But is such a peaceful and just world possible? And if so, how might humanity arrive at this ultimate goal of human history? Kant's philosophy of world history underlies his vision of a pacified world.

Kant's Philosophical History of the World

Kant attached little importance to individual historical events or to particular individuals, and instead he proposed a philosophy of world history. The totality of all deeds past and present reveals a necessary pattern in our history, a lawful development toward transcendental ethical goals. This pattern underlying all contingent acts is a providentially guided plan for the progress of humanity. The "guiding thread of reason" in history propels human events up the road from war to peace, from nature to freedom.

As his works on anthropology, education, politics, history, and other "empirical" subjects make clear, however, Kant by no means believed that the majority of humans lived rational or moral lives. Quite the contrary: Kant's reflections on the past and on his contemporaries reveal a disenchantment with the actual state of the world of his time, one that presaged the much more pessimistic reflections of Schopenhauer (to be discussed later in this chapter) and Max Weber.

Kant's philosophy of history is founded on a teleological theodicy. The "final ends" of history are identical with the "purpose" of human nature and the "purpose of man as a species." And these ends are moral in nature. An obscure supersensible unity of nature and freedom, a unity known only by the name of Providence or through the insight of God, guarantees the positive outcome of history—the cosmopolitan rule of law. At most, history provides us with a "clue to the explanation of the confusing game of human affairs." This is made clear in Kant's essay "Idea for a Universal History from a Cosmopolitan Point of View" (1784), wherein he states:

> However obscure their causes, history, which is concerned with narrating these appearances, permits us to hope that if we attend to the play of freedom of the human will in the large, we may be able to discern a regular movement in it, and that what seems complex and chaotic in the single individual may be seen from the standpoint of the human race as a whole to be a steady and progressive though slow evolution of its original endowment.[95]

In that work, Kant also proposed the following theses:

> Fifth Thesis *The greatest problem for the human race, to the solution of which Nature drives man, is the achievement of a universal civic society which administers law among men* ... The highest purpose of Nature, which is the development of all the capacities which can be achieved by mankind, is attainable....
>
> Eighth Thesis *The history of mankind can be seen, in the large, as the realization of Nature's secret plan to bring forth a perfectly constituted state as the only condition in which the capacities of mankind can be fully developed, and also bring forth that external relation among states which is perfectly adequate to this end.*

But Kant seems at times appears to be ambivalent about this outcome. Will this world, humanity's world, ever fulfill what Kant hoped would be its "ultimate justification," or would that hope be realized only "in another world?"

> Such a justification of Nature—or, better, of Providence—is no unimportant reason for choosing a standpoint toward world history. For what is the good of esteeming the majesty and wisdom of Creation in the realm of brute nature and of recommending that we contemplate it, if that part of the great stage of supreme wisdom which contains the purpose of all the others—the history of mankind—must remain an unceasing reproach to it? *If we are forced to turn our eyes from it in disgust, doubting that we can ever find a perfectly rational purpose in it and hoping for that only in another world?*[96]

The hope for humanity, for the preservation and enlightenment of our world, rests—both for Kant and for us—on the emergence of "perpetual" peace from seemingly unending war and conflict.

Toward a Peaceful World?

Kant elaborated his cosmopolitan proposal for the pacification of our world in his essay "Perpetual Peace: A Philosophical Sketch" (1795). According to Kant, world peace can be achieved only when international relations mirror in important ways the relations between individuals in a just society under a republican constitution.[97] Just as people should neither be objectified nor instrumentalized—neither treated as things nor used as means to another's ends—so, too, should independent and sovereign states not be deployed as mere objects for another state's or empire's usage. And just as individuals must respect others' rights to free self-determination, so, too, "no state shall forcibly interfere in the constitution and government of another state."[98]

Kant argued that since individuals should organize themselves into just and equitable societies, states, considered as individuals writ large, must also arrange themselves into a global federation, a "league of nations."[99] Of course, according to Kant, until a state of perpetual peace is reached, wars will be inevitable, since, in his view, war is "the natural state of affairs."[100]

In Kant's opinion, war is the result of an imbalance or disequilibrium in international relations, a view strikingly consistent with modern strategic "balance of power" theories, although Kant might not have agreed with the view that perpetual war is inevitable without such a "balance of terror" between nation-states, especially in the nuclear age. Even in times of wars, however, Kant argued that certain laws must be respected. For instance, it is never permissible for hostilities to become so violent as to undermine the possibility of a future peace treaty. Although wars are never desirable, they often lead to new conditions in international relations, arrangements that are sometimes more balanced than the previous ones, resulting in a lower probability of new wars. *Overall, although the slow march toward perpetual peace is an often violent process, all the states of the world can slowly work toward international equilibrium, and hence to global peace.*

Kant's optimism about such progress extended *only to the future of the species as a whole, and not to any individual in particular.* And "man," as a species, needs, in Kant's words "a master," since:

> Man is an animal which requires a master for he certainly abuses his freedom with respect to other men, and although as a reasonable being he wishes to have a law which limits the freedom of all, his selfish animal impulses tempt him, where possible, to exempt himself from them. He thus requires a master, who will break his will and force him to obey a will that is universally valid, under which each can be free.[101,102,103,104]

Furthermore, Kant observes, in a tone strikingly familiar to the tenor of our own time, that:

> Although, for instance, *our world rulers at present have no money left over for public education and for anything that concerns what is best in the world, since all they have is already committed to future wars, they will still find it to their own interest at least not to hinder the weak and slow, independent efforts of their peoples in this work.* In the end, war

> itself will be seen as not only so artificial, in outcome so uncertain for both sides, in after-effects so painful in the form of an ever-growing war debt (a new invention) that cannot be met, that it will be regarded as a most dubious undertaking.[105]

Despite the best, and worst, efforts of "our world rulers" to foster bellicosity and to undermine their citizens' interest in peace-making and peace-building, Kant believed in the power of "humanity" as a whole to fulfill nature's "hidden plan." The emergence of reason in the human species, and of the eventual pacification of existence, was to Kant a process that had been preordained by an incomprehensible Providence. And, as Kant declared in "Perpetual Peace":

> by using the common right to the face of the earth, which belongs to human beings generally … publicly established by law … the human race can gradually be brought closer and closer to a constitution establishing world citizenship … The idea of a law of world citizenship is no high-flown or exaggerated notion. It is a supplement to the unwritten code of the civil and international law, indispensable for the maintenance of the public human rights and hence also of perpetual peace. One cannot flatter oneself into believing one can approach this peace except under the condition outlined here.[106]

Alas, the perpetual peace Kant envisioned seems even more distant today than at the time of Kant's death in 1804, the year Napoleon declared himself "Emperor" while waging war against France's neighbors. Today, Kant's moral and political worlds might well appear in retrospect to have been even more idealistic than his metaphysical cosmos.

Kant's Worlds and the Worlds to Come

> If then we call the sight of the starry heaven *sublime, we must not place at the basis of our judgement concepts of worlds inhabited by rational beings,* and regard the bright points, with which we see the space above us filled, as their suns moving in circles purposively fixed with reference to them; but we must regard it, just as we see it, as a distant, all-embracing vault.[107]

Kant proceeds, almost rhapsodically:

> *This world presents to us so immeasurable a stage of variety, order, purposiveness, and beauty, as displayed alike in its infinite extent and in the unlimited divisibility of its parts, that even with such knowledge as our week understanding can acquire of it, we are brought face to face with so many marvels immeasurably great, that all speech loses its force, all numbers their power to measure … and that our judgement of the whole*

resolves itself into an amazement which is speechless, and only the eloquent on that account.[108]

Kant's Impact on the World to Come

Immanuel Kant was, arguably, the greatest Occidental thinker since Aristotle, and the creator of linguistic worlds whose depth and breadth is virtually unrivaled. Kant believed that the universe is inherently purposive and lawful; that humanity is capable of disinterested, practical rationality; that knowledge, truth, and scientific objectivity are both possible and necessary, and that world history is the battleground between liberal, cosmopolitan agencies of reason, and regressive, selfish forces of impulse and superstition. Kant was an epistemological and ethical dualist. But although there are significant tensions within Kant's worldview, Kant's philosophical system allows for greater interaction between its binary oppositions than that of virtually any other Western idealist, especially Plato and Descartes.

According to Michel Foucault, over the course of his intellectual career Kant changed his theoretical emphasis from cosmology to cosmopolitanism.[109] *Kant also thought that this world, the human realm, is a republic to be constructed rather than a natural fact. Human beings become over time citizens of the world (*German: *Weltbürger), and in so doing, they are able to assume the vantage point of the world as a whole.*

Kant's metaphysical construction of moral and metaphysical worlds of divided and ultimately unknowable subjects and objects was jettisoned by most post-18th-century thinkers. Marx, Nietzsche, Kierkegaard, and the existentialists, as well as most positivists, historicists, and analytic philosophers, have tended to renounce speculative metaphysics altogether. But in so doing, they reflect the debt owed to Kant, since their worldviews are inconceivable without his.

Kant's quest to know the phenomenal world we perceive and in which we live, and to elevate the moral world of human behavior from what it is to what it should be, remains a model and a challenge for the thinkers and actors of this, and any possible future, century.

Hegel: The History of the World Is the World's Court of Judgment

In 1821, in his magisterial work on social and political philosophy, whose English title is *Elements of The Philosophy of Right* (abbreviated PR), the German philosopher Georg Wilhelm Friedrich Hegel (1770–1831) made the stunning declaration (derived from Friedrich Schiller's end-of-the-18th-century poem "Resignation") that "*Die Weltgeschichte ist das Weltgericht,*" or, "The History of the World is the World's Court of Judgment." This proclamation is the succinct conclusion of a Hegel's decades-long philosophical trajectory, whose thesis is that "the spirit of the world" [*Weltgeist* in German], which is universal, moves dialectically through human history and "is this spirit which exercises its right—which is the highest right of all—over finite spirits in world history as the world's court of judgment" (German: *Weltgericht*, or the "Verdict" or "Last Judgment" of the world).[110] Hegel's concepts of the world, "inner and outer," "internal and external," "universal and particular," and "historical and spiritual" are linked by the ineluctable, but often contradictory, development of Reason (German: *Vernunft*), Spirit (German: *Geist*), or God, through the cosmos as a totality as well as our particular historical world.

Hegel was probably the most influential and systematic German "absolute idealist" philosopher after Kant. While German idealism also included such eminent thinkers as Johann Gottlieb Fichte (1762–1814) and Friedrich Wilhelm Joseph Schelling (1775–1854), its 19th-century incarnation was Hegel.

Although there are striking, and occasionally, pithy maxims and aphorisms in Hegel's surviving works, few readers have accused Hegel of having been either succinct or clear. There are several different types of work that make up Hegel's published corpus.[111] First, there are Hegel's two major books published during his academic career—*Phenomenology of Spirit* (abbreviation: PS) and *Science of Logic* (abbreviation: SL). Then, there are works that were published as handbooks for use in his university teaching, such as the *Encyclopedia of Philosophical Sciences* and *Elements of the Philosophy of Right* (abbreviation: PR). The third cluster of publications is formed by excerpts from his Berlin courses, including his lectures on the *Philosophy* of *Nature*, *Philosophy of Spirit, Philosophy of History* (abbreviation: PH), *Aesthetics*, *Philosophy of Religion*, and *History of Philosophy*. Finally, there are various miscellaneous essays and short works mostly published during the early part of his career.

My comments are based principally on four English translations of books, the *Phenomenology of Spirit* (*Geist* in German, which may also be translated as "Mind" as well as "Spirit"), the *Science of Logic* ("*Wissenschaft*" in German is much broader than the English word "Science" and may denote knowledge more generally), the *Elements of the Philosophy of Right* ("*Recht*" in German is broader than "Right," and also means "Just," "Lawful," and "Fair"), and the *Philosophy of History*, which has Hegel's most condensed (relatively speaking …) discussion of the world and its history.

Hegel's Metaphysical World as Appearance and Essence (as "In- and For-Itself")

What may constitute the "essence" of Hegel's worldview, his underlying *Weltanschauung*, has been debated for about two centuries, since there seems to be little consensus as to whether his philosophy of politics, for example, is, essentially, "conservative," "liberal," or "radical" (either "far-right," or "far-left."). That is, perhaps ironically, both as it appears to be and as it should be. For Hegel's dialectical approach to understanding nature and the world assumes constant opposition, conflict, and dynamic change, at least until "Reason" (*Vernunft*) has accomplished its providential design for the world. But what, for Hegel, is "the world?"

In his *Science of Logic* (SL) and the *Phenomenology of Spirit* (or *Mind*) (PS), Hegel appears, at times, to be following Kant, in seeking to revive metaphysics and the Kantian ontological distinction between the world as appearance and the suprasensible world, or, as Hegel calls it, the "world-in-and-for-itself" (perhaps equivalent to Kant's "noumenal" world).[112] These "two worlds," which seem to be separate, may in fact, for Hegel, be "negative" "moments" of a "self-identical," "existence" (*Dasein*),[113] or "totality."[114]

In his *Phenomenology of Mind*, in a section called "Virtue and the Way of the World," Hegel provides a striking example of the "antitheses" and "sublated" (German: *aufgehoben*) reconciliation, or unification, between "individuality" and the "new world" of the universal "self-actualization of consciousness":

> In the first shape of active Reason, self-consciousness took itself to be pure individuality, and it was confronted by an empty universality. In the second, the two sides of the anti-thesis each had both moments within them, law *and* individuality; but one side, the heart, was their immediate unity, the other their antithesis. Here, in the relationship of virtue and the "way of the world," the two members are each severally the unity and antithesis of these moments, or are each a movement of law and individuality towards one another, but a movement of opposition. While the initial appearance of the new world is … only the whole veiled in its *simplicity,* or the general foundation of the whole, the wealth of previous existence is still present to consciousness in memory … The general *content* of the actual "way of the world" … is again nothing else but the two preceding movements of self-consciousness.[115]

This passage illustrates the conceptual "movement" of abstractions in Hegel's ontology. But Hegel also applies these abstractions to his analyses of the historical and political worlds, most prominently in his *Philosophy of Right* (PR) and *Philosophy of History* (PH).

The "essence" or underlying metaphysical reality of the world for Hegel is Reason or Mind (*Geist*); and the essence of reason is self-conscious, dialectical negativity, leading fitfully to its substantial incarnation in Consciousness (*Bewusstsein*) writ large, or to freedom (*Freiheit*) in this world in particular. Reason, Mind (or Spirit), and Freedom become what they *essentially* are only by "passing through" intermediate historical stages of "externalization" or "estrangement" (*Entäusserung* and *Entfremdung*) in order to "return" gradually and necessarily to what they are intended to be—the "totality" of reality, the unity of the universal, and the particular. As Hegel put it, in his inimitable stylistic way, in the SL:

> The truth of the unessential world is at first a world in and for itself and *other to it*; but this world is a totality, for it is itself and the first world; both are thus immediate concrete existences and consequently reflections in their otherness, and therefore equally truly reflected into themselves. *"World" signifies in general the formless totality of a manifoldness; this world has foundered both as essential world and as world of appearance; it is still a totality or a universe but* as *essential relation.*[116]

Hegel's descriptions of *Geist* are as "purposive activity;" as the "conscious certainty of being all reality"; as the "law-giver and ruler" (like Kant's); as "spirit consciously aware of itself as its own world and of the world as itself;" as "cunning ... which develops its existence through individual passions," as the "Ethical life and the State," or, simply, as "what is." For Hegel, this *Geist*, this Reason, in its most conscious form, is God. God governs the World; the actual carrying out of his plan is the History of the World.[117] Hegel thereby hypostasizes beneath historical events and individuals an underlying "rational design of the world," a teleology that is the artifice of the "wisdom of Divine Providence," manifested in the historical development of freedom in the world.

Hegel's Historical and Political Worlds

History for Hegel is the stage for the "development of Spirit (*Geist*) in time, the progress of the consciousness of freedom." And since the "essence" (*Wesen*) of reality unfolds by means of such "contradictory oppositions" (*gegenseitige Widersprüche* or *Gegensätze*) as Being (*Sein*) and Non-Being (*Nicht-Sein*)—which are "mediated and transcended" (*aufgehoben*, often rendered into English by the ungainly term "*sublated*") by Becoming (*Werden*)—the development of self-consciousness (*Selbst-Bewusstsein*) is a "negative," or dialectical, succession of immanent changes within history. In the PR, Hegel expresses this as follows:

> Furthermore, it is not just the power of spirit which passes judgement in world history—i.e. it is not the abstract and irrational necessity of a blind fate. On the contrary, since spirit in and for itself is reason, and since the being-for-itself of reason in spirit is knowledge, *world history is the necessary development, from the concept of the freedom of spirit alone, of the moments of reason and hence of spirit's self-consciousness and freedom. It is the exposition and the actualization of the universal spirit.*[118]

For Hegel, the history of Spirit (*Geist*) is the development through historical time of its own self-consciousness through the actions of peoples, states, and "world historical" individual actors who, while motivated by their own interests, are simultaneously the unconscious instruments of the "cunning of Reason" (German: *List der Vernunft*). The actions of great men (and for Hegel, it is such "great *men*" as Julius Caesar and Napoleon) are products of their individual wills and passions, but the historical impact of these actions is really the product, not of individual agents, but of the World Spirit.

Hegel also claimed that in the history of the world there are four world-historical epochs, or "realms," each manifesting a principle of Spirit as expressed through a dominant culture. In the *Philosophy of Right*, Hegel explicitly identifies these realms as "the Oriental, the Greek, the Roman, and the Germanic."[119] The culmination of this history of the world is the modern state.

In his *Philosophy of Right*, Hegel describes the political state as the third "moment" of "Ethical Life," embodying a synthesis between the principles governing the family and civil society. For him, freedom also becomes explicit and objective in the state sphere, which Hegel calls "Mind" (*Geist*) "objectified." Rationality is incorporated in the state insofar as its content embeds the unity of objective freedom (the freedom of the universal or substantial will) and subjective freedom (the freedom of individuals who will particular ends). With its "universal" class of civil servants and an "enlightened" monarch, the Prussian state of his own age, with its rationally distinctive Germanic "People" (*Volk*) and "Spirit of the People" (*Volksgeist*), was apparently viewed by Hegel as the apotheosis of the development of Spirit in human history.[120]

In his *Philosophy of History*, Hegel further illustrates the movement of "Spirit" from "Old" to "New" Worlds, from "East" to "West," and from "ancient" to "modern realms":

> The World is divided into *Old and New*; the name of *New* having originated in the fact that America and Australia have only lately become known to us ... The History of the World travels from East to West, for Europe is absolute the end of History, Asia the beginning ... [and] is the discipline of the uncontrolled natural will [presaging Schopenhauer], bringing it into obedience in a Universal principle and conferring subjective freedom. The East knew and to the present day knows only that *One* is Free; the Greek and Roman world, that *some* are free; the German World knows that *All* are free. The first political form ... we observe in History, is *Despotism*, the second *Democracy* and *Aristocracy*, the third *Monarchy* ... For Spirit as the consciousness of an inner World is, at the commencement, still in an abstract form ... The German Spirit is the Spirit of the New World. Its aim is the realization of absolute Truth as the unlimited self-determination of Freedom.[121]

Unsurprisingly, because of his apparent valorization of "the German World, the German Spirit," and the Prussian Empire, Hegel has been linked to state-centered absolutism, German nationalism, and extreme right-wing political ideologies.

More generally, the rational ideals of the Kant and Enlightenment, and the political ideas of the French Revolution—freedom, self-consciousness, historical progress, and critical rationality, but without the bloody political reality, are carried to their ultimate metaphysical limits in Hegel's philosophy of history and his political theory.[122] But can Hegel's own philosophy be identified with what he defined philosophy in general to be, namely "the spirit of its time comprehended in thought?"[123] The answers to that question are as diverse as the elements of Hegel's conceptual world.

The Philosophical World's Judgments of Hegel's Worldview

Hegel's grand theoretical system simultaneously appears to be an elaborate, rationalist metamorphosis of Christian ontology; a response to and a move beyond Kantian metaphysical idealism; an encyclopedic history and philosophy of the world from its beginning until its "rational completion" in Hegel's own day, and a political ideology resistant to labeling.[124] That is, as Western idealist philosophers from Plato to Kant would say, its "appearance."

Interpretations and evaluations of the "essence" Hegel's system, however, have been as divergent and inconsistent as his own conceptualization of the dialectical, antithetical development of the world itself.[125] His influence was felt most directly soon after his death, from opposing political and religious perspectives, including the more conservative "Right," or "Old," Hegelians. It was, however, from the other end of the political spectrum, beginning with the "Left" or "Young" Hegelians—whose most noted representative was Karl Marx—that Hegel's *Weltanschauung* was simultaneously to be absorbed, opposed, and transformed. For it was Marx and his followers who tried to turn Hegel's idealism on its head and, while retaining such key Hegelian ideas as dialectics, alienation, and the roles of labor and markets within a stratified and polarized capitalist economy, anchored these notions within the framework of historical materialism.

On the other hand, in much academic Anglophonic philosophy, Hegelian idealism seemed to wane dramatically after 1848 and the failure of the revolutionary movements of that year, but it underwent a revival in both Great Britain and the United States in the last decades of the 19th century. In Britain, where such "idealist" philosophers such as T.H. Green (1836–82) and F.H. Bradley 1846–1924) were following in Hegel's footsteps, Hegel's own philosophy came to be one of the main targets of the founders of what would be called the "analytic" philosophical movement—still dominant in Anglophonic universities—Bertrand Russell (1872–1970) and G.E. Moore (1873–1958). For Russell, the revolutionary innovations in logic during the last decades of the 19th century with the work of the German mathematician and philosopher of language Gottlob Frege (1848–1925) had apparently undermined Hegel's metaphysics by overturning the Aristotelian logic on which, as Russell claimed, it was based. Hegel's work then came to be seen within the analytic movement as, generously speaking, of little serious philosophical interest, or, even worse, as wooly obscurantism.

On the "other side of the pond" (the English Channel), Hegel has continued to be quite influential, especially within Existentialism, and also in Marxist-oriented theories after Marx. In France, a version of Hegelianism came to influence a generation of thinkers, including Jean Hyppolite (1907–68), Jean-Paul Sartre (1905–80), and the unorthodox psychoanalyst, Jacques Lacan (1901–81), partially through the lectures on Hegel given by the Russian-born French philosopher Alexandre Kojève (1902–68).

In Germany, interest in Hegel, having receded during the second half of the 19th century, was revived at the turn of the 20th century in the philosophy of history of Wilhelm Dilthey (1833–1911), and important Hegelian elements were incorporated by such members of the Frankfurt School of Critical Theory as Theodor W. Adorno (1903–69), Herbert Marcuse (1898–1979), and later, Jürgen Habermas (1929–), as well as within the Heidegger-influenced hermeneutic approach of Hans-Georg Gadamer (1900–2002). Within English-speaking philosophy, the final quarter of the 20th century saw something of a revival of interest in Hegel's philosophy. By the beginning of this century, even such influential analytic thinkers as Robert Brandon and John McDowell (1942–) began to take Hegel seriously, although they are a distinct minority within the Anglophonic philosophical power structure.

Overall, the dialectic of the philosophical world's "positive" appraisals of Hegel often seems to move as unevenly as Hegel's own "negative" development of the "World Spirit." The jury of World History has yet to pronounce its final verdict on Hegel's worldview.

Schopenhauer's World as Will and Representation

From Nothingness, Through Misery, and Back to Nothingness

The German philosopher Arthur Schopenhauer (1788–1860) was among the first modern thinkers to contend that at its deepest level, the universe is not a welcoming or rational place. Initially inspired by such idealist philosophers as Plato, Kant, and Hegel, but also profoundly influenced by such Indian spiritual traditions as Hinduism and Buddhism, Schopenhauer rejected more "optimistic" philosophies of the world in favor of what has generally been interpreted as a "pessimistic," and ultimately ascetic and aesthetic, renunciation of everything mundane.

Faced with a world filled with seemingly unending frustrations and conflicts, to cope with a perilous and suffering-plagued human condition, Schopenhauer argued that we ought not needlessly to attempt to subdue an omnipotent "will," but instead might consider the cultivation, however fleeting, of an inner world of calm beneficence and measured world-renunciation, and emotional absorption in the distractions of sexual love and aesthetic delight. This *Weltanschauung* was of particular importance to Friedrich Nietzsche (1844–1900, to be discussed later in this chapter) as well as to the German Romantic composer Richard Wagner (1813–83), whose Schopenhaueresque opera *Tristan and Isolde* revolutionized Western music. Subsequently, Schopenhauer's philosophy has had a special appeal for many people pondering "the meaning of life," to such writers as the 20th-century novelist Thomas Mann (1875–1955), as well as to a range of other writers, artists, and philosophers, including such members of the Frankfurt School as Max Horkheimer (1895–1973) and Adorno.[126]

Schopenhauer's *chef d'oeuvre* is *The World as Will and Representation* (German: *Die Welt als Wille und Vorstellung*), initially completed in 1818 in Dresden, Germany. This is a multivolume work (with an abridged one-volume text in English), developed from ideas in a previous book, *The Fourfold Root.*[127] His later works include *Two Fundamental Problems of Ethics*; an accompanying volume to *The World as Will and Representation*, published in 1844 along with the first volume in a combined second edition; a set of philosophical reflections entitled *Parerga and Paralipomena* (from the Greek: "Appendices and Omissions"); and, in 1859, a year before his death, the third edition of *The World as Will and Representation.* After Schopenhauer's death, Julius Frauenstädt published new editions of most of Schopenhauer's works, with six volumes appearing in 1873. In the 20th century, the editorial work on Schopenhauer's previously unpublished manuscripts was done by Arthur Hübscher.[128]

To some degree, Schopenhauer considered himself a successor of Kant, who claimed that the limits of human reason made the inner nature of things and of the world "in-themselves" unknowable to human reason and, consequently, that there are two separate worlds, the "sensory" and the "intellectual." Kant thereby famously left room for "faith" and "divine providence," not knowledge about the meaning and purpose of our existence in this world. *But Schopenhauer. while also claiming that we can have no "certainty" about the world's "inner essence," rejected both Kantian metaphysical dualism (no "unknowable transcendental" world for Schopenhauer) and Kant's "rational faith" in a divine purpose for all existence, and has been considered, respectively, agnostic if not atheistic about God's existence, and "pessimistic" about both our lives in this world and the cosmos as a whole.*

In a passage strikingly similar to remarks later made by Ludwig Wittgenstein in his early work the *Tractatus Logico-Philosophicus* (to be discussed later in this book), Schopenhauer comments on the epistemological limits of his own philosophy:

> The present philosophy ... by no means attempts to say *whence* or *for what purpose* the world exists, but merely *what the world is* ... My philosophy does not presume to explain the ultimate causes of the world. It confines itself to the facts of inner and outer experience ... accessible

> to everybody, and points out the true and intimate connection between these facts, without, however, concerning itself with that which may transcend them. It refrains from drawing any conclusions concerning what lies beyond experience. It merely explains the data of sensibility and self-consciousness, and strives to understand only the immanent essence of the world.[129]

And while acknowledging the historical importance of Descartes in the history of modern philosophy, who posed the human *cogito* ("I think") as the subjective starting point for grounding knowledge of the "external" world and of ourselves while articulating the earliest significant construal of the "mind/body" problem, Schopenhauer nonetheless came to reject both Cartesian rationalism and metaphysical optimism.[130]

But rather than conceiving of two, seemingly autonomous worlds, outer and inner, as Descartes, Kant, Plato, and other Occidental idealists had done, Schopenhauer, like Bishop George Berkeley, rejected the absolute autonomy of the "external world" from the "internal world" of human consciousness and held the "outer" world to be "my idea" and its very existence to depend on our individual perceptions:

> *"The world is my idea:" this is a truth which holds good for everything that lives and knows, although only man can bring it into reflected, abstract consciousness … this whole world—is only object in relation to subject, perception of the perceiver—in a word, idea.*[131]

Not only is the world "my idea" (German: *Idee*), but it is also my "representation" (German: *Vorstellung*), according to Schopenhauer:

> "The world is my representation" is … a Proposition … everyone must recognize as true … To have brought this proposition to consciousness … of the world in the head to the world outside the head, constitutes, together with the problem of moral freedom, the distinctive characteristic of the philosophy of the moderns. For only after thousands of years … *did* [men] *discover that, among the many things that make the world so puzzling and precarious … however immeasurable and massive it may be, its existence hangs nevertheless on a single thread; and this thread is the actual consciousness in which it exists.* This condition [marks] … the existence of the world … with the stamp of *ideality* … and consequently with the stamp of the mere *phenomenon.*[132]

The preceding passage might be recognizable as what Kant and others have called "subjective idealism," in contrast with Kant's own "transcendental idealism," which rejected the epistemological stance that *esse est percipi* ("to be is to be perceived," according to Bishop George Berkeley[133]). Instead, Kant declared the world to be a "transcendental" idea, like God, freedom, and the soul's immortality, something beyond human sense experience and rational proof, but nonetheless as "transcendentally real" as anything material and "empirically real."

But Schopenhauer, in a remarkable passage, eerily similar to strains of mysticism, psychoanalysis, Asian philosophical traditions, and contemporary cognitive neuroscience, but very far from Kant, also declared that:

> *Thus the world must be recognized ... as capable of being put in the same class with a dream.* For the same brain-function that conjures up during sleep a perfectly objective, perceptible, and indeed palpable world must have just as large a share in the presentation of the objective world of wakefulness. *Though different as regards their matter, the two worlds are nevertheless obviously molded from one form ... the intellect, the brain-function.*[134]

In a few brief and strikingly clear (for a German philosopher!) pages, Schopenhauer has managed to put forward as virtually self-evident axioms the following propositions:

1 The existence of the world as an "external" physical totality hangs "like a thread" on the perceptions of each individual human. Hence, "the world is *my* idea."
2 The world is also "my *representation*," so the apparent distinction between "inner and outer worlds" is overcome by the subjectivity of objectivity.
3 But, the world is also "akin to a dream," and hence a transient and illusory, if often powerful and upsetting, perceptual phenomenon.
4 And, furthermore, both the internal, subjective world of ideas and dream states, and the "objective, perceptible, and palpable objective world of wakefulness," are the effects of brain functions!

The last of these four propositions, perhaps ironically, is a premise of virtually all contemporary "materialist" and "neural correlates'" theories of human consciousness, which are epistemologically and metaphysically as far removed from subjective idealism as possible.[135]

As stunning (both substantively and stylistically) as his theory of the world may appear, it is Schopenhauer's doctrine of the "will" that has carved out a unique place in the history of Western ideas. While apparently rejecting Kant's claim that "things-in-themselves" exist, are nonidentical with the phenomenal world, and are, accordingly, unknowable, Schopenhauer also appears, perhaps inconsistently, *to have retained an agnosticism or skepticism regarding humanity's ability to penetrate the deepest mysteries of the world.*

But that is just a preliminary move in his argument. For Schopenhauer then declares that, in fact, *the world's secrets may be unearthed, but only if one abandons hope for any rationality in the world as it really is, which is not as we wish it to be, and instead accepts, however anxiously and grudgingly, that the "inmost nature" of the "world that surpasses all understanding" at bottom is a formless, timeless, unconscious, and unconstrained Will, of which our individual wills and this world as a whole are "manifestations":*

> the inmost nature of the world as will, and all its phenomena as only the objectivity of will; and ... this objectivity from the unconscious working of obscure forces of Nature ... to the completely conscious action of man. Therefore ... with ... the surrender of the will, all those phenomena are also abolished; that constant strain and effort without end and without rest ... *in which and through which the world consists ... the whole manifestation of the will;* and ... also the universal forms of this manifestation, time and space, and ... its last fundamental form, subject and object; all are abolished. *No will: no idea, no world. Before us there is certainly only nothingness. But that which resists this passing into nothing, our nature, is indeed just the will to live, which we ourselves are as it is our world.*[136]

And, as bad as this world may be, and it is something Schopenhauer did not refrain from on occasion calling "the worst of all possible worlds,"[137] our individual wills strive furiously to continue living and to resist annihilation: "That we abhor annihilation so greatly, is simply another expression of the fact that we so strenuously will life, and are nothing but this will, and know nothing besides it."[138]

So what would be the possible prudent and "sage" (as for Chinese thinkers) response to our inescapable condition—fight, flight, surrender, or something else?

Schopenhauer considers such reactions to *Weltschmerz* (the pain felt by being-in-the world) as world-renunciation and self-overcoming, in the spirit of ancient Indian spiritual traditions, as well as flight from this world of suffering into the temporary, but pleasurably distracting realm of the aesthetic in general, and Western classical music in particular:

> But if we turn our glance from our own needy ... condition to those who have overcome the world, in whom the will ... found itself ... in all, and ... freely denied itself, and ... then ... wait to see the last trace of it vanish with the body ... it animates; then, instead of the restless striving and effort ... the constant transition from wish to fruition, and from joy to sorrow, instead of the never-satisfied and never-dying hope which constitutes the life of the man who wills, we shall see that peace which is above all reason, that perfect calm of the spirit, that deep rest, that inviolable confidence and serenity ... the will has vanished.[139]

Schopenhauer continues:

> We look with deep and painful longing upon this state, beside which the misery and wretchedness of our own is brought out clearly by the

> contrast. Yet this is the only consideration which can afford us lasting consolation, when … *we have recognized incurable suffering and endless misery as essential to the manifestation of will, the world; and … see the world pass away with the abolition of will, and retain before us only empty nothingness* … by contemplation of the life and conduct of saints … and … by art, we must banish … that nothingness … which we fear as fear the dark; we must not even evade it …, through myths and meaningless words … as reabsorption in Brahma or the Nirvana of the Buddhists … *what remains after the entire abolition of will is for all … who are … full of will certainly nothing … conversely, to those in whom the will … has denied itself, this our world, which is so real, with all its suns and milky-ways—is nothing.*[140]

In the end, for Schopenhauer, *there is no viable "solution" for us as individuals, for the species as a whole, or for the world at all, to the problem of incarnate existence, with its ineluctable fate that everything that comes into being suffers during its brief term of existence and soon thereafter, cosmically speaking, passes away into nothingness.*

Everything that ever was, is, and ever will be, is ensnared by the will and the universe's omnipotent indifference to our fate. There are only such temporary distractions as sexual love[141] (and Schopenhauer was among the first Occidental thinkers to write extensively, and sometimes glowingly, about the human body and its pleasures),[142] and the aesthetic realm, of which classical music is the highest expression.[143]

But as dire as these reflections may seem, Schopenhauer, like Kant, was as amazed and awe-inspired by the universe as he was dismayed by the phenomenal world in which we struggle futilely and die alone. In this passage from *The World as Will and Representation*, he eloquently expresses the ambivalence of being in this world:

> *If we lose ourselves in contemplation of the infinite greatness of the universe … or if the night sky brings before our eyes countless worlds … we feel ourselves reduced to nothing … All this, however … shows … we are one with the world, and … not oppressed, but exalted by its immensity.*[144]

Schopenhauer's Worldly Influence

Schopenhauer's philosophy has been influential in diverse places, except for most Anglo-American departments and journals of philosophy. His recognition that the universe appears to be a fundamentally irrational place was appealing to a diverse group of 20th-century thinkers, novelists, poets, playwrights, artists (especially Surrealists), and psychologists, who focused on unconscious, irrational feelings, and impulses (Schopenhauer's "will") as motivating much human behavior.[145] Among noted philosophers, it was Friedrich Nietzsche who, perhaps more than any other, appreciated and to some degree incorporated Schopenhauer's views on the meaning of life, the overpowering non-rational will, and the cathartic emotional release of great art and music.

Schopenhauer's ideas about the importance of instinctual urges at the core of daily life also reappeared in Freud's psychoanalytic thought, and his conviction that human history is and cannot be a tale of unending progress, became themes within post-World War II 20th-century French thought, notably in some works of Albert Camus and Jean-Paul Sartre. Finally, although they would seem to be radically different persons and thinkers, anticipating the two dominant and sometimes antagonist contemporary philosophical orientations—"Continental" and "Analytic," respectively—Schopenhauer and Wittgenstein shared important concerns, including the limits of what can be said meaningfully about the world, the constraints of authentic philosophical analyses, and the importance of what cannot be factually stated.[146]

In a nutshell, from a Schopenhauerean perspective, *there is no escape from this world, except to return to the nothingness from which we and everything else have emerged. For Schopenhauer, and for everyone else, there may nonetheless be some consolations to be had while we are in this world, not the least of which are its emerald-like radiance, the few other souls within it to whom we might in some tentative way relate, and its unfathomable mystery.*

Søren Kierkegaard's Existential World

Søren Aabye Kierkegaard (1813–55), had, as far as I can detect, little to say in his published writings about "the world," if, by "the world," one means the "external" "natural," or "outer world" that has been the object of scientific, and scientistic, philosophical, and intellectual discourse since the Ancient Greeks. So what is a comparatively brief section on this remarkably "indirect" and loquacious anti-systematic writer doing in this anything-but-brief, and ambitiously systematic book? Well, as little as Kierkegaard apparently had to say about the outer world we take for granted as the totality of objects that provide the environment in which we find ourselves, it is Kierkegaard's sometimes quirky and iconoclastic way of talking about the *existing subject, and his inner world* of ideas, feelings, phantasies, and anxieties—and Kierkegaard does mean "his," Søren Aabye Kierkegaard's—that thrusts him to the position of forerunner of what would later be called the philosophy of existence, or Existentialism.

But Kierkegaard himself probably would have wanted to have nothing to do with—isms of any kind, especially one of which he would be considered the "father" or forerunner. Instead, he focused almost exclusively on his unique place in his everyday world rather than on some fanciful speculation about his future place in the history of world philosophy or about "the world" outside his personal experience. He, like each one of us, was unique, and his unique inner world, unlike most of the rest of us, deserves a unique place in any history of ideas of the world.

In this regard, Kierkegaard's life may be more relevant to his work than is the case for many other "world-historical" writers.[147] Externally, his life may appear to be quite uneventful compared with those of other influential writers. He rarely left his hometown of Copenhagen, Denmark, and traveled infrequently abroad. Probably, the two most important persons in his life were his father and his onetime fiancée, Regine Olsen, with whom Kierkegaard notoriously terminated their engagement, for reasons that are still the subject of considerable speculation. Not only did Kierkegaard seem to inherit his father's melancholic disposition, his sense of guilt and anxiety, and his pietistic emphasis on the dour aspects of Christian faith, but he also inherited his talents for philosophical argumentation and creative disquisition. In addition, upon his father's death in 1838, Kierkegaard was bequeathed enough of his father's material assets to allow him to pursue his life as a freelance writer.

Kierkegaard studied theology, philosophy, and literature at the University of Copenhagen. For a time, he considered becoming a pastor in the Danish People's Church. But in 1840, just before he enrolled at the Pastoral Seminary, he became engaged to Regina Olsen. This engagement was to form the basis of a great literary love story, propagated by Kierkegaard through his published writings and his journals. It also provided an occasion for Kierkegaard to define himself further as an outsider. The theme of a young woman being the vehicle for a young man to become "poeticized" recurs in Kierkegaard's writings, as does the theme of the sacrifice of worldly happiness for a higher (religious) purpose. Kierkegaard's infatuation with Regine, and the sublimated libidinal energy it may have lent to his poetic production, were crucial for setting his life course. The breaking of the engagement allowed Kierkegaard to devote himself semi-monastically to his quest for faith, as well as to establish his status as an outsider. It also freed him from close personal entanglements with women, possibly leading him to objectify them as ideal creatures, and to reproduce the patriarchal values of his church and father. Nevertheless, whatever one's life circumstances, social roles, and gender, Kierkegaard regarded everyone as equal before God from the standpoint of eternity.

Kierkegaard's engaged on a sustained attack on the Church of Denmark (Lutheran) in his newspaper articles and in a series of self-published pamphlets, which are now included in Kierkegaard's *Attack Upon Christendom.* In these polemics, Kierkegaard argued that the idea of congregations keeps individuals as children since Christians are disinclined from taking the initiative to take responsibility for their own relation to God. He stressed that "Christianity is the individual, here, the single individual."[148] Furthermore, since the Church was controlled by the Danish state, Kierkegaard believed the Church's real mission was to increase membership and not to promote the welfare of its members. According to Kierkegaard, this mission was contrary to Christianity's true doctrine, which, for him, was the spiritual importance of the individual. However, the state-church political alliance was detrimental to individuals, since anyone can become "Christian" without understanding the real meaning of Christianity. The worldly Church also reduced Christianity to a fashionable tradition adhered to by unbelieving "believers," a "herd mentality," remarkably similar to Nietzsche's depiction of Christian mass behavior, though without, of course, Nietzsche's proclamation of the "death of God."

One day in 1855, Kierkegaard collapsed on a street in Copenhagen. He stayed in a hospital and refused communion. He was reported to have said to a childhood friend who kept a record of his conversations with Kierkegaard that his life had been one of immense suffering. Kierkegaard died in a Copenhagen hospital, possibly from complications from a fall he had taken from a tree in his youth. It has also been suggested that Kierkegaard died from Pott disease, a form of tuberculosis. He was interred in a cemetery in Copenhagen. At Kierkegaard's funeral, his nephew caused a disturbance by protesting Kierkegaard's burial by the official Church. His nephew maintained that Kierkegaard would never have approved of this interment, had he been alive, as he had broken from and denounced the Church. For his efforts, in an ironic twist characteristic of his ironist-uncle, Søren, Kierkegaard's nephew was later fined for his "disruption" of a funeral.[149]

Kierkegaard's Singular Literary and Paradoxically Absurd Worlds

In his writings, Kierkegaard used numerous pseudonyms, one of which, "Johannes Climacus," says of Socrates that "his whole life was personal preoccupation with himself, and then Governance comes and adds world-historical significance to it."[150] Kierkegaard may have seen himself as a modern Socratic "gadfly" stinging the pompous preachers and writers of Denmark and Germany with targeted barbs. More abstractly, he may, in his own terms, have himself as a "singular universal," whose self-preoccupation was transfigured by divine "governance," or the grace of God, into universal significance.

Kierkegaard's distinctive literary conceits comprise an array of different narrative points of view, personas, and disciplinary topics, including aesthetic novels, works of psychology, Christian dogmatics and polemics, satirical prefaces, philosophical "edifying discourses and postscripts," literary reviews, and self-interpretations. His arsenal of rhetoric includes irony, satire, parody, humor, polemic, and a "dialectical" method of "indirect communication"—all designed to deepen the reader's subjective engagement with ultimate existential issues. Among the proto-existential terms and literary, psychological, philosophical, and theological categories Kierkegaard invented or played with are "fear and trembling" (anxiety), "the sickness unto death" (despair), melancholy, inwardness, irony, existential stages and choices, inherited sin, the "teleological suspension of the ethical," paradoxes (especially of being a Christian), defining choices and relations, "the self," sacrifice, love as a duty, seduction, the demonic, and, most importantly for such 20th-century "Existentialist" writers as Albert Camus, "the absurd."[151]

In his *Journals and Notebooks*, Kierkegaard illustrates his struggle to understand what God "wished" him to do as well as his choice to "construct a world" in which he might live, and not one that would "only be held up to the view of others":[152]

> The thing is to understand myself, so see what God really wishes me to do ... to find a truth which is true for *me, to find the idea for which I can live and die* ... what good would it do me to be able to develop a theory of the state, and combine all the details into a simple whole, *and so construct a world in which I do not live, but only held up to the view of others.*

This "constructed world" was Kierkegaard's inner world of "the Absurd":[153]

> What is the Absurd? The Absurd, or to act by virtue of the absurd, is to act upon faith ... I must act, but reflection has closed the road so I take one of the possibilities and say: This is what I do, I cannot do otherwise because I am brought to a standstill by my powers of reflection.

And in *Concluding Unscientific Postscript*, once again Kierkegaard asks:

> What, then, is the absurd? The absurd is that the eternal truth has come into existence in time, that God has come into existence, has been born, has grown up. etc., has come into existence exactly as an individual human being, indistinguishable from any other human being, in as much as all immediate recognizability is pre-Socratic paganism and from the Jewish point of view is idolatry.

How can this paradox of absurdity be held or believed? Kierkegaard says:

> The absurd is not the absurd or absurdities without any distinction (wherefore Johannes de Silentio [one of Kierkegaard's numerous pseudonyms]: "How many of our age understand what the absurd is?") ... The absurd is a [Christian] category, the negative criterion, of the divine or of the relationship to the divine. When the believer has faith, the absurd is not the absurd—faith transforms it, but in every weak moment it is again more or less absurd to him. *The passion of faith is the only thing which masters the absurd*—if not, then faith is not faith in the strictest sense, but a kind of knowledge. The absurd terminates negatively before the sphere of faith, which is a sphere by itself. To a third person the believer relates himself by virtue of the absurd; so must a third person judge, for a third person does not have the passion of faith. Johannes de Silentio has never claimed to be a

> believer; just the opposite, he has explained that he is not a believer—in order to illuminate faith negatively.

Absurdity is not just an abstract "category" for Kierkegaard, but an anguished and authentic, passionately "faithful" way of life. And, like his role models Socrates and Jesus Christ, *Kierkegaard viewed the manner in which one lives one's life to be the prime criterion for an honest and authentic existence in this world.*

In addition to being dubbed "the father of Existentialism," Kierkegaard is also philosophically renowned for having been a critic of Hegel and Hegelianism. As he said, indirectly as usual, regarding "speculative" (i.e., Hegelian, philosophical systems):

> we must criticize modern speculation … because it has become comical through having forgotten, in a fit of world-historical absentmindedness what it is to be human, and not what it is to be human in general … but what it is for you or me or him or her to be human, each of us for ourselves and on our own.[154]

And in the guise of his literary persona Johannes Climacus, Kierkegaard distinguished his personalistic brand of unphilosophical philosophizing from Hegelian and other "Systems" of philosophy:

> Those who believe that philosophy has never been closer to solving its problems may well find it strange, vain, or even ridiculous that I choose the narrative form rather than helping, in my humble way, to put the final coping stone on the System. On the other hand, those who are convinced that philosophy, in spite of all its definitions and distinctions, has never been more preposterous or confused … someone who listened to all of them would imagine that nature herself had become confused and the world would not last till Easter.[155]

For Kierkegaard, "*The essential task for an existing human being … is not to speculate philosophically about absolute knowledge* [Hegel], *but to become himself … this is a task involving risky choices, choices that must be made without the comfort of objective certainty.*"[156] For the individual, *living in this uneasy subjectivity is living in the truth, and subjectivity is, typically paradoxically, both truth and untruth.*

"Take away the paradox from a thinker and you have a professor …" declared Kierkegaard in a journal.[157] But Kierkegaard was as far removed from a "professorial" persona as is possible for a thinker who has been taken so seriously, at least in "Continental" philosophical traditions in particular and in theological discussions more generally. It would be refreshing if more professors, especially in philosophy departments, would take these words to heart …

Kierkegaard's polemical attitude, quirky lifestyle, and skeptical, subjectivistic, inward-directed worldview, which has sometimes been regarded as "irrationalist," created problems for him and for the authority figures he savaged. But as Kierkegaard saw it, this was his calling, for "I conceived it as my task to create difficulties everywhere."[158]

In "creating difficulties," Kierkegaard emphasized what he called Christianity's "inverted dialectic," which requires that Christians exercise a kind of "double vision," that is, paradoxically, *to see in worldly conditions their spiritual opposites*, such as hope in hopelessness, strength in weakness, and opportunity in adversity. The purpose of so doing, according to Kierkegaard, *is to call into question and undermine Christians' focus on worldly goals in order to redirect their attention to what is infinitely more worthwhile—the quest for such "other-worldly" goals as faith and salvation.* This redirection of the faithful's attention from the things of this world, such as donations to the Church, literary fame, and material success, to immaterial pursuits, was, of course, not appreciated by the local clerisy or tabloids, and Kierkegaard became not infrequently an object of their anger, scorn, and derision.

In his book *The Sickness Unto Death*, Kierkegaard *redirected philosophy away from "knowing the world," in some kind of unachievable "impersonal and objective" way, and instead toward the subjective—the psychology of "the self"—together with an often unsettling account of such inner emotional states as "despair" and "anxiety"* (which had been the central theme of a previous book, *The Concept of Anxiety*). *Passionate and unremitting reflection on the inner world of the individual had for Kierkegaard replaced the dispassionate analysis of the outer world of the physical cosmos as the central focus of philosophical analysis.* For him, the lifelong struggle for self-knowledge, together with the often anxiety-ridden hope for God's forgiveness of our sins, can bestow meaning on one's life. This is the task of "becoming," or "willing to be a self," according to Kierkegaard.

In Kierkegaard's account, willing to be a self is a form of despair. But so are not willing to be a self and not being aware of the possibility of being a self. The only antidote to despair is Christian faith, an awareness of God as the ultimate ground of one's selfhood. Only in faith can God's grace be received and accepted. But God is an absolutely unknowable otherness, on whom we are completely dependent for the final judgment of our actions and selves. Neither we, as individual selves, nor society, as collectively objectified in its norms and laws, can distinguish with any certainty the differences between "good and evil" deeds or persons. Only God can. And God is silent. Still, we must decide for ourselves how to live, and in the process hope for God's ultimate gift, the "miracle" of faith.

According to Kierkegaard, faith is the ultimate gift for which a human being may hope, because only on the basis of faith does an individual have a chance to become a true self. Becoming this true self is the individual's life-project, whose ultimate success or failure is judged by God, whose edict determines our eternal fate. Individuals are thereby shouldering an enormous responsibility, for upon our existential choices hang our eternal salvation or damnation. Anxiety or dread (Danish: *Angest*) accompanies our awareness of this responsibility since we must make ultimate existential choices. One the one hand, we feel anxious about making choices that will bind us for eternity. On the other hand, we exult in the freedom of choosing to make our own selves. Ultimate choices occur during the paradoxical moment at which our point in time and timeless eternity intersect, because an individual creates through temporal choices a self that will be judged by God for eternity. And, for Kierkegaard, the most ultimate choices, the ones that may be forever rewarded or punished by the unseen God, are the repeated choices that avow our faith in the God who does not appear to us, at least not in this world of sin and the anxiety-inducing freedom to choose our own path through the forest of life.

For Kierkegaard, another key paradox is that the eternal, infinite, transcendent God of Christianity became incarnated as a temporal, finite, historical human being—Jesus Christ. Regarding this pillar of Christianity, either we can have faith, or we can doubt. What we cannot do, according to Kierkegaard, is believe in this paradox by virtue of reason. *If we choose faith, we must suspend our reason in order to believe in something higher than reason: we must believe because it is paradoxical and absurd.*

Kierkegaard's paradoxical, subjective, absurd, and arational worldview assesses our works and deeds, philosophical and otherwise, largely by the passionate intensity and authenticity of our inner lives. Like such illustrious Christian theological predecessors as St. Augustine and Blaise Pascal, Kierkegaard seemed to believe that it is ultimately God's judgment that matters in the evaluation of the totality of an individual's life on Earth. But we, as frail and fallible terrestrial creatures, cannot hope to know the omnipotent, omniscient, and divine Christian deity of an infinite and timeless realm apart from our spatio-temporal universe. We only can only take an existentially anguished and paradoxical "leap of faith" from our "fearful and trembling" existence in this world to the paradoxically uncertain certainty of God's existence.

Kierkegaard's contributions to Western thought were largely unappreciated until long after his death. While his pamphleteering had little immediate impact on his readers, critics, and peers, his philosophical, literary, psychological, and theological works eventually had a significant impact on some of the most important thinkers of the 20th century, especially the "Existentialist" thinkers Martin Heidegger, Jean-Paul Sartre, and Albert Camus.

Heidegger's most influential book, *Being and Time*, for example, exhibits its Kierkegaardean roots in its author's discussions of the phenomenology of moods and the constitutive role of time in the formation of the subject. Sartre's early masterpiece, *Being and Nothingness*, focuses largely on such themes as freedom, authenticity, anxiety, and choice, all of which were raised by Kierkegaard. And, of course, Camus appropriated the idea of "the absurd" as a key theme in *The Myth of Sisyphus* and other works. Recent

Christian theology, including the works of Paul Tillich and Martin Luther King's "Letter from a Birmingham Jail" and "Pilgrimage to Nonviolence," as well as "Existential Psychology" (notably in the writings of Rollo May), are also indebted to Kierkegaard's ideas in particular and to existential philosophy more generally.[159] And, Nietzsche's agnostic or atheistic critique of Western philosophy and theology complements Kierkegaard's theistic revolt against orthodox Christian doctrines and practices.

But, in the final analysis, Kierkegaard most likely would not have cared how he, as a person and as a writer, came to be received. *For it was his inner world, and God's gift or withholding of grace, that mattered most, and not how the world at large perceived him.*

For Kierkegaard, an authentic self does not concern itself with how one appears to others in one's worldly affairs. It is also not fruitful to "speculate philosophically about absolute knowledge," as Kant and Hegel had done. But rather it is one's life project to "become oneself" and, in so doing, to make the risky choice of the "leap of faith" into living a certain form of life, in imitation of the life of Jesus, and also to trust in the mercy and forgiveness of an unknowable and but beneficent God.[160]

Friedrich Nietzsche's Life-World

Friedrich Nietzsche (1844–1900)—as philosopher, psychologist, human being, and legend—resists "correct" interpretation. His life, at least insofar as one's "life" retrospectively consists of its biographical details, has been amply described by others much more interested in its details than I.[161] Nonetheless, perhaps more than any other modern influential thinker, *Nietzsche's philosophy was his life; and his life was his philosophy in action*—at least until his complete mental and physical breakdown in 1888 in Turin, Italy, commonly interpreted as a lapse into insanity (likely due to a brain tumor), the dimensions of which were also unprecedented for a major Occidental thinker.[162]

Yet, it is an interpretation of Nietzsche—my perspective in reading and reflecting about Nietzsche and his commentators as grounded in my own needs and interests—of course, that follows. I make no claims for its "accuracy." I suspect Nietzsche might not have cared. But what I hope readers might care about in reading what is to come in this section of this book is what these words—Nietzsche's, where cited, and mine otherwise—*might mean to your lives and thoughts*. It is *your* interpretation, your reading of and reacting to the text on these pages that matters. Resisting an interpretation and not having a perspective, especially of a writer and figure of Nietzsche's legendary stature, are as improbable as successfully resisting the temptation to analyze one's own motives and behavior. So indulge, don't futilely resist, that temptation to self-interpretation, as well as the seductively guilty pleasures of reading and interpreting Nietzsche's life and work.

Nietzsche's Life's Work

Friedrich Nietzsche was born in 1844, in Röcken (near Leipzig), Germany, where his father was a Lutheran minister. His father died in 1849, probably of a brain injury, while much later, perhaps in a kind of perverse irony, Nietzsche accused many Europeans of having a "herd mentality and slave morality" and of having a "sickness of the brain," a physical pathology to which he himself would later succumb.[163] The very young Friedrich was clearly deeply affected by this family tragedy, and his family relocated to Naumburg, where he grew up within an otherwise all-female household consisting of his mother, grandmother, two aunts, and his younger sister, Elisabeth.

Nietzsche had a meteoric but short-lived academic career, peaking in 1869, at the age of 24, when he was awarded a chair in classical philology at the University of Basel in Switzerland. Most of Nietzsche's teaching and early publications were in classical (Latin and Ancient Greek) philology, but he was already interested in philosophy, particularly the works of Arthur Schopenhauer and Friedrich Albert Lange (1828–75), a "neo-Kantian" member of the "Marburg School" of German idealism.[164]

Nietzsche's published writings from 1869 to 1876 incorporate his interests in philology, art and aesthetic theory, cultural criticism, and philosophy. He was at that time deeply, intellectually, and emotionally influenced by the person and operas of Richard Wagner. Nietzsche's first published book, in 1872, *Die Geburt der Tragödie aus dem Geiste der Musik*, *The Birth of Tragedy from the Spirit of Music* (abbreviation: BT; in English, *The Spirit of Music* is unfortunately usually omitted), appropriated Schopenhaurean categories of artistically-rarefied individuation and collective emotional catharsis in an elucidation of primordial sublimated and unleashed aesthetic and instinctual drives as represented by the Greek gods Apollo and Dionysus, respectively. This book also includes a Wagnerian formula for cultural flourishing: society must cultivate and promote its highest and noblest spirits—artistic geniuses. In the Preface to a later edition of this work, Nietzsche expressed regret for having attempted to elaborate a "metaphysics of art." Nietzsche's second book-length project, on which he worked from 1873 to 1876, is *Untimely Meditations* (German: *Unzeitgemässe Betrachtungen*, also translated as *Unfashionable Observations* and *Thoughts Out Of Season*). This consists of four essays of acerbic cultural and historical criticism, adulatory reflections on Wagner and Schopenhauer, and, most notably, an essay called "On the Use and Abuse of History for Life" (*Vom Nutzen und Nachteil der Historie für das Leben*), an often disparaging critique of the historical scholarship of his day, which Nietzsche, characteristically, claimed did a disservice to "life" (*das Leben*).[165]

Nietzsche's health, always fragile, plus his hostility to the Philistine academic culture of his day, induced him to resign his professorship in 1879 and become an independent, free-thinking intellectual. He gradually developed a distinctive naturalistic critique of traditional morality and culture—an orientation encouraged by his friendship with Paul Rée, who traveled with Nietzsche to Italy, and who, together with Lou Andreas Salomé, formed a Platonic (at least for Nietzsche) ménage à trois. Nietzsche was ultimately resigned to the termination of these once-intense friendships, another bitter disappointment for him in life and love, but one that served as intellectual fodder for Lou, who spurned Nietzsche's marriage proposals but later wrote a book about him.[166]

In 1878, Nietzsche published *Human, All Too Human: A Book for Free Spirits* (German: *Menschliches, Allzumenschliches: Ein Buch für freie Geister,* abbreviation: HH). A second part, *Assorted Opinions and Maxims* (*Vermischte Meinungen und Sprüche*), was published in 1879, and a third part, *The Wanderer and his Shadow* (*Der Wanderer und sein Schatten*), followed in 1880. In this work, Nietzsche savaged conventional pieties in the pithy aphoristic style that would characterize his later works. When he sent the book to the Wagner early in 1878, it effectively ended their friendship.

For the rest of his working life, Nietzsche published a book almost every year, commencing with *The Dawn of Day,* or *Dawn,* or *Daybreak Thoughts on the Prejudices of Morality* (German: *Morgenröte—Gedanken über die moralischen Vorurteile*, 1881), which consists largely of critical observations on conventional morality and its underlying psychology. This was followed in rapid order by the works for which Nietzsche is best known: *The Gay Science* (German: *Die fröhliche Wissenschaft*; abbreviation: GS), occasionally translated as *The Joyful Wisdom* or *The Joyous Science* (1882, second expanded edition, 1887; abbreviation: JS); *Thus Spoke Zarathustra*: *A Book for All and None* (German: *Also sprach Zarathustra: Ein Buch für Alle und Keinen*, 1883–5; abbreviation: Z); *Beyond Good and Evil*: *Prelude to a Philosophy of the Future* (German: *Jenseits von Gut und Böse: Vorspiel einer Philosophie der Zukunft*, 1886; abbreviation: BGE); *On the Genealogy of Morality*: *A Polemic* (German: *Zur Genealogie der Moral: Eine Streitschrift*, 1887; abbreviation: GM), and in the last year of his productive life *Twilight of the Idols, or, How to Philosophize with a Hammer* (German: *Götzen-Dämmerung, oder, Wie man mit dem Hammer philosophiert*, 1888; abbreviation: TI), *The Case of Wagner* (German: *Der Fall Wagner*; abbreviaiton: CW), *The Antichrist* (German: *Der Antichrist*; abbreviation: AC), and his intellectual "biography," *Ecce Homo, How One Becomes What One Is* (German: *Ecce homo: Wie man wird, was man ist*; abbreviation: EC) which was published posthumously in 1908.

In his later working years, Nietzsche moved frequently in an effort to find a climate that might improve his health, spending winters near the Mediterranean Sea (usually in Italy) and summers in Sils Maria, Switzerland. His symptoms included intense headaches, nausea, and trouble with his eyesight. Recent research (Huenemann 2013) has convincingly argued that Nietzsche *probably suffered from a retro-orbital meningioma, a slow-growing tumor on the brain surface behind his right eye.* In January 1889, Nietzsche collapsed on a street in Turin, and when he regained consciousness wrote a series of increasingly deranged letters. After unsuccessful attempts at treatment in Basel and Jena, he was released into the care of his mother, and later his sister, eventually lapsing entirely into silence. He lived on until 1900, when he died of a stroke complicated by pneumonia.

During his illness, his sister, Elisabeth, assumed control of his literary legacy, and she eventually published *The Antichrist* and *Ecce Homo*, as well as a selection of writings from his notebooks for which she used the title *The Will to Power* (abbreviation: WP), following Nietzsche's remark in *On the Genealogy of Morality* (*GM*, Book III, 27) that he planned a major work by that title. Her edited version did not precisely match Nietzsche's surviving plans in his notebooks for that book and was also marred by Elisabeth's vehemently anti-Semitic prejudices, which had been quite distressing to Nietzsche himself. As a result, *The Will to Power* may not be a reliable indicator of his later philosophy.

Much later, during the 1920s and 1930s, Elizabeth, following her husband's (Bernhard Förster) radical right-wing proclivities, became an ardent Nazi supporter and even welcomed Hitler to the Nietzsche home in Weimar, Germany, turning it into a Nietzsche "shrine" for a growing Nietzsche "cult" of admirers.[167] Accordingly, there has been an ongoing debate regarding the degree to which, if any, Nietzsche himself, or at least some of his phrases and concepts most susceptible to distortion or sloganeering—notably "The Will to Power" and "The Overman"—presaged or in some stronger way contributed to Nazi Aryan ideology, or even if he had an explicit social and political philosophy that might have lent itself to Nazism.[168] It is far from clear that Nietzsche himself was in any meaningful way a "Proto-Nazi," and, of course, he died decades before Hitler's accession to power. The case of Nietzsche was therefore unlike that of Martin Heidegger (who wrote volumes about Nietzsche), who was, at least in the early 1930s, an active Nazi and never resigned his membership in the Nazi Party. But some of Nietzsche's harsh phrases regarding people and groups he didn't like do lend support to a critical view of Nietzsche's political and racial sentiments by those "hard" interpreters who decry his implicit or explicit distasteful inclinations.

My own view is closer to the "soft" interpretation of someone like Walter Kaufmann (whom I had the pleasure of once meeting) and lays most of the blame for "Nietzschean Nazism" on his sister, his brother-in-law, and Nazi ideologists. They cherry-picked selections from Nietzsche's work for their own Teutonic nationalistic, (notwithstanding the fact Nietzsche tended to despise both Germans and nationalism), anti-Semitic, misogynistic, homophobic, and militaristic purposes.[169] That said, it does not take much effort to find many passages in Nietzsche's works that lend themselves to *both* "hard" and "soft" readings, or, indeed, *to virtually any interpretation* that matches the reader's projections and desires. Ambiguity, elusiveness, and apparent inconsistencies abound in Nietzsche's texts. And being "understood" was not something he seems to have desired:

> One does not only wish to be understood when one writes; one wishes just as surely *not* to be understood. … [Some authors] did not wish to be understood by "just anybody."
>
> (Nietzsche (*GS*, 381))

Nietzsche's Textual Worlds

Nietzsche's texts, as stylistically scintillating as many of them are, depart dramatically from conventional philosophical writing. Most academic philosophers write scholarly books or articles that present a precisely articulated thesis for which they present a sustained and well-defended argument. Kant, Hegel, and even Schopenhauer wrote long tomes with systematic arguments for the positions they espoused and against those they critiqued. The presentation and refutation of "arguments" in an orderly, logical, systematic, and often boringly abstruse way has been the norm for "philosophical" writing since Aristotle.

Nietzsche's books, however, are nothing like that. Many are divided into short, numbered sections, which only intermittently have obvious connections to nearby sections. While the "books" or "parts" within sections are often thematically related (e.g., *GS* Book II or *BGE* Parts I, V, and VI), they do not typically fit together into a single overall argument. Nietzsche himself noted the fleeting character with which he discussed his concerns, declaring:[170]

> I approach deep problems like cold baths: quickly into them and quickly out again. That one does not get to the depths that way, not deep enough down, is the superstition of those afraid of the water, the enemies of cold water; they speak without experience. The freezing cold makes one swift.

Some of Nietzsche's books (notably *BT*, GM, and the *AC*) do linger longer in the "cold water" by offering sustained argumentation, but even in those Nietzsche frequently abandons one thread of reasoning to raise apparently unrelated points, leaving the reader to piece together if and how the disparate elements of his text fit together (e.g., *GM* Book II, virtually all of Z, and many of his last works). This mode of writing is often classified as "aphoristic," and, together with such French *moralistes* as Michel de Montaigne (1533–92), Blaise Pascal (1623–62), and François de La Rochefoucauld (1613–80), Nietzsche was a distinguished practitioner of that form of discourse.

A number of Nietzsche's pithy aphorisms consist of psychological pronouncements, like "Egoism is the law of perspective applied to feelings: what is closest appears large and weighty, and as one moves further away size and weight decrease" (*GS*, 162), while others encapsulate a point Nietzsche has been developing through the preceding section of the text (e.g., "We are always only in our own company," *GS*, 166). Walter Kaufmann suggested that Nietzsche created his aphorisms in the service of an "experimentalist" mode of philosophizing.[171] But not every Nietzschean aphorism is an experiment, and not every short section is an aphorism. Many sections of his books *build up* to an aphorism, as a kind of summative conclusion, rather than experiment. In any event, Nietzsche's literary style is an indispensable, and multifaceted, component of his distinctively aesthetic way of philosophizing "with a hammer" in an anti-philosophical way.

Nietzsche's Anti-Philosophical Hammer Pummels the "Idols of the World"

In his very late work, Twilight of the Idols or How One Philosophizes with a Hammer, which he had planned to call *Müssiggang eines Psychologen* (The Idleness of a Psychologist), Nietzsche declared:[172]

> *There are more idols than realities in the world: that is my 'evil eye' for this world* ... also my 'evil *ear*.'... to pose questions here with a *hammer* ... what a delight ... for me, an old psychologist and pied piper ... a *great declaration of war* ... the idols ... are not just idols of the age, but *eternal* idols ... here touched with the hammer as with a tuning fork.

One of the first and most striking "hammer blows" landed by Nietzsche's psychological anti-philosophical, or at least philosophically unsystematic, dissection, and rejection of conventional and "eternal idols" is what he calls in *The Will to Power* (German: *Der Wille zur Macht*, or abbreviation: *WP)* "the moral interpretation of the world," stemming, in his view, mainly from Christian moral values and Occidental religious and intellectual traditions more generally:[173]

> one cannot find a greater contrast than ... between a *master morality* and the morality of *Christian* value concepts ... *Master morality is ... the ... language ... of ascending life, of the will to power as the principle of life*. Master morality *affirms* as instinctively as Christian *negates. The former gives to things out of its own abundance—it transfigures, it beautifies the world and makes it more rational—the latter impoverishes, pales, and makes uglier the value of things, it negates the world. 'World' is a Christian term of abuse.*

And, in *Ecce Homo*, Nietzsche continued his assault on "idols":[174]

> Overthrowing idols ... ("ideals") ... part of my craft. One has deprived reality of its value, its meaning, its truthfulness, to precisely the extent *to which one had mendaciously invented an ideal world. The 'true world' and the 'apparent world'—means ... the mendaciously invented world and reality.*

And so from the ashes of the Christian "negation" of the world and his own exposure of the "mendaciously invented apparent world," Nietzsche hoped *there might arise the "affirmation, transfiguration, rationalization, and beautification" of the new world brought into being by the "master morality" of the select few who exercise the will to power as the "principle of life."* This new, post-Christian world seems also to entail what for Abrahamic religions and conventional Occidental rationalism and optimism might be deemed "nihilism:"[175]

> The end of the *moral* interpretation of the world … leads to nihilism. "Everything lacks meaning." … What does nihilism mean? *That the highest values devaluate themselves* … the answer is lacking to our "Why?"

For Nietzsche, we humans, who are the products of "the modern age" with all its skepticisms, idols and ideals, illusions and distractions, discover that what we have held to be our core values—including our beliefs in God, the immortality of the soul, the virtues of compassion and "good deeds," the values of justice and equality, and the existence of "another, better world"—are mere illusions and worthless "idols" (*Götzen*) manufactured by priests and "philosophers" for popular consumption. *But the end of our "metaphysical faith" in "another world," also entails the loss of our faith in science, of a defensible belief in a rational world order set in place by a transcendent God and measurable by scientific tools.* As Nietzsche proclaims in his *Genealogy of Morals*, the man of science, or:[176]

> The truthful man … *affirms another world than that of life, nature and history … that "other world" [*denies*]… its antithesis this world, our world?… a metaphysical faith* … underlies our faith in science—and … the Christian faith … that God is truth … But what if this belief is becoming … lies—if God himself turns out to be our *longest lie*?

And furthermore in *Ecce Homo,* Nietzsche declares:[177]

> The concept of "God" invented as a counter concept of life … hostility unto death against life … *the "beyond," the "true world" invented … to devaluate the only world there is* … The concept of … the "immortal soul," invented … to despise the body, to make it sick … what is most terrible of all—the concept of the *good* man signifies … all that is weak, sick, failure, suffering … an ideal … fabricated … against the proud … human being who says Yes … and … now called evil—all this was believed as morality! *Ecrasez l'infame!*[178]

However, *this "real, godless world," the only world, fails, by its own "nature," to provide us with meaning and purpose, and our restless, unending quest for the pleasures of this world do not provide us with lasting satisfaction or happiness. And so we feel lost and dispirited, without a secure moral compass to find our way in an evanescent world and seemingly trapped in lives without satisfaction or redemption.*

Under these existential conditions, what are we to believe and value, if anything? Is nihilism the "solution" or the "problem," a stark reflection of the times in which "God is dead" and during which we have murdered Him, or an eternal moral void at the core of the cosmos? And, if so, how are we to live without either the hope of redemption in a "better" world or a feeling of fulfillment in this world of pain and suffering? Finally, is there anything that can replace the discredited values that have anchored Occidental civilization for millennia? Can something new and enduring be discovered or created that will assuage our angst and "transvalue" or revalue the valueless world in which we are adrift?

Nietzsche does not frequently provide direct or consistent "answers" to these pressing questions but instead appears to supply just more questions and intellectual quagmires. The closest things to answers, or "solutions," to the riddles of worldly existence lie in such phrases as *amor fati* (Latin: "love of fate," recalling the pre-Christian, "hard world of struggle" Nietzsche seems to have valorized in his persona as a classical philologist), "the will to power" (*der Wille zur Macht*), the "overman" (*der Übermensch*), "perspectivism," the "endless recurrence"—the return of all things (recalling ancient Greek and Indian theories of the repetitive cycles of everything past, present, and future)—and, ultimately, the aesthetic dimension of human life.[179]

The World Is the Will to Power, and We Have Created It!

Nietzsche's demystification of the origins of human values challenges the commonsensical, but false, view that values either just exist somewhere outside ourselves—i.e., created by God, found in nature, or revealed by scripture—and are there for us somehow to internalize and obey. Furthermore, Nietzsche subverts virtually all binary oppositions, or dualisms, including "good and evil," and the widespread reflexive assumption that knowledge in general and human values in particular are either "objective" and "unchanging" (e.g., Plato, Kant, and most Occidental religions) or are "subjective" and "relative" (for many people during this "Postmodern Age"). Nietzsche challenges these basic beliefs and rejects such facile dualisms as objective and subjective when he declares in *The Birth of Tragedy*, *The Gay Science*, and elsewhere that *"life" and "the world" are "beyond good and evil," are hence "amoral," and, accordingly, that value is "created" rather than discovered, just as we have "created" the world to which we ascribe, rather than find, value*:

> We [contemplatives] … are those who really continually *fashion* something that had not been there before: *the whole eternally growing world of valuations, colors, accents, perspectives, scales, affirmations, and*

> *negations ... Whatever has* value *in our world now does not have value in itself, according to its nature—nature is always value-less—but has been* given *value at some time, as a present—and it was* we *who gave and bestowed it. Only we have created the world* that concerns man*!*
>
> (*GS*, 301)

Given this, Nietzsche declares in the *Twilight of the Idols* that:[180]

> There are no moral facts at all. Moral judgment has this in common with religious judgment, that it believes in realities which do not exist.

But what, if anything, did Nietzsche himself value? The closest Nietzsche comes to answering this question systematically is the stress he puts on the importance of power, especially if this is taken together with related ideas about strength, health, and "life." Consider, for example, this passage that appears in the second part of his late (1888) work, *The Antichrist*:[181]

> What is good? Everything that heightens the feeling of power in man, the will to power, power itself.
>
> What is bad? Everything that is born of weakness.
>
> What is happiness? The feeling that power is *growing*, that resistance is overcome.

The phrase the "will to power" is usually viewed as one of Nietzsche's central ideas. It can be generously interpreted to mean that creatures like us, or, more broadly, all life, or even all things, aim at the enhancement of their power—and that this fundamental truth about the nature of things entails the additional truth that enhanced power is good for us. Furthermore, by implication, given Nietzsche's doctrine of the eternal recurrence of all things, this is also the case for everything everywhere that has ever lived and ever will live. For Nietzsche, *"the feeling of power," which is also a Lebensgefühl (feeling for life, vitality), should replace otherworldly measures of value.*

But no otherworldly measures of anything exist for Nietzsche. Yet, this absence of a transcendental measure of value, belief, and action does not necessarily imply that all expressions of power are qualitatively the same. For example, a man such as Julius Caesar, who, from Nietzsche's point of view, practiced a kind of Machiavellian *virtù*, can be powerful, shrewd, and achieve conventional success in this world. But for Nietzsche, the most important *aspect of "the will to power" lies in self-mastery and self-directed action in the world.*

In the chapter, "Of Self-Overcoming" from *Thus Spake Zarathustra*, all living creatures are said to be obeying something, while "he who cannot obey himself will be commanded. That is the nature of living creatures." Either one can be a "master" or a "slave," in one's mentality and in the world of "eternally changing and flooding play of forces." *So what is "the world" and how does Nietzsche conceive "life in this world?"* In aphorism § 1067 of *The Will to Power*, Nietzsche reveals what he regards as *the essence of "the world:"*

> And do you know what *"the world" is to me? ... This world: a monster of energy, without beginning, without end ... as force throughout, as a play of forces ... waves of forces ... a sea of forces flowing and rushing together,* eternally changing and eternally flooding back with tremendous years of recurrence ... out of the play of contradictions back to the joy of concord, still blessing itself as that which must return eternally, as a becoming that knows no satiety, no disgust, no weariness; *this my Dionysian world of the eternally self-creating, the eternally self-destroying, this mystery world of the two-fold voluptuous delight, my "beyond good and evil," without goal, unless the joy of the circle is itself a goal ... This world is the will to power—and nothing besides! And you yourselves are also this will to power—and nothing besides!*

Furthermore, in *Beyond Good and Evil*, Nietzsche asks us to suppose that the only "given" is:[182]

> *our world of desires and passions ... [*no other*] "reality" besides the reality of our drives—* ... thinking is merely a relation of these drives to each other ... the so called mechanistic (or "material") world ... "mere appearance" ... an "idea" but ... *a more primitive form of the world of affects in which everything still lies contained ... as ... instinctive life ...* a *preform of life.*

And what is "life" in this "Dionysian world?" For Nietzsche, life, as the incarnation of the "will to power," is, or could be, self-overcoming, a self-empowering instinctual struggle for growth and survival, as well as kind of intellectual war against Christianity-inspired Western cultural and moral "decadence" and "nihilism." As he says in *The Antichrist*:[183]

> Life itself appears to me as an instinct for growth, for survival, for the accumulation of forces, for *power*: whenever the will to power fails there is disaster. My contention is that all the highest values of humanity have been emptied of this will—that the values of *décadence*, of *nihilism*, now prevail under the holiest names.

To live life fully and powerfully, according to Nietzsche in *The Gay Science,* is to affirm the kind of cosmology implied by his doctrine of "the eternal recurrence of all things," that is to "live dangerously" and to "love fate" (*amor fati*), without guilt, shame, or remorse for who one is and how one lives.

Nietzsche's Aestheticization of the World?

Countless readers of Nietzsche's works, including many distinguished philosophers, have spent unending hours laboring over and struggling to "understand" and "interpret" what Nietzsche "really meant" by his linguistic constructions. Perhaps I will disappoint some of you by disclaiming any "special knowledge or insight" into what Nietzsche "intended to say or meant," and instead leave it to you readers to provide you own interpretations. Is this a metaphysical cop-out? Perhaps. But I, too, cannot entirely refrain from a bit of textual explication of Nietzsche's worldview. And, so I will continue …, but not for too much longer☺.

In what has come to be called *Also Sprach (Thus Spoke) Zarathustra (AZ),* Nietzsche's eponymous prophetic anti-prophet without a home in this world exhibits the "Weariness that wants to reach the ultimate with one leap, with one fatal leap, a poor ignorant weariness that does not want to want any more: this created all gods and other worlds."[184]

But is there at least some possible relief from this weariness, if not a remedy for despair? In his first major work, *The Birth of Tragedy*, Nietzsche seems to reject the "optimistic, superficial, cheerful," and "scientific" "logicizing" of the world, with its attendant "rationality, utilitarianism, and democracy" and the "doltification of existence," in favor of an "amoral" life lived "beyond good and evil" and as *"an aesthetic phenomenon."* For him, art, not morality, is "the truly metaphysical activity …" And "the *existence of the world is justified only as an aesthetic phenomenon.*"[185] This book was written during Nietzsche's (temporary …) infatuations with Wagner and his music as well as with Schopenhauer's philosophy, and his longer-lasting attraction to classical Greek tragedy.[186] Much later, in *The Will to Power*, Nietzsche is said to have declared, "We possess art least we perish of the truth" (Fragment #822).

In tragedy, both theatrical and personal, Nietzsche discerned an alternation between rapturous Dionysian collective abandonment and expressive vitalism on the one hand, and Euripidean and Apollonian individualizing detachment on the other hand. What Nietzsche called the "sweeping opposition of styles" between these competing classical tragic visions of existence also presents the alternative between two worlds, similar to Wagnerian opera, especially in *Tristan and Isolde*. There "… the Apollonian state of dreams, the world of the day becomes veiled, and a new world, clearer … more moving than the everyday world … more shadowy, presents itself … in continual rebirths."[187]

In *The Birth of Tragedy* and his later works, Nietzsche also posed, if somewhat indirectly, this question: *If the "everyday" world of toil and suffering is fundamentally meaningless, what, if anything, makes our lives worth living? Not "another life, the life of the world to come,"* via, as Kierkegaard would argue, our "defining relation" with God, since both this "other, metaphysical, transcendent world" and God are life-denying fictions, according to Nietzsche. Not morality, whether conventional

Judeo-Christian, Kantian, or Utilitarian. And, not the lures and snares of such deindividualizing escapes from the tragedy of existence as "Romantic love and family values"—for which Nietzsche probably longed but never achieved. And, certainly not such political fictions as "democracy, equality, and justice." For there is only the "equality" of the tragic condition for self-reflective human beings.

What remains? *Only, apparently, the ceaseless struggle of living the hard life to the fullest in the present while anticipating and welcoming its "eternal return," and the revaluation of all values such that the aestheticization of this world as a whole, and of our lives individually, might replace the moral and religious idols Nietzsche's linguistic hammer has pulverized.* The construction of our own lives as tragic–comic works of art, and the temporary "loss" of our anxiety-plagued selves in great literature, art, and music, may provide a distraction from, if not a solution to, the loss of hope in a better world, which is never to come.

The Riddle of the World and the Puzzle of Nietzsche's Inner World

And finally, who was Friedrich Nietzsche? A prophet without a home, church, or a following? A heretic whose works should be condemned and banned? A lost, tragic soul, whose unswerving and perhaps maniacal pursuit of the truths of things dislodged previously accepted verities and contributed to unhinging his own mind? Or someone who, in the section "Why I Am a Destiny" from *Ecce Home,* himself declared:[188]

> I know my fate … I am no man. I am dynamite … I want no "believers"… I am too malicious to believe in myself … Perhaps I am a buffoon … my truth is terrible; for so far one has called *lies* truth. *Revaluation of all values*: that is my formula for an act of supreme self-examination … of humanity, become flesh and genius in me. *It is my fate that I have to be the first decent human being—I was the first to discover the truth by being the first to experience lies as lies—My genius is in my mouth* … all power structures of the old society will have been exploded—all … based on lies … And whoever wants to be a creator in good and evil, must first be an annihilator and break values' … *I am the first immoralist: that makes me the annihilator par excellence.*

How is one to interpret such stunning pronouncements? There are as many readings of such apparently grandiose proclamations regarding one man's place in this world as there or responses to Nietzsche's works. For during the century and a quarter that has passed since Nietzsche's mind retreated from his outer world, we have witnessed a virtually unending stream of interpretations of Nietzsche's published works as well as speculations about the inner world of the mind that produced them. I will not add to them anything more than the interpretative "reading" of Nietzsche I have just written.

Except for this: The distinguished American "Pragmatist" philosopher/psychologist William James (1842–1910) proclaimed: "A philosophy is the expression of a man's intimate character." And, as Walter Kaufmann glossed, "Nietzsche would add, of the philosopher's moral notions."[189]

Nietzsche's philosophy, like virtually everyone else's, is chosen as a (partial?) reflection of that person's character and moral sentiments. And what was *his philosophy*? Perhaps his own words might best serve as both his epitaph and an epigraph:

> The very best of all things is completely beyond your reach: not to have been born, not to *be,* to be *nothing.* But the second best thing for you is—to meet an early death.[190]

The world that Nietzsche dissected with unprecedented acuity has been dramatically transformed since his physical demise. Nietzsche's textual worlds, however, have lived on long since his departure from this worldly vale of tears at the dawn of the 20th century.

Notes

1 For Aristarchus, see www.greeka.com/eastern-aegean/samos/history/aristarchus/.
2 For Seleucus, see https://en.wikipedia.org/wiki/Seleucus_of_Seleucia.
3 For Aryabhata, see www.sciencedirect.com/topics/mathematics/aryabhata.
4 For Abu Mashar, see https://encyclopedia2.thefreedictionary.com/Ja%27far+ibn+Muhammad+Abu+Ma%27shar+al-Balkhi.
5 For Ptolemy, see https://en.wikipedia.org/wiki/Ptolemy.
6 For Ali Qushji, see https://muslimheritage.com/ali-al-qushji-and-his-contributions-to-mathematics-and-astronomy/.
7 For Al-Ghazali, see Chapter 1 of this book. For Al-Kindi, see Peter Adamson, "al-Kindi," *The Stanford Encyclopedia of Philosophy* (Spring 2020 Edition), Edward N. Zalta (ed.), https://plato.stanford.edu/archives/spr2020/entries/al-kindi/. For Saadya, see Sarah Pessin, "Saadya [Saadiah]," *The Stanford Encyclopedia of Philosophy* (Fall 2008 Edition), Edward N. Zalta (ed.), https://plato.stanford.edu/archives/fall2008/entries/saadya/.
8 For Tycho Brahe, see https://en.wikipedia.org/wiki/Tycho_Brahe; for Nilakantha_Somayaji, see https://en.wikipedia.org/wiki/Nilakantha_Somayaji.
9 See https://islamsci.mcgill.ca/RASI/BEA/Ibn_al-Shatir_BEA.htm.
10 Nicolas Copernicus, *De Revolutionibus,* available at: www.geo.utexas.edu/courses/302d/Fall_2011/Full%20text%20-%20Nicholas%20Copernicus,%20_De%20Revolutionibus%20(On%20the%20Revolutions),_%201.pdf. For the history and influence of "The Copernican Revolution," see Thomas Kuhn, *The Copernican Revolution Planetary Astronomy in the Development of Western Thought* (Cambridge, MA: Harvard University Press), 1985. A classic synopsis of the history of Occidental planetary astronomy during this period is Alexandre Koyré, *From the Closed World to the Infinite Universe* (Baltimore, MD: The Johns Hopkins University Press, 1957).

11 For Thomas Digges, see https://mathshistory.st-andrews.ac.uk/Biographies/Digges.
12 For Bruno, see https://en.wikipedia.org/wiki/Giordano_Bruno.
13 See Johannes Kepler, *Harmonices mundi libri V*: Trans. E.J. Aiton, A.M. Duncan, J.V. Field, *The Harmony of the World*. Philadelphia, PA: American Philosophical Society (Memoirs of the American Philosophical Society, 1997); and Daniel A. Di Liscia, "Johannes Kepler," *The Stanford Encyclopedia of Philosophy* (Fall 2019 Edition), Edward N. Zalta ed., https://plato.stanford.edu/archives/fall2019/entries/kepler/.
14 Norman Melchert and David R. Morrow, *The Great Conversation A Historical Introduction to Philosophy*, Eighth Edition (New York and Oxford: Oxford University Press, 2019), 358.
15 Galileo, as cited in Melchert and Morrow, *op. cit.*, 355.
16 For Galileo, see *Dialogue Concerning the Two Chief World Systems*, S. Drake (trans.), (Berkeley: University of California Press, 1967); *Dialogues Concerning Two New Sciences*, H. Crew and A. de Salvio (trans.) (Dover Publications, Inc., New York, 1954, 1974); and Peter Machamer, "Galileo Galilei," *The Stanford Encyclopedia of Philosophy* (Summer 2017 Edition), Edward N. Zalta, ed., https://plato.stanford.edu/archives/sum2017/entries/galileo/. Here is a good account of Galileo's forced recantation of his defense before the Inquisition of the Copernican "world system:" www.smithsonianmag.com/smithsonian-institution/378-years-ago-today-galileo-forced-to-recant-18323485/.
17 A good summary of Newton's laws of motion is provided by Wikipedia: these "are three physical laws that, together, laid the foundation for classical mechanics. They describe the relationship between a body and the forces acting upon it, and its motion in response to those forces. More precisely, the first law defines the force qualitatively, the second law offers a quantitative measure of the force, and the third asserts that a single isolated force does not exist. These three laws … can be summarized as follows:

First law: in an inertial frame of reference, an object either remains at rest or continues to move at a constant velocity, unless acted upon by a force.
Second law: in an inertial frame of reference, the vector sum of the forces F on an object is equal to the mass *m* of that object multiplied by the acceleration a of the object: F = *m*a. (It is assumed here that the mass *m* is constant).
Third law: when one body exerts a force on a second body, the second body simultaneously exerts a force equal in magnitude and opposite in direction on the first body." https://en.wikipedia.org/wiki/Newton%27s_laws_of_motion.

18 See Charles Webel, "Modernist Creativity: The Construction of Reality in Einstein and Kandinsky," *World Futures* 63, Fall 2007, 1–33.
19 Isaac Newton, *The Mathematical Principles of Natural Philosophy*, trans. Andrew Motte (New York: Daniel Adee, 1846), aka *Principia*, esp. 511–13, available at: www.maths.tcd.ie/pub/HistMath/People/Newton/Principia/Bk1Sect1/PrBk1St1.pdf.
20 For the "Rules," see https://apex.ua.edu/uploads/2/8/7/3/28731065/four_rules_of_reasoning_apex_website.pdf/. According to Alan Shapiro, Newton refrained

from using the term "experimental philosophy," widely used in Restoration England at the start of his career, until 1712, when he added a passage to the "General Scholium" of the *Principia* that briefly expounded his anti-hypothetical methodology. Newton introduced the term for polemical purposes to defend his theory of gravity against the criticisms of Cartesians and Leibnizians but, especially in the *Principia*, against Leibniz himself. "Experimental philosophy" has little directly to do with experiment, but rather more broadly designates empirical science. Newton's manuscripts provide insight into his use of "experimental philosophy" and the formulation of his methodology, especially such key terms as "deduce," "induction," and "phenomena," in the early 18th century. See Alan Shapiro, "*Newton's 'Experimental Philosophy,'*" Early Science and Medicine Vol. 9, No. 3, Newtonianism: Mathematical and 'Experimental' (2004), pp. 185–217.

21 See https://en.wikipedia.org/wiki/Hypotheses_non_fingo.

22 Newton:

> This most beautiful system of the sun, planets, and comets, could only proceed from the counsel and dominion of an intelligent and powerful Being … This Being governs all things, not as the soul of the world, but as Lord over all; and on account of his dominion he is wont to be called Lord God.

And:

> When I wrote my treatise about our system, I had an eye upon such principles as might work with considering men for the belief of a Deity, and nothing can rejoice me more than to find it useful for that purpose.

(www.christianitytoday.com/history/issues/issue-30/newtons-views-on-science-and-faith.html)

23 For Swedenborg, see www.britannica.com/biography/Emanuel-Swedenborg; for Lambert, see https://en.wikipedia.org/wiki/Johann_Heinrich_Lambert; for Wright, see https://en.wikipedia.org/wiki/Thomas_Wright_(astronomer); and for Laplace, see https://mathshistory.st-andrews.ac.uk/Biographies/Laplace/.

24 For the independent cosmological contributions of Laplace and Kant, see www.encyclopedia.com/science/encyclopedias-almanacs-transcripts-and-maps/laplace-theorizes-solar-system-originated-cloud-gas.

25 Good summaries of Descartes' contributions to mathematics, science, and philosophy include: Gary Hatfield, "René Descartes," *The Stanford Encyclopedia of Philosophy* (Summer 2018 Edition), Edward N. Zalta, ed., https://plato.stanford.edu/archives/sum2018/entries/descartes/; Norman Melchert and David R. Morrow, *The Great Conversation A Historical Introduction to Philosophy*, Eighth Edition (New York and Oxford: Oxford University Press, 2019 (Ch. 17), 360–401; Stephen Gaukroger, *Descartes an Intellectual Biography* (Oxford: Oxford University Press), 2004; and Daniel Garber, *The Cambridge History of Seventeenth-Century Philosophy: Volume I.* (Cambridge; Cambridge University Press, 1998).

26 Melchert and Morrow, *op. cit.*, 361.

27 In Descartes own words,

> But after I had spent some years pursuing these studies in the book of the world, and trying to gain some experience, I made a decision one day

> to undertake studies within myself too and to use all the powers of my mind in choosing the paths I should follow. This has worked better for me, I think, than if I had never left my country or my books.

(René Descartes, *Discourse on the Method of Rightly Conducting One's Reason and Seeking Truth in the Sciences,* Part 1, ed. Jonathan Bennett, 2017, p. 5, available at: www.earlymoderntexts.com/assets/pdfs/descartes1637.pdf)

28 Descartes describes his project in this way:

> But I wanted … to be free to say what I thought … without having to follow or to refute the opinions of the learned. My plan for doing this was to leave our world wholly to the learned folk to argue about, and to speak solely about what would be the case in a new world that would exist if God now created somewhere in imaginary spaces enough matter to compose a world; variously and randomly agitated the different parts of this matter so as to create as confused a chaos as any poets could dream up; and then did nothing but allow nature to unfold in accordance with the laws he had established.
>
> (Discourse on Method, *op. cit.*, 19)

29 Descartes, *The World and Other Writings*, trans. & ed. Stephen Gaukroger (Cambridge, Cambridge University Press 2004). Also see https://en.wikipedia.org/wiki/The_World_(Descartes).

30 Stephen Gaukroger, *The World and Other Writings*, *op. cit.*, Introduction xvi–xvii.

31 Descartes, *The World*, *op. cit.*, 21–2.

32 As Descartes puts it in Chapter 7, "*The Laws of Nature of this new world*:"

> Take it then, first, that by "Nature" here I do not mean some deity or other sort of imaginary power. Rather, I use the word to signify matter itself, in so far as I am considering it taken together with the totality of qualities I have attributed to it, and on the condition that God continues to preserve it in the same way that He created it.
>
> *The World, op. cit.*, 24–5

33 *The World*, *op. cit.*, 23.

34 See *The Oxford Handbook of Descartes and Cartesianism*, ed., Steven Nadler, Tad M. Schmaltz, and Delphine Antoine-Mahut (Oxford: Oxford University Press, 2019).

35 See the various works of Patricia and Paul Churchland, as well as Charles Webel and Anthony Stigliano, "Beyond Good and Evil?" Radical Psychological Materialism and the 'Cure' for Evil," *Theory & Psychology*, Vol. 14 (1) Feb. 2004, 81–103.

36 According to Peter Markie in the SEP,

> The dispute between rationalism and empiricism concerns the extent to which we are dependent upon sense experience in our effort to gain knowledge. Rationalists claim that there are significant ways in which our concepts and knowledge are gained independently of sense experience. Empiricists claim that sense experience is the ultimate source of all our concepts and knowledge. Rationalists generally develop their view in two ways. First, they argue that there are cases where the content of our concepts or knowledge outstrips the information that sense experience can

provide. Second, they construct accounts of how reason in some form or other provides that additional information about the world.

(See Peter Markie, "Rationalism vs. Empiricism," *The Stanford Encyclopedia of Philosophy* (Fall 2017 Edition), Edward N. Zalta, ed., https://plato.stanford.edu/archives/fall2017/entries/rationalism-empiricism/)

37 For overviews overviof Spinoza's life and work, see Steven Nadler, "Baruch Spinoza," *The Stanford Encyclopedia of Philosophy* (Summer 2020 Edition), Edward N. Zalta ed., https://plato.stanford.edu/archives/sum2020/entries/spinoza/ and https://iep.utm.edu/spinoza/.

38 Baruch Spinoza, *Ethics*, trans, from the Latin by R.H.M. Elwes, The Project Gutenberg EBook: www.gutenberg.org/files/3800/3800-h/3800-h.htm.

39 Spinoza, *Ethics*, *op. cit.*, *passim.*

40 *Ibid.*

41 See Antonio Damasio, *Descartes' Error: Emotion, Reason, and the Human Brain* revised edition (London: Penguin, 2005); *The Feeling of What Happens: Body and Emotion in the Making of Consciousness* (New York: Harcourt, 1999); *Looking for Spinoza: Joy, Sorrow, and the Feeling Brain*, (New York: Harcourt, 2003); *Self Comes to Mind: Constructing the Conscious Brain* (New York, Pantheon, 2010; and *The Strange Order of Things: Life, Feeling, and the Making of Cultures* (New York: Pantheon, 2018).

42 Aquinas treats the "Best of all possible worlds" problem in the *Summa Theologica* (1273).

43 Proceeding from Leibniz, contemporary "analytic" philosophers, logicians, and semanticists define basic modal concepts—necessity, contingency, possibility, and impossibility—in non-modal terms:

a *Possibility*: a proposition is possible if and only if it is true in some possible world. A being is possible if and only if it exists in some possible world.

b *Contingency*: a proposition is contingently true if and only if it is true in this world and false in another world. A proposition is contingent if its contrary does not imply a contradiction.

c *Necessity*: a proposition is necessarily true if and only if it is true in every possible world.

d *Impossibility*: a proposition is impossible if and only if it is not true in any possible world.

For overviews of modal logic and possible worlds theories, see Christopher Menzel, "Possible Worlds," *The Stanford Encyclopedia of Philosophy* (Winter 2017 Edition), Edward N. Zalta, ed., https://plato.stanford.edu/archives/win2017/entries/possible-worlds/. For Kripke, see Saul Kripke, "Semantical Considerations on Modal Logic," *Acta Philosophica Fennica*, 16: 83–94, 1963; and *Naming and Necessity* (Cambridge, MA: Harvard University Press, 1980). For Lewis, who, curiously, in *On the Plurality of Worlds*, refrains from a sustained discussion of Leibniz, see David Lewis, "Counterpart Theory and Quantified Modal Logic," *Journal of Philosophy*, 65: 113–26, 1968; "General Semantics," *Synthese*, 22: 18–67; 1970; *Counterfactuals* (Cambridge, MA: Harvard University Press, 1973); and, most importantly, *On The Plurality of Worlds* (Oxford: Blackwell, 1986).

44 *Theodicy*, *op. cit.*, 281.

45 According to Leibniz,

> the world is the creature of God; the whole system is pre-established by him who … perfectly independent of the world … *the first reason of things* … one must seek the reason for the existence of the world, which is the whole assemblage of *contingent* things … in the substance which carries with it the reason for its existence, and which … is *necessary* and eternal. Moreover, this cause must be intelligent: for this existing world being contingent and an infinity of other worlds being equally possible, and holding … equal claim to existence with it, the cause of the world must needs have had … reference to all these possible worlds … if there were not the best (*optimum*) among all possible worlds, God would not have produced any

Leibniz continues:

> I call "World" the whole succession and … agglomeration of all existent things … and … there is an infinitude of possible worlds among which God must … have chosen the best … Some adversary … will perchance answer the conclusion by … saying that the world could have been without sin and without sufferings; but I deny that then it would have been *better*. For … all things are *connected* in each one of the possible worlds: the universe … is all of one piece, like an ocean … if the smallest evil that comes to pass in the world were missing in it, it would no longer be this world; which … was found the best by the Creator … one may imagine possible worlds without sin and without unhappiness … but these same worlds … would be very inferior to ours in goodness … I resort to my principle of an infinitude of possible worlds … where all conditional futurities must be comprised.

Furthermore,

> one must think of the creation of the best of all possible universes … God is not forced, metaphysically speaking, into the creation of this world … as soon as God has decreed to create something there is a struggle between all the possibles, all … laying claim to existence, and … those which … produce most reality, most perfection, most significance carry the day … God is bound … to make things … that there can be nothing better … the place our world holds in the universe, nothing can be done … better than … God does. He makes the best possible use of the laws of nature … he has established and … the laws that God has established in nature are the most excellent … it is possible to conceive.
>
> (*Ibid.*, 252–3, and 340)

46 The American philosopher and psychologist William James used the term "multiverse" in 1895, but in a different context than cosmology. The term was first used in fiction and in its current astrophysical context by Michael Moorcock in his 1963 science fiction adventures novella *The Sundered Worlds.* See https://en.wikipedia.org/wiki/Multiverse.

47 Leibniz, *Theodicy*, *op. cit.*, 269.

48 For Plantinga, see https://en.wikipedia.org/wiki/Alvin_Plantinga.

49 Bertrand Russell, *A Critical Exposition of the Philosophy of Leibniz* (London: George Allen & Unwin, 1900).

50 For Calvin, see https://iep.utm.edu/calvin/.

51 For the Encyclopedists (les Encyclopédistes in French), see https://en.wikipedia.org/wiki/Encyclopédistes; for Rousseau, see Christopher Bertram, "Jean Jacques Rousseau," *The Stanford Encyclopedia of Philosophy* (Winter 2020 Edition), Edward N. Zalta, ed., https://plato.stanford.edu/archives/win2020/entries/rousseau/.

52 For Romanticism, see www.britannica.com/art/Romanticism.

53 For the distinction between the "Western world, or the "Occident," and the "Non-Western world," especially the "Orient," see www.sciencedaily.com/terms/western_world.htm.

54 For the "Old World" and "New World," see Aaron M. Shatzman. *The Old World, the New World, and the Creation of the Modern World, 1400–1650: An Interpretive History* (New York: Anthem Press, 2013), available through *JSTOR*, www.jstor.org/stable/j.ctt1gsmzp1. Accessed Oct. 3, 2020; and https://en.wikipedia.org/wiki/Old_World.

55 Sean Gaston's book *The Concept of World from Kant to Derrida* (London: Rowman & Littlefield, 2013), usefully synopsizes philosophical conceptions of the world in such major thinkers as Kant, Hegel, Husserl, Heidegger, Derrida, and some recent fiction. However, it largely omits Schopenhauer, Wittgenstein, Sartre, Merleau-Ponty, and the analytic philosophical tradition. For the "analytic/continental" split in more detail, see https://philosophynow.org/issues/74/Analytic_versus_Continental_Philosophy.

56 Kant had earlier posed such questions in the *Critique of Pure Reason, op. cit.*, A805, B 833, p. 635: "All the interests of my reason, speculative as well as practical, combine in the three following questions:

1 What can I know?
2 What ought I to do?
3 What may I hope?"

Also see Kant's *Philosophical Correspondence*, 1759–89, trans. Arnold Zweig (Chicago, IL: University of Chicago Press, 1986), 50 and 214; and Charles Webel, *The Politics of Rationality*, *op. cit.*, Chapter 3, pp. 102–51.

57 While researching and writing this book during the COVID-19 pandemic, I did not have access to the standard editions of most of Kant's works, including (in German), the official Deutsche Akademie edition and most of the English translations in the Cambridge edition of Kant's works. I have used the translation by Norman Kemp Smith for the unabridged edition of the *Critique of Pure Reason* (New York: St. Martin's Press, 1965). While I will provide references to passages from the *Critique of Pure Reason's* 1781 and 1787 editions in the standard A (1781 edition) and B (1787 edition) format (thus "A805/B833" refers to page 805 in the 1781 edition and page 833 in the 1787 edition), I will cite mostly online editions of Kant's other major works.

Useful online overviews of Kant's life and work include: Michael Rohlf, "Immanuel Kant," *The Stanford Encyclopedia of Philosophy* (Fall 2020 Edition), Edward N. Zalta, ed.), https://plato.stanford.edu/archives/fall2020/entries/kant/; Nicholas F. Stang, "Kant's Transcendental Idealism," *The Stanford Encyclopedia of Philosophy* (Winter 2018 Edition), Edward N. Zalta, ed., https://plato.stanford.edu/archives/win2018/entries/kant-transcendental-idealism/; Robert Johnson and Adam Cureton, "Kant's Moral Philosophy," *The Stanford Encyclopedia of Philosophy* (Spring 2019 Edition), Edward N. Zalta, ed., https://plato.stanford.edu/archives/spr2019/entries/kant-moral/;

and https://iep.utm.edu/kantview/. For readers who have access to good research collections, the following primary English-language translations of Kant's works are recommended:

The Cambridge Edition of the Works of Immanuel Kant, ed. and trans. Paul Guyer and Allen W. Wood (Cambridge: Cambridge University Press, 1999).

Practical Philosophy, ed. Mary Gregor (Cambridge: Cambridge University Press, 1996), contains most of Kant's ethical writings, including *Groundwork for the Metaphysics of Morals, Critique of Practical Reason,* and *Metaphysics of Morals.*

Critique of the Power of Judgment, trans. Paul Guyer and Eric Matthews (Cambridge: Cambridge University Press, 2000).

Theoretical Philosophy 1755-1770, ed. David Walford (Cambridge: Cambridge University Press, 2002), contains most of Kant's "precritical" writings in theoretical philosophy.

Theoretical Philosophy after 1781, eds. Henry Allison and Peter Heath (Cambridge: Cambridge University Press, 2002), contains Kant's later works in theoretical philosophy, including *Prolegomena to Any Future Metaphysics* and *Metaphysical Foundations of Natural Science.*

History, Anthropology, and Education, eds. Günter Zöller and Robert Louden (Cambridge: Cambridge University Press, 2007), contains, among other writings, *Anthropology from a Pragmatic Point of View.*

The scholarly and "popular" secondary literature on Kant is, unsurprisingly, immense. Some recommended books include: Ernst Cassirer, *Kant's Life and Thought*, trans. James Haden (New Haven, CT: Yale University Press, 1983); Manfred Kuehn, *Kant: A Biography*. (Cambridge: Cambridge University Press, 2002); Paul Guyer, *Kant and the Claims of Knowledge* (Cambridge: Cambridge University Press, 1987); Henry Allison, *Kant's Transcendental Idealism: An Interpretation and Defense,* Second Edition (New Haven, CT: Yale University Press, 2004); Graham Bird, *The Revolutionary Kant: A Commentary on the Critique of Pure* Reason (Chicago, IL: Open Court Press, 2006); Allen Wood, *Kant's Ethical Thought* (Cambridge: Cambridge University Press, 1999); Christine Korsgaard, *Creating the Kingdom of Ends* (Cambridge: Cambridge University Press, 1996), and Onora O'Neill, *Constructions of Reason: Explorations of Kant's Practical Philosophy*, (Cambridge: Cambridge University Press, 1990).

58 Kant's interest in physics, cosmology, and natural philosophy began early. His first published work, *Thoughts on the True Estimation of Living Forces* (1749) was an inquiry into some foundational problems in physics, and it entered into the "vis viva" ("living forces") debate between Leibniz and the Cartesians regarding how to quantify force in moving objects. One of Kant's most lasting scientific contributions came from his early work in cosmology, *Universal Natural History and Theory of the Heavens* (1755), in which Kant proposed a mechanical explanation of the formation of the solar system and the galaxies in terms of the principles of Newtonian physics. He claimed that at the beginning of creation, all matter was spread out more or less evenly and randomly in a kind of nebula (a gaseous interstellar cloud). This has come to be what would later be known as the Kant-Laplace theory. A year later, Kant wrote the *Physical Monadology* (1756), which dealt with other foundational questions in physics and, notably, used Leibniz's key idea of the "monads." Kant's more developed philosophy of science is contained in *Metaphysical Foundations of Natural Science* (1786), whose central concern, according to Kant "is with

Nature—the whole world that we can know about through our senses," Kant, *Metaphysical Foundations of Natural Science,* ed., Jonathan Bennett, 2017, available at: www.earlymoderntexts.com/assets/pdfs/kant1786.pdf, p. 1. In his *Critique of Pure Reason*, Kant elaborated on the relationship between nature and the world:

> We have two expressions, world and nature, which sometimes coincide. The former signifies the mathematical sum-total of all appearances and the totality of their synthesis, alike in the great and in the small … This same world is entitled nature when it is views as a dynamical whole … the unity in the existence of appearances.
>
> (*Op. cit.*, 392 (A 419, B 447))

59 For Berkeley, see https://en.wikipedia.org/wiki/George_Berkeley.

60 As Melchert puts it, "the rational mind has a certain structure, and whatever is knowable by such a mind must necessarily be known in terms of that structure. This structure is not derived from the objects known; it is imposed on them—but not arbitrarily, because the very idea of an object not so structured makes no sense. This is Kant's Copernican revolution in philosophy," *op. cit.*, 467.

61 Kant, *Critique of Pure Reason*, *op. cit.*, 93 (B 75, A 51).

62 For the categories, see the *Critique of Pure Reason*, *op. cit.*, 113–15 (A 80–3, B 106–9).

63 Plato, for example, believed that the supersensible, purely intelligible world of the Forms is not only knowable but also more intelligible and more real than the world of experience. See Melchert, *op. cit.*, 465–6. But, according to Kant "Plato left the world of the senses, as setting too narrow limits to the understanding, and ventured out beyond it on the wings of the ideas, in the empty state of the pure understanding." *Critique of Pure Reason*, *op. cit.*, 47 (A 5, B 9).

64 In Kant's own words in the *Critique of Pure Reason*, *op. cit.*:

> the concept of appearances … establishes the objective reality of noumena and justifies the division of objects into phaenomena and noumena, and so of the world into a world of the senses and a world of the understanding (*mundus sensibilis et intelligibilis*), and … the distinction [refers] … to the difference in the manner in which the two worlds can be first given to our knowledge … For if the senses represent to us something merely as it appears, this something must also in itself be a thing [in itself], and an object of nonsensible intuition, that is, of the understanding … objects will be represented *as they are,* whereas … things will be known only as they a*ppear*.
>
> (266–7 (A 249–50));

> The thinking subject is the object of *psychology*, the sum total of all appearances (the world) is the object of c*osmology*.
>
> (323 (A 334, B 391));

and

> The sensible world contains nothing but appearances, and these are mere representations which are always sensibly conditioned; in this field things in themselves are never objects to us.
>
> (*Ibid.*, 415 (A 563 B 591))

65 As Kant says,

> Nothing has of itself come into the condition in which we find it to exist, but always points to something else as its cause … The whole universe must thus sink into the abyss of nothingness, unless, over and above this infinite chain of contingencies, we assume something to support it, something which … as the cause of the origin of universe secures also … its continuance … We are not acquainted with the whole content of the world … But … we cannot, as regards causality, dispense with an ultimate and supreme being.
>
> (*Ibid.*, 519 (A 622–3 B 650–1))

66 Kant, *Critique of Pure Reason*, *op. cit.*, 29 (B xxx). Gaston focuses on Kant's "regulative ideas" and "as if" declarations. *Op. cit.*, 18–20.

67 *Ibid.*, 517 (A 619).

68 *Ibid.*, 415 (A 563, B 591).

69 *Ibid.*, 639 (A 811, B 839).

70 I will refer to the following editions of these two works: Immanuel Kant, *Groundwork of the Metaphysics of Morals,* trans. and ed., Mary Gregor with an Introduction by Christine Korsgaard (Cambridge: Cambridge University Press, 2006); and the *Critique of Practical Reason*, trans. Philip McPherson Rudisill. Available at: https://kantwesley.com/Kant/CritiqueOfPracticalReason.pdf.

71 Kant, *Critique of Practical Reason, Conclusion*, 1.1–1.6, *op. cit.*, 199.

72 Immanuel Kant, *Critique of Judgement,* trans. with Introduction and Notes by J.H. Bernard (2nd ed. revised), (London: Macmillan, 1914), 197 and 201. Available at: https://oll.libertyfund.org/titles/1217.

73 See Charles Webel, *The Politics of Rationality*, *op. cit.*, 136–7 and 150–1 for an elaboration of this point and also for numerous references to places in Kant's texts that are relevant for my discussion, and Christine Korsgaard, Introduction, in Kant, *Groundwork of the Metaphysics of Morals*, *op. cit.*

74 Regarding these "two worlds," Kant says:

> This must yield a distinction, although a crude one, between a *world of sense* and the *world of understanding,* the first of which can be very different according to the difference of sensibility in various observers of the world while the second, which is its basis, always remains the same. Even as to himself, the human being cannot claim to cognize what he is in himself through the cognizance he has by inner sensation … and thus as regards mere perception and receptivity to sensations he must count himself as belonging to the *world of sense,* but with regard to what there may be of pure activity in him (what reaches consciousness immediately and not through affection of the senses), he must count himself as belonging to the *intellectual world*, of which however he has no further cognizance.
>
> (Groundwork, *op. cit.*, 56)

75 *Ibid.*, 57.

76 *Ibid.*

77 Kant, *Critique of Practical Reason*, *op. cit.*, 38–9.

78 *Ibid.*, 47.

79 As Kant argues, "In a word, the moral law demands compliance for the sake of duty, and not for the sake of a liking, which we cannot and are not supposed to presuppose at all." *Critique of Practical Reason*, *op. cit*, 8.4, p. 194.

80 For "moral luck" in Aristotle's ethics, and in the work of such recent ethicists as Bernard Williams, see Dana K. Nelkin, "Moral Luck," *The Stanford Encyclopedia of Philosophy* (Summer 2019 Edition), Edward N. Zalta, ed., https://plato.stanford.edu/archives/sum2019/entries/moral-luck.

81 For the contrast between deontological ethics (such as Kant's) and consequentialist ethics (including the British Utilitarians, notably Jeremy Bentham and John Stuart Mill), see Larry and Michael Moore, "Deontological Ethics," *The Stanford Encyclopedia of Philosophy* (Winter 2020 Edition), Edward N. Zalta, ed., https://plato.stanford.edu/archives/win2020/entries/ethics-deontological.

82 The Kingdom (or "Realm") of Ends" (German: *Reich der Zwecke*), mentioned by Kant in the *Groundwork of the Metaphysics of Morals* (*op. cit.*, 4:439, p. 41), is a moral world in which all human beings are treated as ends, that is, as if they and their well-being are the goal, and not the self-centered inclinations of any particular individuals who treat other people as mere means for their purposes. A Kingdom of Ends is composed entirely of rational beings, who follow the categorical imperative and who therefore choose to act by laws that imply an absolute necessity. This "kingdom" or "realm" denotes the systematic union of different rational beings under common laws. These common laws, in line with the categorical imperative, especially Kant's second formulation of it, are the standard used to evaluate the worthiness of an individual's actions. When all the realm's members are rational beings who live by the categorical imperative, they have created the "Kingdom of Ends." See https://en.wikipedia.org/wiki/Kingdom_of_Ends.

Regarding freedom, the (moral) laws, and the intelligible world, Kant says: "But instead of a perspective being the foundation to these laws, it appeals to the concept of their existence in the intelligible world, i.e., freedom." *Critique of Practical Reason*, *op. cit.*, 11.2, p. 60.

83 Gaston, *op. cit.*, 20.

84 Immanuel Kant, *Anthropology from a Pragmatic Point of View,* ed. Robert B. Louden and Manfred Kuehn (Cambridge: Cambridge University Press, 2006), Preface, 3–4.

85 Kant, *Religion within the Limits of Bare Reason*, Copyright Jonathan Bennett 2017, available at: www.earlymoderntexts.com/assets/pdfs/kant1793.pdf.

86 For the "evil" in "human nature," see Kant, *Religion*, *op. cit.*, esp. 8–19; Sharon Anderson Gold and Pablo Muchnik, ed., *Kant's Anatomy of Evil* (Cambridge: Cambridge University Press, 2010), available at: https://philpapers.org/archive/MUCAAP.pdf; and https://iep.utm.edu/rad-evil.

87 Kant, *Religion*, *op. cit.*, 9.

88 I*bid.*, Gold and Muchnik, *Introduction*, *op. cit.*, 5.

89 Kant, *Religion*, *op. cit.*, 5.

90 I*bid.*, 11–15.

91 *Kant's Political Writings*, ed. with an Introduction and Notes by Hans Reiss, and trans. H.B. Nisbet (Cambridge: Cambridge University Press, 1970), 179–80.

92 *Ibid.*

93 Kant, *What is Enlightenment?*, available at: https://docs.google.com/viewer?a=v&pid=sites&srcid=aGFuYWxhbmkub3JnfGhvbm9ycy1oaXN0b3J5LW9mLXdlc3Rlcm4tY2l2aWxpemF0aW9ufGd4OjJjMTZhZTFkZmUxNWE2Y2E.

94 On Rousseau's republicanism, see Annelien de Dijn, "Rousseau and Republicanism," *Political Theory*, Volume: 46, Issue: 1, 59–80, first published October 9, 2015. Available at: https://journals.sagepub.com/doi/abs/10.1177/0090591715609101?journalCode=ptxa.

95 Kant, "Idea for a Universal History from a Cosmopolitan Point of View," in Immanuel Kant, *On History*, trans. Lewis White Beck (Indianapolis, IN: The Bobbs-Merrill, 1963), 1–2.

96 *Ibid.*, 6, 11, and 14.

97 In "Perpetual Peace," under the "First Definitive Article for Perpetual Peace," Kant declares:

> *The Civil Constitution of Every State Should Be Republican* … The only constitution which derives from the idea of the original compact, and on which all juridical legislation of a people must be based, is the republican. This constitution is established, firstly, by principles of the freedom of the members of a society (as men); secondly, by principles of dependence of all upon a single common legislation (as subjects); and, thirdly, by the law of their equality (as citizens).
>
> (Kant, "Perpetual Peace: A Philosophical Sketch," available at: http://fs2.american.edu/dfagel/www/Class%20Readings/Kant/Immanuel%20Kant,%20_Perpetual%20Peace_.pdf, p. 4)

98 *Ibid.*

99 Kant elaborates on the idea of a federation of autonomous states, or a "league of nations," in the following passages from "Perpetual Peace:"

> Peoples, as states, like individuals, may be judged to injure one another merely by their coexistence in the state of nature (i.e., while independent of external laws). Each of them may and should for the sake of its own security demand that the others enter with it into a constitution similar to the civil constitution, for under such a constitution each can be secure in his right. This would be a league of nations, but it would not have to be a state consisting of nations … The practicability (objective reality) of this idea of federation, which should gradually spread to all states and thus lead to perpetual peace, can be proved. For if fortune directs that a powerful and enlightened people can make itself a republic, which by its nature must be inclined to perpetual peace, this gives a fulcrum to the federation with other states so that they may adhere to it and thus secure freedom under the idea of the law of nations. By more and more such associations, the federation may be gradually extended.
>
> (*Op. cit*, 6–7)

100 As Kant says in Section II of "Perpetual Peace,"

> The state of peace among men living side by side is not the natural state *(status naturalis);* the natural state is one of war. This does not always mean open hostilities, at least an unceasing threat of war. A state of peace, therefore, must be *established,* for in order to be secured against hostility it is not sufficient that hostilities simply be not committed; and, unless this security is pledged to each by his neighbor (a thing that can occur only in a civil state), each may treat his neighbor, from whom he demands this security, as an enemy.
>
> (*Op. cit.*, 3)

101 Kant, *Perpetual Peace, op. cit.*, 8–9.

102 Kant, *Critique of Judgment, op. cit.*, 4.

103 Kant, *Critique of Pure Reason, op. cit.*, 519.

104 Kant, *On History, op. cit.*, 12.

105 *Ibid.*
106 Kant, *Perpetual Peace*, *op. cit.*, 8–9.
107 Kant, *Critique of Judgment*, *op. cit.*, 4.
108 Kant, *Critique of Pure Reason*, *op. cit.*, 519.
109 Foucault, in Gaston, *op. cit.*, 24.
110 Georg Wilhelm Friedrich Hegel, *Elements of the Philosophy of Right* (PR), trans. H.B. Nisbet and edited by Allen W. Wood (Cambridge: Cambridge University Press 1991), Section 341, p. 372. *Gericht* in German can also mean the "verdict," "final," or "last judgment," with a possible religious implication for the Last Judgment. For a commentary, see Michael Rosen, "Die Weltgeschichte ist das Weltgericht." In *Internationales Jahrbuch des deutschen Idealismus*, ed. F. Rush: 256–72. Berlin: De Gruyter. Available at: https://dash.harvard.edu/handle/1/32186260.
111 For a succinct if somewhat "analytically" oriented overview of Hegel's life and work, see Paul Redding, "Georg Wilhelm Friedrich Hegel," *The Stanford Encyclopedia of Philosophy* (Spring 2020 Edition), Edward N. Zalta, ed., https://plato.stanford.edu/archives/spr2020/entries/hegel/. A somewhat narrower overview is David Duquette, *Hegel Social and Political Thought,* available at: https://iep.utm.edu/hegelsoc. Important secondary literature on Hegel in particular and German Idealism more generally includes the following books by Frederick C. Beiser, *The Cambridge Companion to Hegel,* (Cambridge: Cambridge University Press, 2014), *German Idealism: The Struggle Against Subjectivism* (Cambridge, MA: Harvard University Press, 2002), *Hegel* (New York and London: Routledge, 2005), *The Cambridge Companion to Hegel & Nineteenth Century Philosophy* (Cambridge: Cambridge University Press, 2008), and *After Hegel: German Philosophy, 1840–1900* (Princeton, NJ: Princeton University Press, 2014); as well as Herbert Marcuse, *Reason and Revolution Hegel and the Rise of Social Theory,* 2nd ed. (London: Routledge, 1986); Dean Moyar, *The Oxford Handbook to Hegel* (Oxford: Oxford University Press, 2017); Robert Stern (ed.), *G.W.F. Hegel: Critical Assessments*, 4 volumes (London: Routledge, 1993); and Charles Taylor, 1975, *Hegel* (Cambridge: Cambridge University Press, 1975).
112 In this typically dense, and possibly opaque, text, Hegel elaborates on this point:

> 1. The concrete existing world tranquilly raises itself to a kingdom of laws; the null content of its manifold determinate being has its subsistence in an other; its subsistence is therefore its dissolution. In this other, however, that which appears also *comes to itself*; thus appearance is in its changing also an enduring … Thus appearance reflected-into-itself is now a *world* that *discloses* itself above the *world of appearance* as *one which is in and for itself*. The kingdom of laws contains only the simple, unchanging but diversified content of the concretely existing world.—This world which is in and for itself is also called the *suprasensible world*, inasmuch as the concretely existing world is characterized as *sensible*, that is, as one intended for intuition, which is the immediate attitude of *Appearance*.—The suprasensible world likewise has immediate, concrete existence, but reflected, essential concrete existence … The world which is in and for itself is the totality of concrete existence; outside it there is nothing. But, within it, it is absolute negativity or form, and therefore its immanent reflection is *negative* self-reference. It contains opposition, and splits internally as the world of the senses and as the

world of otherness or the world of appearance. For this reason, since it is totality, it is also only *one side* of the totality and constitutes in this determination a self-subsistence different from the world of appearance. The world of appearance has its negative unity in the essential world to which it founders and into which it returns as to its ground. Further, the essential world is also the positing ground of the world of appearances; for, since it contains the absolute form essentially, it sublates its self-identity, makes itself into positedness and, as this posited immediacy, it is the world of appearance.

(G.W.F. Hegel, *The Science of Logic*, trans. and ed. George de Giovanni [Cambridge: Cambridge University Press, 2010], 443–8)

113 *Dasein* is the German term that Hegel's translators have often rendered in English as "Existence." This foreshadows the 20th-century German philosopher Martin Heidegger's appropriation of *Dasein*, or "Being-There," as "Human Existence."

114 For "Left-Hegelian" elaborations by two leading Frankfurt School critical theorists of the "negativity" at the core of Hegelian ontology, see Herbert Marcuse, *Negations Essays in Critical Theory* (Boston, MA: Beacon Press, 1969), and Theodor W. Adorno, *Negative Dialectics*, trans. E.B. Ashton (New York: Continuum Books, 2007).

115 Hegel, *Phenomenology of Spirit*, tr., A.V. Miller and Foreword J.N. Findlay (Oxford: Oxford University Press, 1977), 228–38.

116 Hegel, *The Science of Logic, op. cit.*, 448.

117 See Hegel, *The Philosophy of History*, tr. J. Sibre (New York: 1956/2004), 13–17, 19, 33, 42, 59, 72–8; Hegel, *The Philosophy of Right*, tr. T.M. Knox (Oxford: Oxford University Press, 1967), 13, 31, 156, 176, and 279; and *The Phenomenology of Mind,* tr. J.B. Baillie (New York: Dover, 1967/2003), 83, 243, and 445–57.

118 Hegel, *Philosophy of Right, op. cit.*, 372.

119 Regarding world-historical individuals, or "subjectivities," and cultures, or "realms," Hegel says:

> At the forefront of all actions, including world-historical actions, are individuals as the subjectivities by which the substantial is actualized. Since these individuals are the living expressions of the substantial deed of the world spirit and are thus immediately identical with it, they cannot themselves perceive it and it is not their object and end ... In accordance with these four principles, the world-historical realms are four in number: 1. the Oriental, 2. the Greek, 3. the Roman, 4. the Germanic ... The Oriental Realm. The world-view of this first realm is inwardly undivided and substantial, and it originates in the natural whole of patriarchal society ... The Germanic Realm ... Having suffered this loss of itself and its world and the infinite pain which this entails (and for which a particular people, namely the Jews, was held in readiness), the spirit is pressed back upon itself at the extreme of its absolute negativity. This is the turning point which has being in and for itself. The spirit now grasps the infinite positivity of its own inwardness, the principle of the unity of divine and human nature and the reconciliation of the objective truth and freedom which have appeared within self-consciousness and subjectivity. The task of accomplishing this reconciliation is assigned to the Nordic principle of the Germanic peoples.
>
> (Hegel, *ibid.*, 375–7)

120 As Hegel says in the PR:

> the state as freedom, which is equally universal and objective in the free self-sufficiency of the particular will; this actual and organic spirit of a people actualizes and reveals itself through the relationship between the particular national spirits … and in world history as the universal world spirit whose right is supreme … But the state emerges only at the third stage, that of ethical life and spirit, at which the momentous unification of self-sufficient individuality with universal substantiality takes place. The right of the state is therefore superior to the other stages: it is freedom in its most concrete shape, which is subordinate only to the supreme absolute truth of the world spirit.
>
> (*Ibid.*, 62–3)

121 Hegel, *Philosophy of History*, *op. cit.*, 80, 103–4, 109, and 341.

122 As Hegel says in his *Philosophy of History*, "It has been said, that the French Revolution resulted from Philosophy, and it is not without reason that Philosophy has been called '*Weltweisheit*' [World Wisdom]; for it is not only Truth in and for itself, as the pure essence of things, but also Truth in its living form as exhibited in the affairs of the world," PH, *op. cit.*, 446.

123 Hegel, PR, *op. cit.*, 21.

124 See Charles Webel, *The Politics of Rationality*, *op. cit.*, 155–5.

125 For markedly different appraisals of Hegel's influence, see Redding, *Stanford Encyclopedia of Philosophy*, *op. cit.*, and Marcuse, *Reason and Revolution*, *op. cit.*

126 A useful summary of Schopenhauer's life, works, and influence is: Robert Wicks, "Arthur Schopenhauer," *The Stanford Encyclopedia of Philosophy* (Spring 2019 Edition), Edward N. Zalta, ed., https://plato.stanford.edu/archives/spr2019/entries/schopenhauer. Another online overview is available at: https://iep.utm.edu/schopenh/. Among the recommended secondary sources are Bernardo Kastrup, *Decoding Schopenhauer's Metaphysics* (Alresford: John Hunt Publishing, 2020); Brian Magee, *The Philosophy of Schopenhauer* (Oxford: Oxford University Press, Clarendon Press, 1983), and Christopher Janaway, *Schopenhauer A Very Short Introduction* (Oxford: Oxford University Press, 2002).

127 Because of research limitations during the pandemic, I had only intermittent access to the complete English edition of Schopenhauer's *The World as Will and Representation*, Two Volumes, trans. E.F.J. Payne (New York: Dover Publications, Inc., 1966), abbreviated as *WWR Long*. Accordingly, I will also refer to the abridged, one volume edition of this work, ed. David Berman, trans. Jill Berman (London: Everyman, 1995), abbreviated as *WWR Short*.

128 See Wicks, *The Stanford Encyclopedia of Philosophy*, *op. cit.*

129 Schopenhauer, *WWR Long*, Vol. 1, 82 as cited in Magee, *op. cit.*, 36.

130 *Ibid.*, Vol. 2, 4–5.

131 Schopenhauer, *WWR Short*, Book One, *op. cit.*, p. 3.

132 *WWR Long.*, Vol. 2, *op. cit.*, 3–4.

133 For Berkeley's subjective idealism, previously mentioned in this chapter, also see www.britannica.com/topic/esse-est-percipi-doctrine.

134 *WWR Long*, Vol 2, *op. cit.*, 3.

135 There is a huge literature on "consciousness" and the brain. A useful summary is by Christof Koch, *"What Is Consciousness?" Scientific American,* 318, 6, 60–4 (June 2018), www.scientificamerican.com/article/what-is-consciousness/.

For "materialist," or "physicalist," philosophical "identity theory" approaches to the brain and consciousness (e.g., "consciousness is what the brain does") see the many books by Patricia Churchland and a brief, accessible overview of her theory: www.prospectmagazine.co.uk/magazine/out-of-mind-philosopher-patricia-churchlands-radical-approach-to-the-study-of-human-consciousness. For alternative positions to neurophilosophical reductionisms of conscious minds to physical brains, see the many works of David Chalmers, summarized here: https://blogs.scientificamerican.com/cross-check/david-chalmers-thinks-the-hard-problem-is-really-hard/, as well as Webel and Stigliano, "Beyond Good and Evil? Radical Psychological Materialism and the 'Cure' for Evil," *op. cit.*

136 *WWR Long*, Vol. 1, *op. cit.*, 531.

137 As Berman says in his "Introduction" to the abridged edition of Schopenhauer's *The World as Will and Representation*:

> "The world is something we should be sorry about, for at bottom it is something that ought not to be ... This ... is the worst of all possible worlds." [vs. Leibniz] ... the cause of our discontent lies in the substance of the world—in its being will to life. Life ... (1) is morally wrong, (2) has no meaning or purpose, and (3) will always have more pain than pleasure. So there is ultimately nothing we can do to alter this worst of all possible worlds—except to end it by strangling all desire, thereby achieving what the Buddhists call Nirvana, which is that state closest to pure nothingness.
>
> (David Berman, *WWR Short*, *op. cit.*, xxxii–xxxiii)

138 *WWR Long*, Vol. I, *op. cit.*, 526–32.

139 *WWW Short*, Book 4, *op. cit.*, 262.

140 *Ibid.*

141 Schopenhauer on sex, egoism, and death could not be more un-Kantian:

> The sexual impulse ... the strongest affirmation of life ... the ultimate purpose, the highest goal of life ... Every knowing individual ... is ... the whole will to live, or the inner being of the world itself, and also as the integral condition of the world as idea ... every individual, though infinitesimally small and reduced to insignificance in the boundless world, nonetheless makes itself the center of the world, considers his own existence and well-being before everything else ... is ready to sacrifice everything else for this—... to annihilate the world ... to safeguard his own self ... Everyone looks upon his own death as ... the end of the world. Every person also accepts the death of ... acquaintances as a matter of comparative indifference ... egoism ... must have reached its highest grade, and the conflict of individuals ... must appear in its most terrible form.

Donald Trump might be recognizable in this passage. *WWR Short*, Par. 81, *op. cit.*, 209–10.

142 Some comments by Schopenhauer on the body and the sexual impulse remarkably presage Freud, who deemed Schopenhauer (and Nietzsche) to be among the very few philosophers of interest to him. For example, Schopenhauer called the body

> nothing but objectified will ... become idea ... body is the only real individual in the world ... the world ... mere idea, object for a subject ... will

> … simply what this world is … the world as idea, both in whole and in parts, the *objectification of will* … means the world become object, i.e. idea.
> (*WWR Short*, *op. cit.*, pp. 33, 36, and 97)

143 Schopenhauer rhapsodizes about (Romantic) music in the following passages:

> we must attribute to music a far more serious, deeper significance for the inmost nature of the world and our own self … the phenomenal world, or nature, and music [are] two different expressions of the same thing … Music … if regarded as an expression of the world, is … a universal language … The actual world … the world of particular things, provides the objects of perceptions … both to the universality of concepts and the universality of melodies.
> (*WWR Short*, Par. 12, *op. cit.*, 163, 169–71)

144 *Ibid.*, Book 3, Par. 40, p. 129.
145 See Wicks, *The Stanford Encyclopedia of Philosophy*, *op. cit.*
146 Magee claims:

> Like a number of other great philosophers, he [Schopenhauer] was convinced that he had finally solved the fundamental problems of philosophy—indeed, as he put it: "Subject to the limitation of human knowledge, my philosophy is the real solution of the enigma of the world." He believed that he had unveiled the mystery of the world.

Wittgenstein in his "Preface" to *Tractatus Logico-Philosophicus* [wrote]

> the *truth* of the thoughts that are here set forth seems to me unassailable and definitive. I therefore believe myself to have found, on all essential points, the final solution of the problems.
> (*Op. cit.*, 19–20)

But Schopenhauer himself was at times much more modest in his philosophical ambitions:

> Philosophy can never do more than interpret and explain what is given … The world as idea is a mirror which reflects the will … in itself, is without knowledge … is merely a blind, irresistible impulse … Will is the thing-in-itself, the inner content, the essence of the world.

For Schopenhauer, "Life, the visible world, the phenomenon" only "mirror" the world. *WWR Short*, *op. cit.*, Book 4, par. 53–4, pp. 176–7. And the world is "… so rich in content that not even the profoundest investigation of which the human mind is capable could exhaust it." *WWR Long*, *Vol 1*, *op. cit.*, 273.
147 My primary source of Kierkegaard's works is Jane Chamberlain and Jonathan Rée, *The Kierkegaard Reader* (Oxford: Blackwell, 2001). The most comprehensive English translation of Kierkegaard's works, unavailable to me during the pandemic, is: *Kierkegaard's Writings* Volumes 1-XXVI, ed. & trans. H.V. Hong, et al. (Princeton, NJ: Princeton University Press, 1978–2000).

Two good overviews of Kierkegaard's life and works are by William McDonald, "Søren Kierkegaard," *The Stanford Encyclopedia of Philosophy* (Winter 2017 Edition), Edward N. Zalta, ed.), https://plato.stanford.edu/archives/win2017/entries/kierkegaard/, and also McDonald's article on Kierkegaard for the *Internet Encyclopedia of Philosophy*, https://iep.utm.edu/kierkega/.

Interesting books about Kierkegaard's life, works, and influence include: Theodor W Adorno, *Kierkegaard: Construction of the Aesthetic*, Robert Hullot-Kentor (trans.) (Minneapolis, MN: University of Minnesota Press, 1989); Harold Bloom (ed.), *Søren Kierkegaard* (New York: Chelsea House Publishers, 1989); Joachim Garff, *Søren Kierkegaard: A Biography*, trans. Bruce H. Kirmmse (Princeton, NJ: Princeton University Press, 2005); John Lippitt and George Pattison (eds.), *The Oxford Handbook of Kierkegaard* (Oxford: Oxford University Press, 2013); Walter Lowrie, *Kierkegaard*, 2 Volumes (New York: Harper & Brothers, 2002); Jonathan Rée & Jane Chamberlain (eds.), *Kierkegaard: A Critical Reader* (Oxford: Blackwell, 1998); Michael Weston, *Kierkegaard and Modern Continental Philosophy* (London: Routledge, 1994); and Merold Westphal & Martin J. Matustik, *Kierkegaard in Post/Modernity* (Bloomington & Indianapolis, IN: Indiana University Press, 1995).

148 This quote and some of the details of Kierkegaard's life are drawn from https://en.wikipedia.org/wiki/Søren_Kierkegaard.

149 As Kierkegaard says in *The Concept of Irony*:

> irony exhibits itself most nearly as conceiving the world, as attempting to mystify the surrounding world not so much in order to conceal itself as to induce others to reveal themselves. But irony may also manifest itself when the ironist seeks to lead the outside world astray respecting himself … just as a sexton regards suicide as a disgraceful stratagem designed to sneak oneself out of the world—in our time … it may occasionally seem necessary for a person to play false.
>
> (Kierkegaard in Chamberlain and Rée, *op. cit.*, 34)

150 Kierkegaard, as cited in William McDonald, "Søren Kierkegaard," *The Stanford Encyclopedia of Philosophy*, *op. cit.*

151 Kierkegaard's alter ego Johannes Climacus makes a distinction between two types of religiousness: "Religiousness A" and "Religiousness B." The former is the pagan conception of religion and is characterized by intelligibility, immanence, and recognition of continuity between temporality and eternity. Religiousness B is "paradoxical religiousness" and ostensibly represents the essence of Christianity. *It posits a radical divide between immanence and transcendence, a discontinuity between temporality and eternity, yet also claims that the eternal came into existence in time. This is a paradox and can only be believed "by virtue of the absurd."* See Kierkegaard in Chamberlain and Rée, *op. cit.*, 290–379.

152 Kierkegaard, *Journals and Notebooks*, in Chamberlain and Rée, *op. cit.*, 15.

153 Quotes by Kierkegaard taken from Chamberlain and Rée, *op. cit.*, 22, and also from *Absurdism.* https://en.wikipedia.org/wiki/Absurdism.

154 Kierkegaard, *Concluding Unscientific Postscript*, in Chamberlain and Rée, *op. cit.*, 259.

155 Kierkegaard, *Johannes Climacus, or De Omnibus Dubitandum Est A Narrative*, in Chamberlain and Rée, *op. cit.*, 346.

156 Melchert and Morrow, *The Great Conversation*, *op. cit.*, 536–7.

157 Kierkegaard, *Journals and Notebooks*, *op. cit.*, 22.

158 Kierkegaard, *Concluding Unscientific Postscript*, trans. David F. Swenson and Walter Lowrie (Princeton, NJ: Princeton University Press, 1944), 166, cited in Melchert and Morrow, *op. cit.*, 537.

159 For the influence of Kierkegaard, Hegel, and Existentialism on the theological and political writings of Martin Luther King Jr., see www.critical-theory.com/martin-luther-king-jr-hegel/.

160 See Melchert and Morrow, *op. cit.*, 532–7.

161 As in Walter Kaufmann's extremely, perhaps excessively, detailed account, *op. cit.* Between 1954 and 1974, all but three of Nietzsche's major works appeared in translations, edited by Walter Kaufmann, who also contributed introductions and notes. My main source for primary texts in English is Kaufmann's *Basic Writings of Nietzsche (BWN),* Introduction by Peter Gay (New York: The Modern Library, 2000), which contains *The Birth of Tragedy (BN); Seventy Five Aphorisms from Five Volumes (APH); Beyond Good and Evil (BGE); On the Genealogy of Morals (GM); The Case of Wagner (W); and Ecce Homo (EH).* Other primary sources in English translations by Kaufmann are: *Thus Spoke Zarathustra (Z)* (New York, Viking, 1966); *The Will to Power (WP)* (New York: Vintage Books, 1968); and *The Gay Science (GS)*, (New York: Vintage Books, 1974).

Reliable online overviews of Nietzsche's life and work include: R. Lanier Anderson, "Friedrich Nietzsche," *The Stanford Encyclopedia of Philosophy* (Summer 2017 Edition), Edward N. Zalta, ed., https://plato.stanford.edu/archives/sum2017/entries/nietzsche/; and Dale Wilkerson, "Nietzsche," *Internet Encyclopedia of Philosophy*, https://iep.utm.edu/nietzsch/.

The immense and still expanding secondary literature on Nietzsche, in addition to Kaufmann, includes: David Allison, (ed.), *The New Nietzsche* (Cambridge, MA: MIT Press, 1977); Keith Ansell-Pearson, *An Introduction to Nietzsche as Political Thinker: the Perfect Nihilist* (Cambridge: Cambridge University Press, 1994); Maudemarie Clark, *Nietzsche on Truth and Philosophy* (Cambridge: Cambridge University Press, 1990); Arthur Danto, *Nietzsche as Philosopher* (New York: Columbia University Press, 1965); Gilles Deleuze, *Nietzsche and Philosophy*, trans. Hugh Tomlinson (New York: Columbia University Press, 1983); Jacques Derrida, *Spurs: Nietzsche's Styles*, trans. Barbara Harlow (Chicago, IL: University of Chicago Press, 1978); Michel Foucault, "Nietzsche, Freud, Marx," in Gayle Ormiston and Alan Schrift, eds., *Transforming the Hermeneutic Context* (Albany NY: State University of New York Press, 1990), 59–67, and "Nietzsche, Genealogy, History," in *Language, Counter-memory, Practice: Selected Essays and Interviews*, trans. Donald Bouchard and Sherry Simon (Ithaca, NY: Cornell University Press, 1977); Ken Gemes and John Richardson, eds., *The Oxford Handbook of Nietzsche* (Oxford: Oxford University Press, 2013); Ronald Hayman, *Nietzsche: a Critical Life* (Oxford: Oxford University Press, 1980); Martin Heidegger, *Nietzsche*, 2 vols., trans. David Farrell Krell, (New York: Harper Collins, 1991); Karl Jaspers, *Nietzsche: an Introduction to the Understanding of his Philosophical Activity*, trans. Charles Wallraff and Frederick Schmitz (Baltimore, MD: Johns Hopkins University Press, 1965); Joshua Landy and Michael Saler, eds., *The Re-Enchantment of the World: Secular Magic in a Rational Age* (Stanford, CA: Stanford University Press, 2009); Karl Löwith, *Nietzsche's Philosophy of the Eternal Recurrence of the Same*, trans. J. Harvey Lomax (Berkeley, CA: University of California Press, 1997); Alexander Nehamas, *Nietzsche: Life as Literature* (Cambridge, MA: Harvard University Press, 1985); Robert B. Pippin, *Nietzsche, Psychology, and First Philosophy* (Chicago, IL: University of Chicago Press, 2010); Bernard Reginster, *The Affirmation of Life: Nietzsche on Overcoming Nihilism* (Cambridge, MA: Harvard University Press, 2006); John Richardson, *Nietzsche's System*, (Oxford: Oxford University Press, 1996) and *Nietzsche's New Darwinism*

(Oxford: Oxford University Press, 2004); Rüdiger Safranski, *Nietzsche: a Philosophical Biography*, trans. Shelley Frisch (New York: Norton, 2003); Richard Schacht, *Nietzsche* (London: Routledge, 1983); Robert Solomon and Kathleen Higgins, eds., *Reading Nietzsche* (Oxford: Oxford University Press, 1988); Tracy Strong, *Friedrich Nietzsche and the Politics of Transfiguration* (Urbana, IL: University of Illinois Press: 2000); Charles Webel, *The Politics of Rationality, op. cit.*, 159–62; Charles Webel and Anthony Stigliano, "Beyond Good and Evil," *op. cit.*; Rex Welshon, *The Philosophy of Nietzsche* (Montreal: McGill University Press: 2004); and Julian Young, *The Death of God and the Meaning of Life* (London: Routledge, 2003).

162 After more than a century of speculation and armchair "diagnoses" of Nietzsche's mental malady, the following article might resolve, for now, the issue by pointing to an organic cause, ironically, a "brain disease" (but not of the kind Nietzsche accused his opponents of having): Charlie Huenemann, "Nietzsche's Illness," in Gemes and Richardson, *op. cit.*, 63–80.

163 For the putative cause(s) of Nietzsche's father's death and its traumatizing effects see www.dartmouth.edu/~fnchron/sidelights/KLdeath.html.

164 For Lange, see Nadeem J.Z. Hussain and Lydia Patton, "Friedrich Albert Lange," *The Stanford Encyclopedia of Philosophy* (Winter 2016 Edition), Edward N. Zalta, ed., https://plato.stanford.edu/archives/win2016/entries/friedrich-lange/.

165 Friedrich Nietzsche, *Untimely Meditations*, ed., Daniel Breazeale, trans. R.J. Hollingdale (Cambridge: Cambridge University Press, 2007).

166 See https://en.wikipedia.org/wiki/Lou_Andreas-Salomé/.

167 I visited the Nietzsche house in 2019 and, although there are striking busts of Friedrich and plenty of memorabilia, it did not strike me as a sanctuary.

168 Regarding "the Aryans," in section 9 of *BT*, Nietzsche was contemptuous of Germans [as he seemed to have been of the vast majority of humans), since, to an "Aryan," the

> contradiction at the heart of the world reveals itself ... as a clash of different worlds ... of a divine and a human one, in which each ... has right on its side ... In the heroic effort of the individual to attain universality ... to become *one* world-being, he suffers ... the primordial contradiction ... concealed in things.
>
> (*BWN*, 79)

Regarding "German virtue," on the other hand, Nietzsche declares: "One should counter the moral arrogance of the Germans with one little word, *schlecht*" (bad, poor). *Seventy-Five Aphorisms, ibid.*, 170. For an interpretation of Nietzsche's overall social and political views, see Brian Leiter, "Nietzsche's Moral and Political Philosophy," *The Stanford Encyclopedia of Philosophy* (Spring 2020 Edition), Edward N. Zalta (ed.), https://plato.stanford.edu/archives/spr2020/entries/nietzsche-moral-political/. For the appropriation/distortion of Nietzsche by the Nazis, see Max Whyte, "The Uses and Abuses of Nietzsche in the Third Reich: Alfred Baeumler's 'Heroic Realism'," *Journal of Contemporary History* Vol. 43, No. 2 (Apr., 2008), 171–94, www.jstor.org/stable/30036502.

169 For a "hard" interpretation of Nietzsche's remarks on Jews, see Robert C. Holub, "Nietzsche and the Jewish Question," *New German Critique* 66 (1995): 94–121. doi:10.2307/48858, For his (misogynistic?) views on women,

see https://en.wikipedia.org/wiki/Friedrich_Nietzsche%27s_views_on_women. For what Nietzsche thought about nationalism in general and its German incarnation, see Carol Diethef, "Nietzsche and Nationalism," *History of European Ideas*, Vol. 14, Issue 2, March 1992, 227–34; and for his views on war and militarism, see Rebekah S. Peery, *Nietzsche on War* (New York: Algora Publishing, 2009).

170 Nietzsche, GS, *op. cit.*, 381.

171 See Kaufmann, BWN, *op. cit.*, 72–94, and Anderson, SEP, *op. cit.*

172 Nietzsche, *WP*, as cited in Kaufmann, BWN, *op. cit.*, 112–13.

173 Nietzsche, *Epilogue* to *The Case of Wagner*, in BWN, *op. cit.*, 646.

174 Nietzsche, *Ecce Homo (EH)*, in *BWN*, *op. cit.*, 674.

175 Nietzsche, *The Will to Power (WP)*, as cited in Kaufmann, *BWN*, *op. cit.*, 122.

176 Nietzsche, *On the Genealogy of Morals (GM)*, Third Essay, Section 24, in *BWN*, *op. cit.*, 588.

177 Nietzsche, *EH*, *ibid.*, 790–1.

178 Nietzsche cites in French Voltaire's motto "Crush the infamy!" in his savaging of the Catholic church.

179 According to Melchert and Morrow, paraphrasing Nietzsche,

> What the thought of eternal recurrence teaches ... is that the meaning of life cannot be sought in anything beyond it. Eternal recurrence is the ultimate denial of all "real worlds": there is just this world ... With "real worlds" gone, life must justify itself as it is, or it cannot be justified again ... With the disappearance of "real worlds," any contrast between distinct ultimate realities also vanishes. Every aspect of human life ... virtues, vices, values, science—has to be accounted for in terms of the same fundamental reality. And what is that? Will to power.
>
> (*Op. cit.*, 590)

Regarding Nietzsche's perspectivism: "There is *only* a perspectival seeing, *only* a perspectival 'knowing.'" (*GM*, III, 12 in *BWN*, *op. cit.*, 555)

180 Nietzsche, *Twilight of the Idols*, 33, cited in Melchert and Morrow, *op. cit.*, 574.

181 In *The Antichrist* (*AC*), Nietzsche vents his spleen against Christianity, "the religion of pity" in particular, and against "humanity" in general, to wit:

> It is a painful and tragic spectacle that rises before me: I have drawn back the curtain from the rottenness of man. This word, in my mouth, is at least free from one suspicion: that it involves a moral accusation against humanity. It is used ... without any moral significance: and this is so far true that the rottenness I speak of is most apparent to me precisely in those quarters where there has been most aspiration, hitherto, toward "virtue" and "godliness" ... I understand rottenness in the sense of décadence: my argument is that all the values on which mankind now fixes its highest aspirations are décadence-values. I call an animal, a species, an individual corrupt, when it loses its instincts, when it chooses, when it prefers, what is injurious to it. A history of the "higher feelings," the "ideals of humanity" ... would almost explain why man is so degenerate.
>
> (*AC*, Par. 6, trans. with an introduction by L. Mencken (New York: Alfred A. Knopf, 1924), available at: www.gutenberg.org/files/19322/19322-h/19322-htm)

182 Nietzsche, *BGE*, in *BWN*, *op. cit.*, 237–8.

183 Nietzsche, *AC*, *op. cit.*, par. 6.
184 Nietzsche, *Thus Spake Zarathustra (Z),* 1 3, cited in Kaufmann, *op. cit.*, 125.
185 Nietzsche, *The Birth of Tragedy (BT)*, in *BWN*, *op. cit.*, 21–2.
186 In this context, Nietzsche seems approvingly to cite Schopenhauer's *The World as Will and Representation*, especially his "tragic spirit" and "resignation," *ibid.*, 24.
187 I*bid.*, 66.
188 Nietzsche, *Ecce Homo*, in *BWN*, *op. cit.*, 782–3.
189 William James, *A Pluralistic Universe*, p. 20, cited in Walter Kaufmann, *Nietzsche Philosopher, Psychologist, Antichrist*, *op. cit.*, 79.
190 Nietzsche, *The Birth of Tragedy*, trans. Douglas Smith (Oxford: Oxford University Press, 2000), 27, cited in Melchert and Morrow, *op. cit.*, 564.

3 Existential and Phenomenological Words and Worlds

Since Nietzsche's death in 1900, the world and its conceptualizations have emerged in ways unthinkable to millennia of previous thinkers, scientists, and political actors. Two philosophical and intellectual traditions that have developed since the turn of the 20th century are "Continental" (Western European) and "Analytic" (Anglophonic, but sometimes with a German accent) philosophy. Although they frequently "talk past" each other, in doing so they use language. And the analysis of language, as well as the consciousness, bodies, brains, and perceptions from which words emerge, is a common focus for both traditions, which differ in many other important ways.

Husserl's Phenomenological Worlds

Edmund Husserl (1859–1938) is known today as having been the "father of phenomenology," one of the most important philosophical traditions of the 20th century, particularly in the German and French-speaking worlds.[1] He was born in Prossnitz (Moravia), at that time part of the Austro-Hungarian Empire and now called Prostějov, a city in the Czech Republic. While Husserl's parents were non-orthodox Jews and he was raised Jewish, Husserl and his wife eventually converted to Protestantism. One of his sons, Wolfgang, died in World War I during the battle (1916) of Verdun, France. For a year thereafter, Husserl was in mourning and kept professional silence.

During his student days, Husserl studied mathematics, physics, astronomy, and philosophy in Leipzig, Germany. There, he also attended lectures given by Wilhelm Wundt (1832–1920), who founded what is considered the first institute for experimental psychology. At that time, Husserl was also mentored by Tomáš Masaryk (1850–1937), who would later become the first president of Czechoslovakia. At Masaryk's advice, following his military service, Husserl went to Vienna to study with the one-time Catholic priest Franz Brentano (1838–1917), author of *Psychology from an Empirical Standpoint* (1874). Brentano's lectures on logic and psychology, and his conceptions of "intentionality" and a "scientific philosophy," had a lasting impact on Husserl, as well as on Sigmund Freud (1856–1939).

DOI: 10.4324/9781315795171-4

In his first published monograph, *Philosophy of Arithmetic* (1891), Husserl attempted to provide a psychological foundation for mathematics. However, the eminent logician and mathematician Gottlob Frege (1848–1925) criticized this book for what Frege called its underlying "psychologism," a critique Husserl took very seriously. (Frege would also have enormous influence on the founders of what has since come to be called "analytic philosophy," especially Bertrand Russell, Rudolf Carnap, and Ludwig Wittgenstein.) About a decade later, Husserl published his first explicitly "phenomenological" work, the two-volume *Logical Investigations*. The first volume contains a devastating critique of psychologism, especially as represented in Brentano's works, and the second volume consists of six "descriptive-psychological" and epistemological investigations into, among other topics, expression and meaning, universals, and the interrelations among truth, intuition, and cognition. The *Logical Investigations* revived a kind of philosophical Platonism.[2]

During the early 1900s, while teaching in Germany, Husserl created what he called "transcendental phenomenology," in developing the "phenomenological method." In his second major book, *Ideas* (1913), Husserl argued that this method enables the phenomenologist to formulate an essentially unprejudiced view of the world and its "essences." In such later works as *On the Phenomenology of the Consciousness of Internal Time* (1928), *Formal and Transcendental Logic* (1929), *Cartesian Meditations* (1931), and *The Crisis of European Sciences and Transcendental Phenomenology* (often referred to as *The Crisis*, which was published posthumously in 1954), Husserl developed his theories of time, as well as the constitutional roles of intersubjectivity, the "life-world" (*Lebenswelt*), and the "surrounding environment" (*Umwelt*) in the construction and development of prescientific, scientific, and everyday forms of knowledge, worldly experience, and perceptual reality.

When Husserl formally retired from teaching in 1928, his successor for the chair in philosophy at the University of Freiburg was his former assistant Martin Heidegger, whose major work *Being and Time* had been published in Husserl's *Yearbook* in 1927. The following year, Husserl held lectures in Paris that were later published as the *Cartesian Meditations*. In 1931, Husserl gave a number of talks on "Phenomenology and Anthropology," which were critical of Heidegger's ideas as well as those of the German phenomenologically – oriented philosophical anthropologist Max Scheler (1874–1928).

Adolf Hitler came to power in Germany in 1933. Husserl received and rejected an opportunity to move to Los Angeles, a decision he may have later regretted because of the ill-treatment he received during Nazi rule due to his Jewish ancestry. In 1935, Husserl gave lectures in Prague, eventually resulting in his last major work, *The Crisis.* After a brilliant academic career, during which he taught and mentored several of the 20th-century's leading philosophers, including Heidegger—by whom Husserl may have felt betrayed—he died of pleurisy in Freiburg, Germany, on the eve of the World War II.

Phenomenological Worlds as Bracketed, Intended, and Constituted

Although he did not coin the term, "phenomenology" is most closely associated with Husserl's teaching and publications. The noun "phenomenology" means the "study or science of phenomena," and is derived from the Latin noun *phaenomenon* ("appearance"), whose root is the Ancient Greek substantive *phainómenon*, "a thing appearing to view."[3] In the early 18th century, the Latin term *phenomenologia*, was used by the German philosopher and theologian Christoph Friedrich Oetinger (1702–82) to refer to the epistemological analysis of sense appearances. Subsequently, the Swiss polymath Johann Heinrich Lambert (1727–77), in his book the *New Organon*, called phenomenology "the doctrine of appearance," specifically referring to sensory, psychological, and moral appearances. Immanuel Kant occasionally used the term, *Phänomenologie*, as did the German idealist philosopher Johann Gottlieb Fichte. In 1807, Hegel wrote the *Phänomenologie des Geistes* (the *Phenomenology of Spirit*). Toward the end of the 19th century, Brentano used the term to characterize "descriptive psychology." Edmund Husserl adapted *Phänomenologie* to denote his new "science of consciousness," which he famously declared, in his *Logical Investigations*, was a methodological return back "to the things themselves" (German: *zu den Sachen selbst*).[4]

Phenomenology is the thus the study of phenomena: literally, of appearances as opposed to reality, as exemplified in the parable of the cave in Plato's *Republic,* in which the eternal reality of the world of the forms is contrasted with the mere appearances of things in the visible world. Today, phenomenology usually refers to the field of knowledge (*Wissenschaft* in German) that in an unprejudiced way investigates individual phenomena as distinct from being as a whole (ontology). It also has a more technical meaning in recent philosophy of science.[5]

Phenomenology also refers to *a movement* in 20th-century philosophy, stemming from Husserl and including, to varying degrees, such distinguished philosophers as Heidegger, Jean-Paul Sartre, Maurice Merleau-Ponty, Jacques Derrida (1930–2004), and Emmanuel Levinas (1906–95). It has also influenced such hermeneuticians as Hans-Georg Gadamer (1900–2002) and Paul Riceour (1913–2005), as well as numerous psychologists, sociologists, anthropologists, literary theorists, and theologians. Accordingly, "phenomenological" refers to both a way of doing philosophy and to the corresponding movement that claims to utilize the phenomenological method for understanding the structural features of experience and of things as experienced and perceived by human beings. Phenomenology is conceived by its proponents as a *descriptive* study of human consciousness and related phenomena and is undertaken in a way ostensibly independent of empirical, including causal, accounts of consciousness, and experience. Beginning with Husserl, the philosophical phenomenological tradition has focused on such topics as the nature of intentionality, perception, time-consciousness, self- and bodily awareness, consciousness of others, and the world as the "horizon" of human life and experiences.[6]

Husserl claimed that only by suspending or bracketing the "natural attitude" could philosophy become its own distinctive and rigorous science, and he insisted that phenomenology is a *science of consciousness* rather than of empirical things. Husserl also famously argued that human *consciousness is intentional, that each act of consciousness is a consciousness of something,* that is, intentional, or directed toward something. *Phenomenology is therefore a systematic description of intentional states of consciousness (such as "I desire, wish, plan, or hope for x") as well as a rigorous study of the internal objects, or qualities, of specific conscious experiences* (such as "I am now in pain").

Husserl's Natural and Arithmetical Worlds

In the first book of *Ideas*, Husserl described in detail the everyday, seemingly commonsensical "natural attitude" most people take for granted in their view of "the natural world" as ordinarily conceived and perceived, and contrasted the natural world with the "arithmetical world":

> We begin … as human beings living naturally, objectivating, judging, feeling, willing "in the natural attitude" … with the world as existing in the order of the spatial present, which … is also … with respect to its order *in the sequence of time. This world, on hand for me now and manifestly in every waking Now, has its two-sidedly infinite temporal horizon, its known and unknown, immediately living and lifeless past and future … The complexities of my … changing spontaneities of consciousness then relate to this world, the world in which I find myself and which is … my surrounding world (Umwelt)* … Living along naturally, I live continually in this *fundamental form* of "active" *living … The arithmetical world is there for me only if, and as long as, I am in the arithmetical attitude. The natural world, however, the world in the usual sense of the world is, and has been, there for me continuously as long as I go on living naturally* … the background for my act-consciousness, but it is *not a horizon with which an arithmetical world finds a place. The two worlds simultaneously present are not connected* … To cognize "the" world more comprehensively … than naïve experiential cognition can, to solve all the problems of scientific cognition … within the realm of the world, that is the aim of *the sciences belonging to the natural attitude.*[7]

In addition to distinguishing between "the natural world" we seldom call into question, and the "arithmetical" world of essential spatial and numerical essences, Husserl's phenomenology, "the science of the essences of consciousness" as he called it in *Ideas I* (1913), also proceeds from a critique of both naturalism and psychologism, or what Husserl considered the illegitimate reduction of logical propositions to psychological attitudes.[8]

Naturalism, according to Husserl and its other critics, is the claim that everything in the natural world, including human beings, can and should be studied by the methods of the "hard" sciences, especially physics and biology. Husserl argued that the study of human consciousness must differ from the study of the natural world. For him, phenomenology, unlike natural science, does not proceed from the collection of large amounts of empirical data and subsequently to a general theory, as in scientific induction. *Rather, it aims to describe intentional states and conscious experiences without theoretical presuppositions in order to uncover the essential features of these experiences.*[9]

Husserl's Transcendental and Transcendent Worlds

In order to reveal the underlying structures of conscious experience, what Husserl called "transcendental phenomenology" resembles Kant's "transcendental" *a priori* inquiry into the possibility of knowledge and of experience in general. But Husserl also proposed a "suspension" or "bracketing" of the natural attitude via what Husserl calls its transcendental "*epoché*" (from the Ancient Greek, referring to Greek skeptics' notion of abstaining from belief). *To do phenomenology, Husserl argued, one must "bracket" the natural world around us and thereby "suspend" or "parenthesize," but not "negate" our belief in its existence.* By doing so, we may then turn our attention, in reflection, to the structure of our own conscious experience. As Husserl describes it:

> *We put out of action the general positing ... the essence of the natural attitude ... thus the whole natural world* which is continually "there for us," "on hand," and which will always remain there ... as an "actuality" even if we choose to parenthesize it ... If I do that ... then I *am not negating* this "world" ... I am not doubting its factual being; ... rather I am exercising the phenomenological *epoché* which also *completely shuts me off from any judgment about spatiotemporal factual being.*[10]

Husserl also distinguished between the "transcendent world" of possible perceptual objects and individual selves, which conceal a virtually inexhaustible number of as-yet-unperceived features, from the "whole psychical world" which, together with the "whole physical world," is constituted, or "posited," by our acts of consciousness.[11]

Husserl's World as a Whole, as Form, as a Concept, and as a Community of Incarnate Egos

In suspending any judgment about the immediately given "natural world" about us, by performing this "phenomenological reduction," we may, in Husserl's view, simultaneously explore in a non-prejudicial way both the ontological properties of that world as well as the essential features of "another world," the "world as *Eidos*" (the ancient Greek term for "Form," as in Plato's "world of the Forms"), or "the sphere of general essences." *Only in so doing, it is possible to have a perspective on "the world as a whole" and not just on the natural world of "things" in the world. We are, accordingly, both "in the world" as naturally perceiving and experiencing beings, and are also able, as conscious actors with a transcendental perspective, to construct an essentially unified world.*[12]

Husserl also aimed to provide both a history of the concept of the world and a demonstration that the world has a history. In so doing, he hoped, in part, to avoid what he considered to be the "modern distortions" of our relation to the "prescientific world," beginning, in Husserl's view, with Galileo, Thomas Hobbes, and empiricism. As Husserl declared:[13]

> As each new form of world is constituted and ... suspended, the history of the concept of world ... serves to affirm the freedom of the philosopher to stand above and beyond its own worlds: "I stand above the world, which has now become for me ... a phenomenon."
>
> (*The Crisis*, 152)

And:

> The world of things in which spirits [*Geister*] live is an objective world constituted out of subjective surrounding worlds and is the objectively determinable surround world of the spirits.
>
> (*Ideas II*, 292)

In his later works, especially in the *Cartesian Meditations* and *The Crisis*, Husserl elaborated on the "subjective surrounding worlds," including the "world of spirits" (German: *Geister*, which can also mean "minds" or "intellects"), and also distinguished among "prescientific, scientific, and arithmetical or mathematical worlds." And, *crucially, Husserl considered the intersubjective lifeworld (*German: *Lebenswelt) as the "ground" or "horizon" of these previously articulated and differentiated "worlds."* He also rooted this "ground of the world" in our personal egos, as well as—unlike most of his philosophical predecessors and successors (except for Merleau-Ponty, who will be discussed later in this chapter)—*in the lived body.*

For Husserl,

> The Body is ... the *medium of all perception* ... on this original foundation, all that is ... real in the surrounding world of the Ego has its relation to the Body ... the constitutive role of the sensations, is of *significance for the construction of the spatial world.*[14]

Our bodies provide for us, for our individual egos (or "solipsistic subjects"), according to Husserl, *an orienting stability in "the true world," the "one objective world," on the basis of its spatiotemporal grounding.*[15] *When a "community of Egos," comes together, they constitute a common world, a lifeworld, of incarnate beings in the world.*

Husserl's Lifeworld and Environmental World

The notion of the lifeworld was introduced in the posthumously published second volume of Husserl's *Ideas*, under the heading of *Umwelt*, the "surrounding world," or the "environment." In that work, Husserl characterized the *Umwelt* as a world of entities that are "meaningful" to us in that they exercise a "motivating" force on us and present themselves to us.[16] And, according to Husserl, intersubjective experience plays a fundamental role in our constitution of ourselves as objectively existing subjects, of other experiencing subjects, and of the objective spatiotemporal world. From a first-person, or ego-centered, point of view, intersubjectivity arises when we experience instances of *empathy*. It occurs in the course of our conscious attribution of intentional acts and experiences to other subjects. Empathy signifies how we may envision another ego's point of view. To do so, however, implies the existence of a preexisting, pregiven common world in which we perceive our own bodies, intuit our own selves, and interact with other incarnate egos.

A fuller elaboration of Husserl's concept of our intersubjective "living together" in this world—the "pregiven world," our common lifeworld (*Lebenswelt*)—appears in *The Crisis* (1936). There, he argued that the very possibility of science posing and answering questions required returning to "the world common to us all," the "pregiven world," or "horizon" within which we coexist with our fellow humans.[17]

And, furthermore, Husserl attempted to ground the "objective-scientific world" in the "soil" of the "concrete lifeworld," with the "interrelationships" between these two worlds being "paradoxical," in a manner recalling the difficulties such previous idealists as Plato, Descartes, and Kant had in explaining the interactions between physical and mental planes of being:

> two different things, life-world and objective-scientific world, although ... related ... The knowledge of the objective-scientific world is "grounded" in the ... the unity of the life-world ... the grounding soil of the "scientifically true" world and at the same time encompasses its own universal significance ... The paradoxical interrelationships of the "objectively true world" and the "life-world" make enigmatic the being of both.[18]

Also in *The Crisis*, Husserl elaborated more fully on the distinctions and interactions among "the world," the "pregiven world," and the life world:

> *the world is the universe of things, which are distributed within the world-form of space time and are "positional," ... according to spatial position and temporal position ... the life-world, for us who ... live in* it, is

> always already there, existing in advance for us, the "ground" of all praxis whether theoretical or extratheoretical. *The world is pregiven to us ... as the universal field of all actual and possible praxis, as horizon. To live is always to live-in-certainty-of-the-world,* being ... "conscious" of the world and of oneself as living in the world, experiencing and ... effecting ... the world. The world is pregiven ... in such a way that individual things are given. But there exists a fundamental difference between the way we are conscious of the world and the way we are conscious of things or objects ... Things, objects ... are "given" ... in such a way that we are conscious of them as ... within the world-horizon ... *The world, on the other hand, does not exist as an entity, as an object, but exists with such uniqueness that the plural makes no sense when applied to it* ... presupposes the world-horizon ... *World is the universal field into which all our acts ... are directed.*[19]

For Husserl, the lifeworld signified the dynamic "horizon" of all our experiences, the background upon which all things appear to us, in which we *live*, and that "lives with us." *Our lifeworld is the universal horizon of existing objects and subjects. Each of us belongs to this world and we live together in it. As subjective and conscious beings, we co-exist in this pre-given lifeworld.*

Husserl's notion of lifeworld can also be considered as the environment (*Umwelt*), or "homeworld," underlying a person's "natural attitude." An individual person's lifeworld consists of the beliefs they hold about themselves and the world as a whole, as well as the system of meanings constituting their common language, or "form of life," in Wittgenstein's sense of that term (to be discussed later in the next chapter of this book).

For Husserl, the lifeworld would also seem to signify the ways that the members of different social groups, cultures, and linguistic communities structure and make meaning of the pregiven world in which they find themselves. *It constitutes a "world-horizon" of potential future experiences within the sum total of "possible worlds and environments." The lifeworld is therefore a kind of a priori background that is universally taken for granted by all conscious beings in which they live and interact, irrespective of their culturally ascribed differences.* But while we may take it for granted in our everyday lives and assume it to be permanent, the very being of the world as a whole may be ontologically fragile.

The Non-Being or Annihilation of the World According to Husserl

In *Ideas*, Husserl raised the possibility of the "non-being" or "annihilation of the world." While he provided few details, either theoretical or empirical, he did seem to imply that some kind of "mental processes" would persist even if the "world of physical things" does not:

> the possibility of non-being of everything physically transcendent ... while the *being of consciousness*, of any stream of mental processes

> whatever, *would indeed be necessarily modified by an annihilation of the world of physical things, its own existence would not be touched.* Modified, to be sure. For *an annihilation of the world means … nothing else but that in each stream of mental processes … certain ordered concatenations of experience … would be excluded. But* that does not mean that other mental processes and concatenations of mental processes would be excluded. no real being … *is necessary to the being of consciousness itself* (… the stream of mental processes) … the world of transcendent *"res"* is entirely referred to … actual consciousness.[20]

Husserl seems to be saying that consciousness is a precondition for the existence of the world, but the world, at least in its physical essence, is not a necessary condition for the existence of (some kind of?) consciousness. If so, this might imply—in a manner that recalls ancient, medieval, and early modern claims for the immortality of the soul as well as the Hegelian ontological difference between being and non-being—that while the non-being of the entire physical world cannot entirely be ruled out, our own living being can never be doubt. This ontological difference reinforces "the essential detachability of the whole natural world from the domains of consciousness," and, therefore the coexistence of two world-like spheres, the "whole natural world" and "the domains of consciousness." *The existence of the actual, physical world hence appears not to be necessary for the continuation of consciousness in some form, thus leaving open "a multitude of possible worlds."*

In posing the possibility of the "annihilation" of the world, or the possibility of its "non-being," without the termination or extinction of "consciousness itself," Husserl also referred to something somewhat enigmatically called "immanental being." He may thereby have assumed that consciousness in some way may exist as "immanental" being, independent from both the individual body in which it is incarnate and also from the "world of things" and the lifeworld as a whole. Whether this means that my personal mental states may in some way survive the annihilation of the physical world, or that there may exist some other form of non-incarnate being, possibly divine in nature, that does not require a physical world for its being, seems unclear.

Is there then a kind of imperishable conscious being in no need of a physical or lifeworld? We do not know. But Husserl's most distinguished—and notorious—student and successor, Martin Heidegger, grounded our being firmly in this world, and not in the multiple worlds articulated by his mentor.

Martin Heidegger's Being-In-This-World

Martin Heidegger (1889–1976) is perhaps the most controversial major thinker in the history of Western ideas. This is due in part to his philosophy, which is widely considered as among the most influential in

20th-century Continental thought but has been less acknowledged by much of Anglophonic academic philosophical world. However, the "case of Heidegger" is due less to his thinking than to his public persona in Nazi Germany, especially in the early 1930s, but also during the decades following the end of World War II, when he never clearly decried his participation in Nazi-fostered activities or condemned the Holocaust. This has resulted in a kind of "Heidegger industry," in which numerous scholars have interpreted, defended, or critiqued both Martin Heidegger the man and Heidegger the thinker. While I will touch briefly on the "Heidegger debate," I will focus mainly on what is most relevant for my concerns, Heidegger's intellectual contributions to understanding the world and our place in it.

Regarding humans' being-in-this world, Heidegger had at least as much to say as any major 20th-century Occidental thinker, although, given the density and ambiguity of much of Heidegger's oeuvre, especially what is considered his main work, *Being and Time* (German: *Sein und Zeit*), any exegesis of his texts is fraught with virtually unprecedented interpretative challenges (even if one reads German well, as I do). Accordingly, while I am by no means a "Heidegger-expert," much less a subscriber to or sophisticated opponent of the "cult of Heidegger," I will nonetheless attempt to provide a "reading" or "interpretation" of Heidegger's textual and political being-in-this-world that might contribute to our sense of worldly existence, both individually and collectively.[21]

Heidegger's Life- and Political Worlds

Martin Heidegger was born in Messkirch (German: *Meßkirch*), Germany, in 1889. Messkirch at that time was a conservative, religious rural town, and this region of provincial southwestern Germany was the background for Heidegger's lifeworld, as a person, as a political actor, and as a thinker.

Heidegger's father was a craftsman and a sexton at the local Catholic church, and Catholic theology and ontology would have a lasting influence on Heidegger's life and thought. Following primary school in Messkirch, where a local Gymnasium (an academically oriented high school) is named after him, Heidegger was tutored in Latin by a local priest and subsequently studied at several Gymnasiums, completing his secondary education "*cum laude*" (with distinction) in Freiburg. During this time, he was presented by the future Archbishop of Freiburg with a copy of *On the Manifold Meaning of Being according to Aristotle* by Franz Brentano, who had taught Edmund Husserl. In 1909, Heidegger entered the Jesuit novitiate near Feldkirch, Austria, but he left after two weeks, ostensibly for medical reasons. He then began to study Catholic theology at the Albert-Ludwig University in Freiburg, and began to write articles and reviews for *Der Akademiker* ("*The Academic*," the journal of the German Association of Catholic Graduates), including his first publication, a religious narrative on "All Souls' Day." He also studied Husserl's *Logical Investigations* and switched his academic emphasis from theology to mathematics and philosophy. Heidegger received his doctoral degree in philosophy in 1913 with a dissertation on *The Doctrine of Judgment in Psychologism* and two years later completed his *Habilitation* (postdoctoral monograph) with a work called *The Theory of Categories and Meaning in Duns Scotus* (and on Thomas of Erfurt) under the supervision of neo-Kantian philosopher Heinrich Rickert and deeply influenced by Edmund Husserl's phenomenology.

From 1916 to 1917, Heidegger was a *Privatdozent* (unsalaried academic who is qualified for a university appointment) before serving as a weatherman on the western front in France during the last three months of World War I. In 1917, Heidegger married Elfriede Petri, and by 1919, they had both converted to Protestantism. Heidegger was then employed as an assistant to Edmund Husserl at the University of Freiburg until 1923, when he was appointed to a professorship in philosophy at the University of Marburg. During this time, he built a cabin in Todtnauberg in the Black Forest, a retreat that he would use throughout the rest of his life. In 1923, he became a professor without a chair at the university in Marburg, where he had many notable students, including Hans-Georg Gadamer (who, during a chat with me, referred to himself as "the liberal Heidegger," and is best known for his hermeneutic masterpiece, *Truth and Method*,), Karl Löwith (1897–1973, whose book *Meaning in History* is a classic), Leo Strauss (1899–1973, who would late be enormously influential on generations of "Straussian," mostly conservative, political theorists in the United States), Herbert Marcuse (1898–1979, whom I knew and about whom I coedited my first book[22]), the Czech philosopher Jan Patočka (1907–77), and Hannah Arendt (1906–75, who had a passionate affair with Heidegger and then went on to become of the most significant political theorists following World War II). After publishing his magnum opus *Being and Time* in 1927, he returned to Freiburg to occupy the professorial chair in philosophy that had been vacated by Husserl's retirement.

In 1933, Heidegger became a member of the NSDAP (the National Socialist German Workers Party, aka the Nazi Party) and was soon after appointed rector of the university in Freiburg, a post from which he resigned the following year, but not before he had made incendiary pro-Nazi addresses. Heidegger may have wished to position himself as the philosopher of the Nazi Party, but never fulfilled that ambition, possibly because of the abstract nature of his work and the opposition of the leading Nazi ideologist, Alfred Rosenberg, who eventually played that role.[23] Heidegger's resignation from the rectorate may have been due less to any principled opposition to the Nazis than to his administrative frustrations. In his inaugural address as rector in May 1933, he expressed his call for "a German revolution" and a new historical "awakening" (German: *Aufbruch*), and in an article and a speech to the students later that year, Heidegger also declared his support for Adolf Hitler. In November 1933,

Heidegger signed the *Vow of allegiance of the Professors of the German Universities and High-Schools to Adolf Hitler and the National Socialistic State*.[24]

Heidegger resigned from his position as rector in April 1934, but he remained a member of the Nazi Party until 1945, even though the Nazis eventually prevented him from publishing. In the autumn of 1944, Heidegger was drafted into the *Volkssturm* (the "People's Storm," the German national militia) and was assigned to dig anti-tank ditches along the Rhine. Heidegger's *Black Notebooks*, or *Ponderings*, written between 1931 and 1941 and in the process of being published in English, contain expressions of anti-Semitic sentiments. Donatella Di Cesare, after having analyzed the *Black Notebooks*,[25] asserts in her book *Heidegger and the Jews* that "metaphysical anti-Semitism" and antipathy toward Jews were central to Heidegger's philosophical work.[26]

Heidegger, according to Di Cesare, considered the Jewish people to have been agents of "modernity disfiguring the spirit of Western civilization"; and she considers Heidegger to have deemed the Holocaust (or the *Shoah*, the Hebrew term for "catastrophe") to have been the "logical result of the Jewish acceleration of technology," and thus, Heidegger, in her view, blamed the *Shoah* on its victims themselves. Others, including such distinguished philosophers as Theodor Adorno and Jürgen Habermas,[27] claim to have detected resonances between the content of Heidegger's philosophy and Nazi ideology.

Heidegger's critics, including but not limited to Adorno, Habermas, Hans Jonas, Karl Löwith, Pierre Bourdieu, Maurice Blanchot, Emmanuel Levinas, Luc Ferry, and Jacques Ellul, assert that Heidegger's affiliation with the Nazi Party revealed flaws inherent in his philosophy and probably in his character as well. Heidegger's supporters, including Hannah Arendt, Otto Pöggeler, Jan Patočka, Jacques Derrida, Jean Beaufret, Richard Rorty, Julian Young, and François Fédier, tend to regard Heidegger's involvement with Nazism as a personal "error"—a word Arendt placed in quotation marks when referring to Heidegger's Nazi-era politics—that is irrelevant to his philosophy.[28] The "case" of Heidegger's Nazism is not yet closed.

Following the Allied victory over Nazism, France conducted its *épuration légale* (French for "legal purge") in 1946. These included official trials that followed France's liberation from the Nazi occupation and the fall of the Vichy Regime in occupied France. The French military authorities in occupied southwestern Germany determined that Heidegger should be blocked from teaching or participating in any university activities because of his association with the Nazi Party. The denazification procedures against Heidegger continued until early 1949, when he was judged to have been a Nazi *Mitläufer* (German for "fellow traveler," the second lowest of five categories of "incrimination" by association with the Nazi regime). No punitive measures against him were proposed. This opened the way for his readmission to teaching at Freiburg University. He was granted professor emeritus status and then taught regularly from 1951 until 1958, and, by invitation, until 1967. A few months before his death, Heidegger met with Bernhard Welte, a Catholic priest and Freiburg University professor of Christian theology and philosophy. While the exact nature of their conversations is unclear, what is known is that they discussed Heidegger's relationship to the Catholic Church and his subsequent Christian burial, at which Father Welte officiated. Heidegger died in 1976 and was buried in the Messkirch cemetery.

Heidegger's Philosophical Being-In-The-World

In the recently-published *Black Books (Sketches* or *Ponderings),* dating from the early 1930s to the early 1940s, in tones eerily similar to Kant's orienting questions,[29] Heidegger asked:[30]

> What should we do?
> Who *are* we?
> Why should we *be*?
> What are beings?
> Why does being happen?

While Heidegger spent thousands of pages discussing who and what kinds of "beings" we are as well as the meaning and nature of "Being" *per se*, the omission from at least his major published works of any sustained responses to the questions "*why* Being" happens," and what *should* we be and do? is striking. It might therefore not be amiss to claim that the lack of a humanitarian or humanistic ethics in Heidegger's political life is matched by the absence of an explicitly elaborated Ethics in his philosophical oeuvre.

Accordingly, Heidegger's focus is through and through *ontological*, the study of Being (German: *Sein*) and of human existence as "being-there," as "being-in-the-world" (*Dasein, in-der-Welt-sein*), and not with ethical or even epistemological matters. As he "jots," in his *Ponderings*[31]: *"Every great thinker thinks only one thought; this one is always the unique thought—of being."*

Even in some of his later writings, where he raises issues related to poetry, literature, and architecture, for example, his interest seems less purely aesthetic than ontological or even theological (with frequent references "a god" or "the gods"). This is consistent with Heidegger's earliest intellectual interests.

Heidegger's philosophical development and focus on ontology began when he read Brentano (whom Husserl had criticized) as well as Aristotle and his medieval scholastic interpreters. Aristotle's quest in his *Metaphysics* to know what it *is* that unites all possible modes of Being (or "is-ness") is, in many ways, the question that initiated Heidegger's ontology. Heidegger then intensively studied the Presocratics, Plato, Kant, Kierkegaard, Nietzsche,

Wilhelm Dilthey (whose stress on the role of interpretation and history in the study of human activity also profoundly influenced him) and, above all, his mentor, Edmund Husserl (whose understanding of phenomenology as a science of essences he eventually rejected). Although Heidegger's intellectual and personal relationship with Husserl was complex and occasionally strained, *Being and Time* was initially dedicated to Husserl, "in friendship and admiration."

Being and Time, which Heidegger published in 1927, is often regarded as one of the most significant texts in modern European, or Continental, Philosophy.[32] This book elevated Heidegger to a position of international intellectual distinction and provided the philosophical impetus for a number of later movements and ideas, including Existentialism, Hans-Georg Gadamer's philosophical hermeneutics, Derrida's "deconstruction," and the later work of the American thinker Richard Rorty (1931–2007), who sometimes explored the interface between the contemporary continental and the analytic philosophical traditions.[33]

In *Being and Time*, Heidegger attempted to describe being (*Sein*) by means of a phenomenological analysis of human existence (*Existenz*, or, more often, *Dasein*) in terms of its temporal and historical character. By focusing on the existential–phenomenological issues related to our *Being-In-The-world* (*Sein-in-der-Welt*), Heidegger dramatically called into question the Western philosophical tradition's subject/object division—which stems in large part from Cartesian doubts about our direct knowledge of the "external" world, as well as Descartes' radical cleft between "mental" and "physical substances," especially between individual minds (*res cogitans*) on the one hand, and spatially extended (*res extensa*) bodies, others, and the material world on the other hand.[34] To do so, Heidegger created not just a new and provocative way of conceiving the world and our distinctive "everyday," "engaged," and "practical" ways of being-in-the-world, but also devised an original and often intimidating vocabulary, consisting of many neologisms derived from German and Ancient Greek roots, and that is often deployed in obscure ways of expressing his ideas.[35]

Initially citing Plato (in both Ancient Greek and German), Heidegger began *Being and Time* by citing Plato's "perplexity" about what one means in using the expression "being." He then attempts to raise anew "the question of the meaning of being" (German: *Sinn des Seins*). His "provisional aim" in this book "is the Interpretation of time as the possible horizon for any understanding whatsoever of Being."[36]

As a way to investigate the meaning of "being-in-the-world," Heidegger uses the neologism *Dasein* (derived historically from *Dass-sein,* the "that-it-is of being," and literally meaning "Being-there") to denote *human* being, or our *existence as being there*, as *standing out,* in the world.

In "Division One" of *Being and Time*, Heidegger introduces *Dasein* as an "entity" whose "very Being ... is an issue for it."[37] Heidegger discusses *Dasein's "Existenz"* and the "existentiality" of those entities that exist, that have a "being-in-the world."[38] In his later *Ponderings (Black Notebooks)*, Heidegger visualizes *Dasein*'s relationship to the Earth and world as:[39]

Da-sein
Earth—World
(Event)

The conceptual relations between his idea(s) of the world and the Earth are adumbrated in the *Black Notebooks* in somewhat greater detail than in *Being and Time*.

In the section of *Being and Time* called the "Preliminary Sketch of Being-In-The-World, in terms of an Orientation," Heidegger explores (rather than clearly explicates …) the relations between *Dasein*, *my* being, authenticity/inauthenticity, and Being-In-The-world, as contrasted with, for example, the Cartesian cogito's ("I think") being separate from the physical world—even its own body—in the following way:[40]

> *"Being-in-the-world" … a unitary phenomenon …* [has] *several constitutive items in its structure … the "in-the-world." … inquiring into the ontological structure of the "world" and defining … worldhood as such.*

Then, in the section of *Being and Time* most relevant for my purposes, Heidegger elaborates on his notions of "the world" and "worldhood," for which he intends to give a "phenomenological description":[41]

> *BEING-IN-THE-WORLD shall … be made visible with … its structure … the "world" itself … What can be meant by …"the world" as a phenomenon? It means to let us see what shows itself in "entities" within the world …* houses, trees, people, mountains, stars … *to give a phenomenological description of the "world" will mean to exhibit the Being of those entities which are present-at-hand (Vorhanden) within the world* … Things of Nature, and Things "invested with value" … *Is "world" perhaps a characteristic of Dasein's Being? … does every Dasein "proximally" have its world? Does not "world" thus become something "subjective?" How … can there be a "common" world "in" which … we are? And … what world do we have in view? Neither the common world nor the subjective world, but the worldhood of the world as such.*

Heidegger then "discloses" the "worldhood of the world" as an "ontological concept" constitutive of our Being-in-the-world. And *for him "world" in an "ontological characteristic" of Dasein itself.*[42] There is also a preontological "*existential*" public world of our own "closest environment" that must be "disclosed" and "lit up":[43]

> *world is … something "wherein" Dasein as an entity already was … Being-in-the-world … a … circumspective absorption in references or*

> *assignments ... a totality of equipment ...* In this familiarity *Dasein* can lose itself in what it encounters within-the-world and be fascinated with it.

In having defined "the world" and its "worldliness" in various ways, and by having made the crucial distinction between the "entities," or "things" that are "in" or "within" the world, and "the world" as a kind of ontological "container" for them, Heidegger has provided perhaps the most innovative account of the ontological status of "the world" since the Ancient Greeks.

Reminiscent of the very beginning of *Being and Time*, where Heidegger seeks to inquire into the long-neglected question of the "meaning of being," he then poses four questions regarding the very "phenomenon of the world," a phenomenon that has not been called into question for millennia by the Western ontological tradition and that we in the "modern world" seldom if ever consider, and asks:[44]

1. Why *the phenomenon of the world has been passed over* since the beginning of the Western ontological tradition and keeps being passed over?
2. Why *"entities-within-the-world" have ..."intervened* as an ontological theme?"
3. Why these entities are found ... in Nature?

And

4. Why value has been the recourse "when it has seemed necessary to round out such an ontology of the world?"

Heidegger claims that in answering these questions, the problematic of the world "will be reached for the first time," since the world, *Dasein*, and entities within-the-world are the ontological states that are "closest to us." He also recalls the Cartesian analysis of spatiality as a constituent of entities within-the-world. But although Heidegger provides a sympathetic critique of Cartesian ontology, especially of the spatial "extensionality" of the world, later in *Being and Time* Heidegger will stress *the temporality (Zeitlichkeit) and historicity (Historizität) of Dasein's being-in-the world rather than its (Cartesian) spatiality.*

Prior to that analysis, however, Heidegger focuses *on our absorption in and fascination with the "everyday (alltäglich) world," in which we become "who" we are by being with others in "elementary," environmental encounters and experiences. He stresses that these are both ontological and existential ways of Being-in-the-World, of Being-with-Others, and also that Dasein is a "locative personal designation,"* as in "*I here," a "dwelling place" of our "concern" (Sorge).*[45] For Heidegger, *this existential "care," or "concern," is the very "Being" of Dasein (Sein des Daseins).*[46]

Heidegger also stresses *Dasein*'s often "inauthentic" (*uneigentlich*) "falling" into the world of the "they" (*das Man*) and "idle talk" (*Gerede*) as ways of fleeing from its anxiety (*Angst*) in the face of its finitude. For Heidegger, this "fallenness into the world":[47]

> *means an absorption in Being-with-one-another … guided by idle talk, curiosity, and ambiguity … "Inauthenticity" … a … kind of Being-in-the-world … completely fascinated by the "world" and by the Dasein with Others in the "they"* [das Man]. *Not-Being-its-self … is absorbed in a world.*

When *Dasein* "falls," *it may encounter one of its fundamental ways of being concerned with the world, which is anxiety*, a mood and state of mind also central to Kierkegaard's account of existence. This existential Angst is in the face of a sense of the "nothing" and "nowhere" within-the world as well as a sense that the entities within-the-world are insignificant. For Heidegger, anxiety denotes both a "basic kind" of Being-in-the-world and a feeling of "not at home" due to:[48]

> the uncanniness (*Unheimlichkeit*) … lies *in Dasein as thrown Being-in-the-world … anxiety shows Dasein as factically existing Being-in-the-world. The fundamental ontological characteristics of this entity are existentiality, facticity, and Being-fallen.*

Anxiety is a theme to which Heidegger also returns in the second division of *Being and Time*, when he illuminates the state of mind *Dasein* has in the face of the "nothing" (*das Nichts*) and the insignificance of the world that is thereby disclosed to us:[49]

> *The world in which I exist has sunk into insignificance; and the world … is one in which entities* [have] *… no involvement. Anxiety is anxious in the face of the "nothing" of the world … our concernful awaiting … clutches at the "nothing" of the world … is brought to Being-in-the-world … through anxiety. Being-in-the world, however, is both what anxiety is anxious in-the-face-of and what it is anxious about.*

There may be no clearer, and potentially terrifying, account of existential dread in the face of "nothing" than the preceding paragraph; only the journals and diaries of psychiatric patients and some French novelists (the "early" Sartre and Camus, for example) may come close …

In the final section of the first division of *Being and Time*, Heidegger addresses the related questions of the nature of "reality" and whether and how the existence of the "external world" might be proven. These are questions that have haunted the Western intellectual tradition since its origins in the ancient Greek-speaking world and were put on the agenda of modern philosophy, especially in epistemology, by Descartes, who was initially skeptical about the reality of the external world, but then proposed God as the way to solve the problems related to seeming gap between the non-extended inner world of minds and the material, "external," and spatially extended world.[50] For Heidegger, in contrast, our ontological Being-in-the-world *"is bound up ... in the structural totality of Dasein's Being,"* of which *"care as such"* is *"a totality."*

So, it seems for Heidegger that *the (phenomenon) of the world is part of our, of Dasein's, very Being. This is a remarkable shift from conceiving the world, and its worldhood, as something external and unchanging, to positing the worldliness of the world as an essential ontological feature of the very structure of the totality of our existence. Hence, Heidegger seems to claim that the existential–ontological reality of Dasein's being is to be in-the-world, not apart from it as some disinterested, disembodied observer. Our concern, our care about and for the world, is a core structure of Being-in-the-world.*

In the second division of *Being and Time*, Heidegger elaborates on *Dasein*'s Being-in-the-world by exploring its "Being-towards-death," "resoluteness" (*Entschlossenheit*), "authentic potentiality for Being-a-Whole," "care and selfhood," "everydayness," "temporality," and "historicity." Here, he presents one of the most compelling and potentially anxiety-arousing discussions of *death, the limit and end of our, and Others' Being-in-the-world.*

For Heidegger, death is part of the character of *Dasein*, of our "existentiell being toward Death" (*Sein zum Tode*), of our "going out of the world." Our very Being as Dasein is "Being-towards-death," which is essentially anxiety in the face of our "Being-towards-the-end," and toward which we can authentically anticipate the "resoluteness" of death, or flee into the world of distraction and the 'They" (*das Man*).[51]

On that cheerful note, Heidegger concludes his Kierkegaardesque disquisition on the "fear and trembling" that constitute our awareness of our own, and others' finitude. He then commences a discussion of "conscience" (though not in its ordinary ethical or psychological senses), "disclosedness," "fallenness," "care," and "discourse" as constitutive structures of the "there" of *Da-sein's* "Being-in-the-world." For Heidegger, "This Thrown into its 'there" [*da*]; *Dasein* has been "factically submitted to a definite world-its world."[52]

And so, after almost 350 pages of one of the densest ontological discourses in Western intellectual history, Heidegger has come to the point where he discusses the "there" (*da*), the partially spatial location into which our being (*Sein*) has been thrown into the world and is "disclosed" by our involvements, both authentic and inauthentic, and in the ways our "care" for the world are "projected" and "thrown." And where is that? In a *situatio*n (German: *Lage*), a term Sartre would later emphasize in his own existential ontology), or the Being of the "there" in the existence of its "situation."[53]

So, who am, or rather, what is *Dasein*'s, "I' (*das Ich*), according to Heidegger? And where and when is *Dasein* in the history of the world? Heidegger provides a remarkably succinct (for him ...) account of how "the Self" (*das Selbst*), the "saying 'I'" of *Dasein,* is temporally situated in past, present, and future modes of the world. His account of the "historicality" of *Dasein*, however, is more suggestive than exhaustive.[54]

While Heidegger mentions "world-history' (*Welt-Geschichte*), his brief discussion of this term is as far removed from, say, Hegel's, as is philosophically conceivable. "World-history," in Heidegger's account, is ontological and ontic, not a Hegelian "development of reason, spirit, and freedom in time," much less the saga of world-historical individuals, as it is for Hegel. Instead, Heideggerean history, as temporality, exhibits *Dasein's* "factical existence," which is simultaneously, the "disclosure of the world," and *Dasein's* "transcendence."[55]

In one of the most semantically scrumptious passages in a tome filled with neologistic jargon, Heidegger surpasses much of what has come before in the following statements near the conclusion of *Being and Time*:[56]

> as *Dasein* temporalizes itself, a world is too. In temporalizing itself … to its Being as temporality, *Dasein is essentially "in a world,"* by … the ecstatico-horizonal constitution of that temporality. The world … temporalizes itself in temporality. It "is," with the "outside-of-itself" of the ecstases, "there."

Yup … However, there then follows perhaps the clearest and most striking declaration about *Dasein* and its world in the entire book, one striking in its resemblance to similar pronouncements by subjective idealists in the Western intellectual tradition: *"If no Dasein exists, no world is 'there' either."*[57]

So *Dasein*, or human existence as an "entity" without "humanity," is the precondition for the, for any (?) "world" to "be there?" One might have thought, after having attempted to travel so far with Heidegger, that Being-in-the-world would *require the, or a, world for Being to be "there," rather than the other way around* … Anyway, Heidegger's, and our, labyrinthine and occasionally torturous ontological journey through *Being and Time* is about to reach its … incomplete … end, with "the thesis of *Dasein's* historicality," cloaked in the mantle of quintessential Heideggerese:[58]

> *Dasein's historicality* does not say that the worldless subject is historical, *but that what is historical is the entity that exists as Being-in-the-world … Dasein's historicality is essentially the historicality of the world …*

Heidegger's existential–ontological inquiry into "the meaning of Being" in general, and of *Dasein*'s being-in-the-world, has come to what some might say a "premature," and others a "long-awaited" halt …

The World in Being-and-Time

To summarize, in *Being and Time*, Heidegger, in his own inimitable words, uses the word "world" in the following ways, *inter alia*:[59]

1 *... as an ontical concept ... the totality of ... entities ... present-at-hand within the world.*
2 *... as an ontological term ... the Being of those entities ... "world" ... any realm which encompasses a multiplicity of entities ... the "world" of a mathematician ... signifies the realm of possible objects of mathematics.*
3 *"World" ... in another ontical sense-not ... as those entities which Dasein essentially is not and which can be encountered within-the-world, but ... as that "wherein" a factical Dasein ... can ... live. "World" has ... a preontological existentiell signification ... the "public" we-world, or one's "own" closest (domestic) environment.*
4 *Finally, "world" designates the ontologico-existential concept of worldhood ... may have as its modes whatever structural wholes any special "worlds" may have at the time ... the a priori character of worldhood in general ... "worldly" will ... apply ... to a kind of Being which belongs to Dasein, never ... to entities present-at-hand "in" the world ... these latter entities as "belonging to the world" or "within-the-world."*

Heidegger's grounding of *Dasein in* the world was meant to prepare the way for an answer to the question of the meaning of Being in general. However, *Being and Time* was left unfinished, and the agenda Heidegger set for himself was never fully completed. Regarding the never-completed "second volume" of *Being and Time*, Heidegger said:[60] "People are waiting for the second volume of *Being and Time*; I am waiting for this waiting to cease and for people to finally confront the first volume."

*Heidegger's Post-*Being-and-Time *Worlds*

After *Being and Time*, there was what many of Heidegger's commentators have called a shift in Heidegger's philosophy known as "the turn" (German: *die Kehre*), from about 1930 until the early 1940s. Heidegger himself characterized it not as a "turn" in his own thinking alone but as a *turn in Being*. The core elements of the turn are indicated in Heidegger's *Contributions to Philosophy (Of the Event)* (German: *Beitrage zur Philosophie (Vom Ereignis)*), dating from 1936 to 1937 but not published until quite recently.[61]

On the other hand, in contrast to the alleged "turn," in his thinking, both Heidegger's later works and *Being and Time emphasize language and discourse as the vehicles through which the question of being can be unfolded and our world disclosed.* In *Being and Time*, Heidegger stated:[62]

> *Language is a totality of words ... in which discourse has a "worldly" Being of its own; and as an entity within-the-world ...* Discourse is existentially language ... Being-in-the-world ... has been thrown and submitted to the "world." *As an existential state in which Dasein is disclosed, discourse is constitutive for Dasein's existence.*

And, as Heidegger later famously declared in his "Letter on Humanism" from 1946 to 1947:

> man is not only a living creature who possesses language along with other capacities. *Rather, language is the house of Being in which man ek-sists by dwelling ... he belongs to the truth of Being, guarding it.* So ... *what is essential is not man but Being—as the dimension of the ecstasis of ek-sistence*. However, the dimension is not ... spatial ... Rather, everything spatial and all space-time occur ... in the dimensionality that Being is ...[63]

So, language "houses" Being, and discourse "is constitutive for *Dasein*'s existence." Much of mainstream philosophy of language, linguistics, and "universal pragmatics" with a Habermasian accent couldn't agree more, although with the possible exception of Habermas, they would express it more clearly than Heidegger has done. However, in his later essay "What Calls for Thinking?" from 1952, Heidegger reverts to Heideggerese by stating:[64]

> Language ... once called the "house of Being." ... is the guardian of presencing ... as the latter's radiance remains entrusted to the propriative showing of the saying. Language is the house of Being because ... it is propriation's mode.

Perhaps the "transformation of language" called for by Heidegger lies outside Heidegger's oeuvre, in linguistics and the philosophy of language, especially in semantics and discourse-analysis? But that is another topic for another time ...

In his works after *Being and Time*, Heidegger also frequently turned to the exegesis of historical texts, especially of the Presocratics, but also of Kant, Hegel, Nietzsche, Marx, and the German poet Friedrich Hölderlin (1770–1843), as well as to literature, architecture, technology, and other subjects. Instead of looking for a full clarification of the meaning of being, he tried to pursue a kind of thinking that was no longer "metaphysical." He criticized the tradition of Western philosophy as nihilistic, for, as he claimed, the question of being as such was obliterated by "the tradition." He also stressed the nihilism of modern technological culture. By going to the Presocratic beginning of Western thought, Heidegger wanted both to revive and call into question the early Greek focus on being, so that the West could turn away from the "dead end of nihilism" and begin anew. For a while, it appeared to Heidegger that Nazism might provide a "new era," as his notorious interview in 1966 with the German news periodical *Der Spiegel* ("The Mirror"), published shortly after his death in 1976, makes clear in this excerpt:[65]

> *Yes, I was also convinced of it* [the "greatness and glory of this new era"] At that time I saw no other alternative.

Heidegger had said in the fall of 1933: "Let not doctrines and ideas be the rules of your Being. The Führer, himself and he alone, is today and for the future German actuality and its law." But in the *Spiegel* interview decades later, he added: "When I took over the rectorate [of Freiburg University], it was clear ... *I would not survive without compromises. The sentences ... I would no longer write today.*"[66]

The *Spiegel* interviewer also focused on Heidegger's politics of "this supposedly new era," and referred to some comments made by Heidegger, after he'd relinquished his position as Rector of the University of Freiburg, in a 1935 lecture course he gave that was later published in his 1953 work *Introduction to Metaphysics*, in which Heidegger stated:[67]

> *What today is bandied about as the philosophy of National Socialism ... has absolutely nothing to do with the inner truth and greatness of this movement* (namely, with the encounter between technicity on the planetary level and modern man).

To which Heidegger later appended, and decried, the forms "of planetary technicity" supposedly characteristic of "the communist movement" and also "Americanism."

Thus spoke Heidegger in perhaps his last public statement. The peril of "planetary technicity" is one he seems to have believed Nazism might address and overcome. And, it is unclear if Heidegger himself ever authentically addressed, much less overcame, his own "fallen" political way of being-in-the world. Regarding the Jews and millions of other victims of the Nazi technologies of mass extermination—Heidegger infamously remained silent ...

In his later works, Heidegger also frequently mentioned "the world," sometimes reflecting and in other places augmenting his initial ontological conception of the "worldliness" of "world." In his "Letter on Humanism," for example, he provided perhaps the clearest explication of his concepts of world, and earth, in all his hitherto published works, and also referred to "the gods and god," and "the work-being of the work":[68]

> *World is never an object that stands before us and can be seen. World is the ever-nonobjective to which we are subject as long as the paths of birth and death, blessing and curse keep us transported into Being ... there the world worlds. A stone is worldless. Plant and animal likewise have no world ... The peasant woman ... has a world because she dwells in the overtness of beings ... In a world's worlding ... this doom, of the god remaining absent, is a way in which world worlds ...* The setting up of a world and the setting forth of earth are two essential features in the work-being of the work.

Heidegger follows this introduction of the "world's worlding" and the work's "setting up of a world" in which "the god" remains absent by proposing a basic, non-identical, strife-laden relation between world and earth, both of which are "set up" by "the work":[69]

> *The world is the self-opening openness of the broad paths of the simple and essential decisions in the destiny of a historical people. The earth is the spontaneous forthcoming of that which is continually self-secluding and to that extent sheltering and concealing.*
>
> *World and earth are essentially different from one another and yet are never separated. The world grounds itself on the earth, and earth juts through world … The world, in resting upon the earth, strives to surmount it …* The earth … tends always to draw the world into itself and keep it there. *The opposition of world and earth is strife …* In setting up a world and setting forth the earth the work is an instigating of this strife …

In addition to this almost neo-Heraclitean positing of "strife" as the "opposition of world and earth," Heidegger's meta-poetic discourses on the world are perhaps most remarkably revealed in the recently published *Black Notebooks*, or *Jottings*, where he declares, in a section called "The Concept of the World," how, pace Shakespeare's *Hamlet,* this time, "the world is out of joint":[70]

> To bring the world as a world to a worlding is to venture the gods once again … the bringing to a worlding, as an act of violence … *The "world" is out of joint; it is no longer a world, or … never was a world … Along with losing the gods, we have lost the world; the world must first be* [*sein*] *erected in order to create space for the gods in this work; yet such an opening of the world cannot proceed from, or be carried out by, the currently extant humanity—instead, it can be accomplished only if what basically grounds and disposes the opening of the world is itself acquired—for Da-sein and for the restoration of humanity to Da-sein.*

And, Heidegger then proceeds in his *Black Notebooks*, at time poetically, at times in jargon stemming from *Being and Time*, to depict …[71]

> The world—as empowerment of the "there," this the tethered time, without a flight into empty eternity … World overpowers being, but only in order to be sacrificed to being, never itself coming to presence … The world worlds, whereby being prevails so that beings [*das Seiende*] might be … *The worlding of the world happens in the world-producing, opening, ordaining authority of administration—care.*

For Heidegger in these "Jottings," world is …:[72]

> *not a mantle, not an external enclosure; but also not the soul and something interior—quite to the contrary, the vibrant middle of the "there,"*

a grounded middle that stands in the clutches and joints of time … The world as the abyssal ground and the grounding of what is ungrounded. Dasein inhuman—as the thrown breaking in, which quarrels with—beings …

And Heidegger's concept of the world is:[73]

where the "there" opens up … and history, i.e., a people, becomes itself; history is the venturing of the gods out of a world and for a world … To question the concept of world disclosively … with all possibilities of comportment and attitudes and world pictures [*Weltbilder*] … World—the opening up of the counterplay between remoteness and nearness, been-ness and future: the gods … World: space and time appearing in each other … *World is to be grasped only through art as given to the originary event; not first on the basis of knowledge (thinking) or action … The world is now out of joint; the earth is a field of destruction. What being "means" no one knows … Can we at all know it? And if yes, should we know it? And if yes to that, how must it become knowable?*

So, after decades spent inquiring into "the worldliness of the world," and "the meaning of being," Heidegger seems to be as perplexed, and agnostic, as when he began his questioning. Except, perhaps, for this: "world can be grasped only through art," and "art's essence" is "taken as poetry," coeval with "thinking" and both originating in "speech?"

Perhaps what might also be gleaned is what Heidegger later referred to—in his "Letter on Humanism"—as the "homelessness" (*Heimatlosigkeit*), the loss of that "homey" (*heimisch*) feeling of dwelling comfortably on the earth, that is coming to be the "destiny of the world," in large measure due to the "oblivion of Being."

In the remarkable passage below, Heidegger also, for one of the few times, invokes both Hegel and Marx (who, at first glance, might seem to be as far from Heidegger politically speaking as imaginable, although Heidegger's student Herbert Marcuse, among others, would conceive a kind of "Heideggerian," or "Existential Marxism"). Heidegger, also, in a typically oracular pronouncement, declares that "the world's truth is heralded in poetry":[74]

Homelessness … consists in the abandonment of Being by beings … the symptom of oblivion of Being … the truth of Being remains unthought … Being remains concealed. *But the world's destiny is heralded in poetry … Homelessness is coming to be the destiny of the world …* What Marx recognized in an essential and significant sense, though derived from Hegel, as *the estrangement of man has its roots in the homelessness of modern man … Because Marx by experiencing estrangement attains an essential dimension of history, the Marxist view of history is superior to that of other historical accounts.* But … neither Husserl nor … Sartre

> recognizes the essential importance of the historical in Being, neither phenomenology nor existentialism enters that dimension within which a productive dialogue with Marxism first becomes possible.

Unfortunately, Heidegger himself seems not to have entered into that "productive dialogue with Marxism." But Sartre would later do so, as would other intellectual progeny of Husserlian phenomenology, including Maurice Merleau-Ponty and members of the Frankfurt School of Critical Theory. Instead, Heidegger would both "return to his roots" as the "philosopher of being" and proclaim that the world can only be disclosed, not fundamentally changed, unless a "god" might appear, who could "save" us from the "technicity" we have created to destroy our terrestrial home.

Regarding "Man's ek-sistence" as Dasein, and the world that which we are tasked to "disclose" and "clear," Heidegger states, at times in expressions redolent of his earliest training in Christian theology, that:[75]

> *Man is not the lord of beings. Man is the shepherd of Being* ... In his essential unfolding within the history of Being ... Man is the neighbor of Being ... "being-in-the-world" as the basic trait of the *humanitas* of *homo humanus* does not assert that man is merely a "worldly" creature understood in a Christian sense ... What is really meant ... would be ... "the transcendent" ... supersensible being. This is ... the highest being ... the first cause of all beings. God is thought as this first cause.

Heidegger then recalls "being-in-the-world" but does so in the context of "world," not as signifying "earthly as opposed to heavenly being" nor as "worldly" as "opposed to the spiritual" much less "beings or any realm of beings," but rather "the openness of Being," and "Man" (*Der Mensch*):[76]

> is man ... the ek-sisting one. He stands out into the openness of Being. Being itself, which ... has projected the essence of man into "care," is as this openness. Thrown in such fashion, man stands "in" the openness of Being. "*World" is the clearing* [German*: Lichtung] of Being into which man stands out on the basis of his thrown essence.* "Being-in-the-world" designates the essence of ek-sistence with regard to the cleared dimension out of which the "ek-" of ek-sistence essentially unfolds ... "*world" is ... "the beyond" within existence and for it* ... the essence of man consists in being-in-the-world ... With the existential determination of the essence of man ... nothing is decided about the "existence of God" or non-being.

That's Heidegger's existential ontology in a nutshell. But what about God, or, instead, as Heidegger's preferred term, "a god." Can we be saved, from ourselves, and from the technical world *Dasein* has fabricated to deface and destroy our earthly dwelling and to forget our Being-in-the-world?

Can "only a god save us," Save Heidegger, and Save the World?

In his 1966 *Der Spiegel* interview, Heidegger mused about what he considered the bane of *Dasein's* contemporary Being-in-the-world, what he called "technicity" (German: *die Technik*), and the possibility that "a god" (*ein Gott*) might appear. The interview almost assumes the quality of Heidegger's last will and testament. And in it, Heidegger professed both his consternation about "modern technicity" as a planetary-wide "power whose magnitude in determining history" that "can hardly be overestimated" and his ignorance, perhaps skepticism, about the ability of any political system, including democracy, to adapt to and master "the world of technicity."[77]

Heidegger was also queried by his *Spiegel* interviewer about the role of philosophy—which he, perhaps prophetically, believed was being replaced by cybernetics. And, he demurred about philosophy's ability to "effect any immediate change in the current state of the world," or, indeed, about the power of "all human reflection and endeavor" to do this, either. Instead, Heidegger famously declared that:[78]

> *Only a god can save us … It is not through man that the world can be what it is and how it is—but also not without man …*

And of this Heidegger seemed certain: history will take its revenge upon us if we do not understand it.[79]

Finally, Heidegger expressed a kind of world-resignation, even of fatalism, regarding "the situation of man in the face of planetary technicity," lamenting not over the Nazis' failure in its alleged effort to achieve "a satisfactory relationship to the essence of technicity" but that they were "far too poorly equipped" to have done so! As for our current situation, Heidegger laments:[80]

> *I know of no way to change the present state of the world immediately, [even] assuming that such a thing be at all humanly possible.*

Heidegger's Continuing Virtual Being-In-This-World

Taken as a whole, much of Heidegger's oeuvre, especially after *Being and Time*, is less "philosophy"—if by that term one means what was deemed philosophical in the Western intellectual tradition from Aristotle to Kant and what is now considered philosophy in most academic departments in the Anglophonic world—than poetry. As he himself declared in *Ponderings*, "The truth of a philosophy lies in the allegorical power of its work … What is seen is first visible on the basis of the remote, only so—in such seeing—does the world come to be." [81]

This "allegorical truth" of philosophy in its genesis of the world might, in principle, be to Heidegger's credit, since much recent and contemporary academic philosophy is deadening to read and irrelevant to how one might live. However, Heidegger's often obscurantist pronouncements and ontological speculations on the one hand, and oracular and ambiguous "ponderings" and "jottings" on the other hand, tend to mystify rather than clarify the often-important issues with which he appears to be wrestling. Furthermore, deductive and inductive arguments are noticeable by their absence, and "evidence," as known in the social and natural sciences of the modern era, is inconspicuous. And that's excluding his political words and deeds.

So, why is Heidegger taken seriously enough to have generated a virtually unrivaled academic industry of commentaries, expositions, interpretations, courses, and critiques, both of his life and of his work? Well, his existential–ontological way of considering who and what we are, the structure and limits of our being-in-the-world, and of "disclosing" the nature our minds and importance of our moods, is original and sometimes illuminating—if one can "penetrate" the jargon and ambiguity of much of his writing.

Are his claims about human existence "true?" Possibly the wrong question. Are his ontological and existential views indicative of how Heidegger thought about *his own* being-in-the-world? Quite possibly. Does that make them generalizable to *Dasein per se*, to the human race as a whole? Hard, perhaps impossible to demonstrate, especially if there's no consensually validated tool or instrument at our disposal to do so. So where does this leave us, and Heidegger the philosopher? In the limbo of interpretation, guesswork, and speculation. Just as Nietzsche, about whom Heidegger probably wrote more than any other thinker, would have had it ....

Then, of course, the Heidegger legend and industry have also been buoyed, perhaps somewhat perversely, by his having been a notorious Nazi. Nazism was and is "sexy" and charismatic, in a horrifically entertaining way. And, unlike any other influential philosopher, Heidegger was a card-carrying Nazi—another pathway to the publishing grail of seemingly endless interpretations, speculations, reading between-the-lines, psychologizing, and debate ....

Now, isn't it interesting how a "great philosopher" could have been such a nefarious political actor? Indeed, it is, just as Hitler's rise to and fall from power sells more books and film tickets than, say, the history of post-war Germany, or even the political and personal trajectories of Roosevelt and Churchill on the one hand and, and Stalin and World War II Japanese prime minister Tojo on the other. Collective and individual psychopathology, choreographed and filmed by such Nazi superstars as Minster of Propaganda Joseph Goebbels and the legendary film director Leni Riefenstahl, and leading to one of the greatest mass murders in human history, is the stuff of which cultural and critical legends are made (and as I'm writing this, there's a long segment on an acclaimed American public television show about ... Neo-Nazism in Germany today; no comparable broadcast on the German left or political center is anticipated). And Martin Heidegger *is* the stuff of legends. Only Nietzsche and Wittgenstein (the latter to be discussed later in this book) come close among philosophers who have died or lived since 1900.

In addition, Heidegger's most famous work, *Being and Time*, *does* raise many existentially intriguing issues that go to the heart of what it means for "us," or at least for Heidegger's readers, to "be" "in this world." That Heidegger expounds on these matters, often in what was for his time original ways, demands attention. But the *manner* in which he does so is often so obscure and elusive that one is left asking "What does Heidegger mean by x?" And, his commentators and acolytes are only too happy to rush in and publish lengthy tomes on "this is what Heidegger meant …."

It is not difficult to understand why such logical positivists as Rudolf Carnap (responding to self-declared "nonsensical sentences" in Heidegger's "What is Metaphysics"[82]), many analytic philosophers, and even readers sympathetic to ontological endeavors in the "grand," often Teutonic, manner, such as myself, come away from, even after multiple readings, his texts puzzled and frustrated by these "encounters with a master." It is even less difficult not to sympathize with generations of students and readers without backgrounds in academic philosophy who flee from those texts, as one of my best students did, comparing their most tedious and unrewarding daily chores to what it was like "to read Heidegger" (ugh). For me, Heidegger's "ponderings and jottings" are usually clearer, more succinct, and, yes, more conceptually and personally interesting than his lengthy tomes. For they are closer in style, and often in content, to many of Nietzsche's works—*personal, poetic, and philosophical at the same time*. And like Nietzsche, whose own politics has been the subject of numerous commentaries, *Heidegger is a thinker best taken in small doses*.

That said, Heidegger's works, especially *Being and Time*, have been enormously influential in a variety of fields, including phenomenology (Maurice Merleau-Ponty); existentialism (Jean-Paul Sartre and Albert Camus, Heidegger's one-time friend and colleague, Karl Jaspers, and the Spanish philosopher José Ortega y Gasset); hermeneutics (Hans-Georg Gadamer and Paul Ricoeur); political theory (Hannah Arendt, Herbert Marcuse, and Jürgen Habermas); existential psychiatry and psychology (Medard Boss, Ludwig Binswanger, and Rollo May); philosophical pragmatism, comparative literature, and literary theory (especially the "later" works of Richard Rorty); and existential theology (Ludwig Bultmann, Karl Rahner, Paul Tillich, and, perhaps surprisingly, Martin Luther King Jr.).[83] Heidegger's critique of traditional metaphysics and his opposition to positivism and technological world domination have been embraced by leading poststructuralists and theorists of postmodernity (including Jacques Derrida, Michel Foucault, and Jean-François Lyotard). On the other hand, his involvement in the Nazi movement has catalyzed a stormy debate. Although Heidegger never claimed that his philosophy was directly concerned with politics, in many intellectual and cultural circles, his political allegiance has overshadowed his philosophical work.

And perhaps that is as it should be. *A renowned thinker, even perhaps the most prominent Existentialist philosopher, is not simply a "world unto themselves," but is also an actor in the world in which they exist. The case of Martin Heidegger's being-in-the-world is one of the clearest demonstrations in Western intellectual history that a person's being-in-the-world is not framed simply by how one thinks and writes, but also by how one lives.*

Jean-Paul Sartre's Engagement with the World

Jean-Paul Sartre (1905–80) is probably the best-known existentialist writer and one of the few who explicitly accepted a kind of membership in the "movement" now called Existentialism, most notably in his essay *Existentialism and Humanism* (1946). For some, he is also the most renowned and revered philosopher of the 20th century.

Sartre spent most of his life in Paris, where he was born and died. His major philosophical works are *Being and Nothingness* (French: *L'Être et le néant*, 1943), which is heavily indebted to Heidegger's *Being and Time*, the unfinished multivolume *Critique of Dialectical Reason* (*Critique de la raison dialectique,* 1960) and *Flaubert: The Idiot of the Family (Flaubert, L'Idiot de la famille, 1960–72)*. He also wrote many novels, plays, a very influential short work *Anti-Semite and Jew*, and an autobiography, *The Words* (*Les Mots*, 1963).[84]

Prominent among Sartre's literary work is a four-volume novel the first three parts of which Sartre published under the title *Les Chemins de la liberté* (*The Roads to Freedom*): *L'Âge de raison* (1945; *The Age of Reason*), *Le Sursis* (1945; *The Reprieve*). Four years later, he finished *La Mort dans l'âme* (1949; *Iron in the Soul*, or *Troubled Sleep*). After the publication of the third volume, Sartre changed his mind concerning the usefulness of the novel as a medium of communication and focused on writing plays. The fourth volume, however, *La dernière chance* (i.e., *The Last Chance*), although left incomplete, was reconstructed and was published posthumously in 1981.

What a writer must attempt, according to Sartre, is to show humans as they really are. And, what a person really is may best be revealed when in action, and this is what drama portrays. Accordingly, during and after World War II, Sartre wrote and produced one play after another: *Les Mouches* (produced 1943; *The Flies*), *Huis-clos* (produced 1944, published 1945; *No Exit*), *Les Mains sales* (1948; *Crime passionel,* 1949; U.S. title, *Dirty Hands*; acting version, *Red Gloves*), *Le Diable et le bon dieu* (1951; *Lucifer and the Lord*), *Nekrassov* (1955), and *Les Séquestrés d'Altona* (1959; *Loser Wins*, or *The Condemned of Altona*). While these plays exhibit conflict and hostility between humans, they do not exclude the possibility of a kind of redemption. Sartre also wrote *Baudelaire* (1947), a study of that French writer, a biography of the French poet Jean Genet titled *Saint Genet, comédien et martyr* (1952; *Saint Genet, Actor and Martyr*), and many articles published in *Les Temps Modernes* (*Modern Times*, Sartre's journal). Some of these pieces were later collected in several volumes under the title *Situations*.

Sartre's World of Words

According to Sartre, words saved him during his childhood, since *writing provided him with the escape from a world that had rejected him but that he would refashion according to the dictates of his own imagination*. In this way, *he may have been expressing the view that individual creativity may provide the writer or artist with a substitute gratification, a means of replacing the dreary and intimidating outer world with a lively and self-mastered inner world capable of creating a new and purer aesthetic world.* For his lifelong fabrication of a new world of words, Sartre was awarded the Nobel Prize for Literature in 1964. But he refused to accept it, declaring that a writer must "refuse to let himself be transformed into an institution."[85]

In complete contrast with Heidegger and unlike most academic philosophers, Sartre did not make a living as a university professor and was a committed leftwing political activist, initially engaged with the French Resistance to Nazi occupation during World War II—part of which he spent in a German prisoner-of-war camp—and later in a variety of progressive causes, notably in 1968, when he joined the workers and students who were taking to the streets of cities around the world to decry the Vietnam War and to promote anti-capitalist and anti-imperialist revolutions.

Commencing in 1929, Sartre was also intimately associated with the prominent writer and feminist activist Simone de Beauvoir (1908–86), whom he never married but with whom, in 1945, he founded the distinguished periodical *Les Temps modernes*. Sartre also had on- and off-again professional and personal relations with such acclaimed French intellectuals as Albert Camus, Raymond Aron, Maurice Merleau-Ponty, Simone Weil, Emmanuel Mounier, Jean Hippolyte, and Claude Lévi-Strauss.

During the last decade of his life, Sartre's health deteriorated and he lost his eyesight, which had always been weak. In 1980, he died of a lung tumor. His funeral was attended by tens of thousands of people, including both the Parisian literary and intellectual elites as well as throngs of ordinary people who viewed him as an icon of French culture and political resistance. Perhaps more than any other modern philosopher, Jean-Paul Sartre was revered as both a man of the people and a towering intellectual—to such a degree that the post-war period in France (1945–80) is sometimes referred to as "The Age of Sartre."

From Nausea *to* Being and Nothingness

While teaching at a *lycée* (a French academic high school) in Le Havre, Sartre published a philosophical novel called *La Nausée* (1938; *Nausea*). Written in the form of a diary, *Nausea* narrates the feeling of revulsion, or "nausea," its protagonist, Roquentin, experiences when confronted with the world of matter—not merely the world of other people and of things, but the acute self-awareness of his own body.

Nausea rehearses many of the major themes of Sartre's best-known philosophical book, *Being and Nothingness*, which appeared about five years later. It is an extended meditation on the contingency of our existence and on the psychosomatic experience that captures this sometimes terrifying anxiety. In his reflections by a tree root, Roquentin experiences the brute facticity of its existence and of his own: both are simply there, without justification, in excess (*de trop*). The physicality of this revelatory "sickly sweet" sensation is overpowering. Like the embarrassment felt before the Other's gaze, our bodily intentionality (what Sartre calls "the body as for-itself") exhibits the ontological encounter between humans and the world. This theme would be developed further in Sartre's later works.

Following a year spent studying philosophy in Berlin, Sartre adapted Husserl's phenomenological method and combined it with his appropriation of Hegel's and Heidegger's ontology and Cartesian methodology to compose a number of philosophical works—*L'Imagination* (*Imagination: A Psychological Critique*, 1936): *La Transcendance de l'ego: Esquisse d'une description phénomenologique (The Transcendence of the Ego: Sketch for a Phenomenological Description,* 1936*), in which Sartre situates the human ego within the world0; Esquisse d'une théorie des émotions* (*Sketch for a Theory of the Emotions*, 1939), and *L'Imaginaire: Psychologie phénoménologique de l'imagination* (*The Phenomenological Psychology of Imagination*, 1940). But, it was above all in *L'Être et le néant* (*Being and Nothingness*, 1943, abbreviated BN) that Sartre revealed himself as an original, if sometimes opaque philosopher. In all these books, Sartre *places human consciousness, or no-thingness* (*néant), in opposition to being (être*), *or thingness, at the center of his existential–phenomenological world.*

Being and Nothingness is one of a triumvirate of early-to-mid-20th century tomes penned (literally) by the great triad of "continental" philosophers, Martin Heidegger (his previously discussed *Being and Time*), Maurice Merleau-Ponty (whose *Phenomenology of Perception* will be discussed following this section), and Sartre. Apart from their daunting length, fearsome opacity in many places, and weighty tone, these books form the core of what has been called existential phenomenology (though it's unclear if their authors would have gladly accepted this designation). They might be considered *"phenomenological" in that they all proceed from Husserl's focus on intentionality, the description of consciousness and its intentional states and activities, the key role of intersubjectivity, and a view of the world as both constructed by, and as the horizon or limit of, human awareness. The "existential" dimension comprises the important roles played by death, finitude, freedom, responsibility, contingency, inauthenticity (bad faith), anxiety, choices, situations, historicity, Others, and our individual bodily experiences and perceptions* (at least for Sartre and Merleau-Ponty) *in constituting our being-in-this-world.*

Sartre's immense ontological investigation of what it is to be human is both a continuation and an expansion of core themes in his preceding works. For Sartre as for Heidegger, ontology—the "prejudice-free" description of "pure being"—is the philosophical foundation of everything, and *Being and Nothingness* is Sartre's existential–phenomenological, account of what is, what is not, and how they are connected.

Adopting Hegel's terminology and a modified kind of Cartesian "substance dualism" (mind and body as separate "substances"), and proceeding phenomenologically, that is, starting with the "phenomenon"—the being of appearances and the being of those to whom phenomena appear to be—Sartre distinguishes between two distinct but connected types of being (French: *être*), the *for-itself* (French: *pour-soi*, derived from Hegel's "Being-for-itself," German: *Sein-für-sich*) and the in-itself (French: *en-soi*, derived from Hegel's "Being-in-itself," German: *Sein-an-sich*).[86] On the one hand there is the being of the *transcendent object* of consciousness, the in-itself (the *en-soi*), and on the other is the being of *consciousness* (French: *connaissance*), or the for-itself (the *pour-soi*). A phenomenological description of nothingness (*le néant*), or non-being, reveals consciousness to be characterized by its power to "nihilate" (*neantir*),[87] or via "nihilation," to encase being with nothingness. *This constitutes the existential situation of our being-in-the-world, or what Sartre sometimes calls our individual and collective "human reality" (réalité humaine).*[88]

Sartre's notion of human reality is akin to that of Heidegger's foundational notion of human existence as "being-in-the-world," and he also appropriates Heidegger's somewhat mystifying concept of Nothing/Nothingness/Nihilation. For both Heidegger and Sartre, our being is in a world that is:[89] "*... a synthetic complex of instrumental realities.*"

Sartre's "pursuit of being" also remains, at least initially, faithful to Husserl's vision of phenomenology, in that *consciousness is intentional, it is consciousness of something it is not*. Sartre will eventually postulate such seeming paradoxes as x (consciousness) "is what it is not and is not what it is."

Near the beginning of *Being and Nothingness*, also somewhat paradoxically (or, to be less generous, confusingly ...) Sartre declares consciousness to be a "knowing being in his capacity as being and not as being known." He struggles to overcome the very Cartesian/Husserlian dualisms he seems to take for granted, when, for example, he divides the being of the *cogito* (i.e., of thinking consciousness, or the *pour soi*) from the being of what "transcends" consciousness—the object(s) of consciousness, namely, that which consciousness is not (the *en soi*).[90] *By declaring the "positionality" of consciousness, Sartre hopes to bridge this apparent gap between minds and objects, and thereby to "reestablish" the "true connection" between consciousness, "the knowing being," and the world.*[91]

But what, for Sartre, is "the world" to which consciousness is (somehow ...) connected? Unlike Heidegger, Husserl, Descartes, Kant, and other thinkers to whom Sartre's ontological project is indebted, Sartre does not provide a detailed account of what the world is. Instead, throughout *Being and Nothingness*, there appear somewhat scattered remarks regarding the world, rather than a kind of Kantian transcendental deduction of the world, and it is to those passages I will now turn.

Hazel Barnes, the translator of the edition of *Being and Nothingness* to which I'm referring, in her "Key to Special Terminology," claims that, *for Sartre, world is*

> *The whole of non-conscious Being as it appears to the For-itself and is organized by the For-itself in "instrumental complexes."* Because of its facticity [the *pour-soi*'s necessary connection with the *en-soi*, and thus with the world and its past] *the For-itself is inescapably engaged in the world.*

And then in a declaration regarding the world-constituting power of consciousness that recalls similar claims by Kant, Schopenhauer, Husserl, Heidegger, and others, Barnes views Sartre as holding that "… *without the For-itself, there would be not a world but only an undifferentiated plenitude of Being.*"[92]

For Sartre, our "engagement with the world" underscores both our total freedom (as *pour-soi*) in a godless universe to make of ourselves and of our world what we wish. We are also free to create and live by or violate our own values and thereby to forge our destiny. But for Sartre, this "absolute" freedom imposes on us an "overwhelming" responsibility for those existential and ethical choices.[93]

Sartre (like Merleau-Ponty) *situates our freedom in our ontological relation to the world, the "totality which is man-in-the-world" and is actualized by our conduct in the world.*[94] Then, in perhaps the fullest description of the world in Sartre's ontological project, he defines the person (in words reminiscent of Kierkegaard) as a "free relation to himself" and the world in partially Heideggerian-terms:[95]

> *the world-i.e., the totality of beings … is "that in terms of which human reality makes known to itself what it is."*

Sartre then postulates the existential mutual dependency between the world and the person:[96]

> *Without the world there is no selfness, no person; without selfness, without the person, there is no world … The world … is haunted by possibles … which give the world its unity and its meaning as the world.*

So, for Sartre, as for the existential–phenomenological tradition in general, the world and the person are ontologically intertwined, the former being the precondition for the existence of the latter, and, in Kantian form, the person being the precondition for the world. But the person, for this tradition, has an insurmountable limit—mortality—whereas for Wittgenstein and other thinkers in the "analytic" tradition, the world has as its limit, the limits of language. For Sartre, as for Heidegger, we live and die within the horizon of temporality.[97]

In some of the most challenging passages in a book laden with interpretative issues, Sartre then discusses the relations between the for-itself (*pour-soi*) and the world in whose "midst" it has "fallen," in terms of the past and present.[98] In a passage anticipating the central focus of Merleau-Ponty's phenomenology, Sartre states, but does not flesh out (as it were) the ontological importance of perception. He also discloses the ontological ties between the For-Itself and the "totality" of the "ambiguous character" of the world:[99]

> *revealed ... as a synthetic totality and ... collection of all the "thises." ... the world ... the correlate of a detotalized totality, appears as an evanescent totality ... an ideal limitation ... a collection of thises.*

"This" was quite an ontological mouthful, one worthy of Heidegger. ☺ What does this passage mean? Totally obscure in places, especially as a "detotalized totality," at least to me ...

Moving on, Sartre notes both what is beautiful and imperfect about the world, which is encased and determined by "the shell of nothingness":[100]

> *man realizes the beautiful in the world ... in the imaginary mode ... in the aesthetic intuition ...* the beautiful, like value ... *is revealed implicitly across the imperfection of the world.*

Sartre depicts the "disintegrating" spatial relation of the world. For him, *"this world of tasks" is neither Kantian nor Platonic, but rather the mere totality of phenomenal appearances, without an underlying metaphysical substantial unity or "presence."*[101]

And, in a variation on the Heideggerian theme of inauthenticity and thereby of being "lost-in-the-world" (of *das Man* for Heidegger), Sartre notes that[102] "*... being-in-the-world means radically to lose oneself in the world through the very revelation which causes there to be a world ....*"

Sartre then mentions but does not fully explicate the point of view which the for-itself has on the world, involving consciousness's falling into and flight from the world, and mentions "the time of the world," especially the "probables" of the future.[103]

Sartre next pays homage to Husserl and Heidegger, appropriating the former's view regarding "the Other" in the constitution of the psychophysical self that forms a part of and is contemporary with the world, as well as Heidegger's concepts of "being-in-the-world" and the "being-with" others that characterize the being that is human reality.[104] And, Sartre fills in some of the details of what is it is for me (for the cogito) to be in a world with others who, by "appraising" and looking at me, transform both me and the world in which we coexist, since "*… All … the world's density is necessary*" so *"I may … be present to the Other."*[105]

Sartre's stresses *our "engagement" in "situations" in which both the selfhood of myself and of the other are projected and apprehended, and by which we come to exist, since "*I *exist only as engaged …"*[106]

Sartre then discusses, in perhaps unprecedented detail for a major Western philosopher, the existential reality of the body, of *my* body, in the midst of the world.[107] He moves on to discuss in greater detail the "*facticity" of the "body as being-for-itself" in the world, our consciousness of the world, and how the for-itself "makes there be a world," but only in relation to me, and to "my engagement with the world" through which both my self and the world (for me) are constituted:*[108]

> *The for-itself is a relation to the world … makes there be a world …* is-in-the-world … is consciousness of the world … *the world exists confronting consciousness as an indefinite multiplicity of reciprocal relations … this world cannot exist without … relation to me …* through human reality … there is a world … *It is an ontological necessity.*

The preceding passage is *the core of Sartre's ontology*: *the world and the self* (or, in Sartre's terms, the for-itself) *are co-constituted through my inescapable location in and engagement with the human reality in which I am situated. There is no "there" outside my positionality in the world.*

Sartre then provides an account of the role of sensation as the basis of my knowledge of the external world, and how I "enter into" the world through my body, my being-in-this world, which is "*co-extensive with the world.*"[109]

Sartre depicts our body and sense organs as our "being-in-the-midst-of-the-world." He claims "I have caused the world-to-be-there by transcending being toward myself." And he asserts "It is only in a world that there can be a body," our "contingent point of view on the world."[110] He also stresses the centrality of action for our being-in-the-world as well as the relations between consciousness and the body.

In a quasi-Shakespearean passage, Sartre mentions the "silence" and "nothingness" apart from our bodily consciousness:[111]

> *only in a world … can be a body … as a contingent point of view on the world … what this consciousness is … is not even anything except body. The rest is nothingness and silence.*

Sartre then focuses on how my body is "instrumental" and "alienated," as well as how it is the "facticity of my being-in-the-world" and an object for others-in-the-world. My incarnate consciousness is the ground of the "upsurge in the world," which renders my existence both concrete and contingent, and the me that is my freedom "makes a world, exist" in which others' being "haunts me."[112]

After this stunning set of pronouncements, Sartre then moves on to a depiction of how sexual desire and emotion constitute a "radical alteration of the world." He then highlights "the engagement of the world by the body," the relationship between sex and death, and the world as the ground for explicit relations with others, revealed through the body as the "world of desire."[113]

Following this, Sartre turns, in a quasi-Marxian mode, to the "We-Subject" and the world of "manufactured objects" as related to work, and then to the "undifferentiated look," which, borrowing from Heidegger, he calls the "They".[114]

Sartre's life and work as a whole, and his ontology in particular, are building to the theme with which his philosophy is most identified: *freedom,* which, at least in *Being and Nothingness, is both "situational" and "absolute."* For Sartre, being, nothingness, negation, existence, human reality, and our consciousness are all aspects of freedom, the first condition of action, and *how freedom can "modify the shape of the world." And thus comes the credo of Sartrean existentialism: the for-itself "is what it is not while not being what it is ... existence precedes and conditions essence." Hence, in a godless world, we are absolutely free to create our own lives, our destiny, and because of this, we must take absolute responsibility for our actions:*[115]

> *that man being condemned to be free carries the weight of the whole world on his shoulders ... is responsible for the world and for himself as a way of being ... responsibility* [is] *overwhelming since he is the one by whom it happens that there is a world.*

In choosing ourselves, we also choose the world:[116]

> *if there is a world, it IS because we rise up into the world suddenly and in totality ... we lose ourselves in nihilation in order that a world may exist ...* I *choose myself as a whole in the world ... by denying that we are the world, we make the world appear as world.*

Like Heidegger, by "hammering" home the point that we make ourselves what we are by acting in and engaging with the world in all its specific "thises" and "thats," *Sartre illustrates how we discover who we are by what we do in the world, and simultaneously, the world "reveals itself across our conduct."*[117] For Sartre, *our intentional choices "illuminate" the world.*[118]

Sartre's notion of freedom penetrates, in his words, to the nothingness as the "heart of being," but is always in a situation in which the world reveals its, and my, existence, the contingency of freedom and the world.[119]

Sartre continues his elaboration of the existential/ontological importance of freedom, which is situational and absolute. He also describes our living "in a world haunted by my fellowman."[120]

We then come to the existential themes that have come to be identified, for better or worse, with Existentialism as a movement as a whole, and with the writing of Sartre and Albert Camus in particular, death, absurdity, and the meaning of life:[121]

> *I live as a situation ... in the midst of the world ... Death becomes the meaning of life ... becomes mine ...* which makes of this life a unique life ... *It is absurd that we are born; it is absurd that we die.*

After this series of passages containing some of what might be considered Sartre's, and therefore Existentialism's, "greatest hits," Sartre returns to a discussion of our being-there, of the for-itself's human reality, as a "privileged situation."[122]

For Sartre, at the core, or the "lack," of being, is the fruitless human desire to be:

> the *Man-God ... He* [Man] *is what he is not and he is not what he is; Human reality is the pure effort to become God.* [123]

Accordingly, "man is a useless passion."[124] And our concrete human existence is based on the three big categories, namely "to do, to have" and "to be."

Anticipating his future work on "the serious man," especially on political and social revolutions as analyzed and promoted by Marx and his followers who adopt a materialist" perspective, Sartre says: "*revolutionaries ... know themselves ... in terms of the world which oppresses them, and ... wish to change this world ...*"[125]

And, in his Conclusion of this deep and lengthy inquiry into the ontology of human existence, Sartre declares the for-itself as "a hole of being at the heart of being," the "pure nihilation of the in-itself":[126]

> *If ... a single one of the atoms which constitute the universe were annihilated, there would result a catastrophe which would extend to the entire universe ... the end of the Earth and of the solar system ... This upheaval is the world.*

The "phenomenon of the world" leads Sartre to a brief consideration of the project and problem of action.[127] In a curious appropriation of the fact/value distinction, in which ontology "concerns itself solely with what is," Sartre reaches the end of his magisterial, but often mystifying masterpiece, *Being and Nothingness*, with paean to freedom and a sketch of a theory of ethics, which, unfortunately, would never be fleshed out in detail by Sartre.[128]

Although Sartre repeats that our freedom makes "the world come into being," *Being and Nothingness* ends with more questions than answers, especially about the world, which is appropriate, given that Sartre's existential ontology in particular and philosophy in general engage in ceaseless questioning, both into what is (ontology) and what should be (ethics).

Sartre's Post-War World, Words, and Deeds

Following World War II and the publication of *Being and Nothingness*, Sartre's books and political activities, in conjunction with those of Camus and de Beauvoir, catalyzed a cultural movement in post-war France, especially in Paris, that has come to be identified with what we today call Existentialism. However, Existentialism with a Sartrean face came under fire by Christians, who objected to Sartre's atheism, by Communists and philosophical materialists, who disliked the "Cartesian subjectivity" in Sartre's ontology, and also by those who perceived what they claimed to be the "ugliness" and "pessimism" in his philosophy. Sartre's rejoinder to these charges is his work *Existentialism Is a Humanism*, which based on a lecture he gave in 1946.[129]

In the published version of his public address, Sartre rejected these criticisms and restated, in a clear and quotable fashion, some of the "biggest hits" of *Being and Nothingness*.[130] He claimed that the core principle, of both the religious existentialism of Gabriel Marcel and Karl Jaspers and the atheistic existentialism he viewed Heidegger (who may have spurned this designation as he did existentialist "humanism") and himself as proposing, is that "existence precedes essence," or that we must "begin with the subjective," in complete contrast with "viewing the world from a technical standpoint ... that production precedes existence":[131]

> What do we mean by saying that *existence precedes essence? We mean that man first of all exists, encounters himself, surges up in the world—and defines himself afterwards*. If man as the existentialist sees him is not definable, it is because to begin with he is nothing. He will not be anything until later, and then he will be what he makes of himself.

From this, according to Sartre, it follows that:

> *there is no human nature, because there is no God to have a conception of it. Man simply is. Not that he is simply what he conceives himself to be, but he is what he wills, and as he conceives himself after already existing—as he wills to be after that leap towards existence. Man is nothing else but that which he makes of himself. That is the first principle of existentialism.*

From this "nature-less" existence, it follows for Sartre that we are absolutely free and also absolutely responsible for how we exercise our free choices, not merely for ourselves, but also, and debatably, "for mankind as a whole":

> If, however, it is true that *existence is prior to essence, man is responsible for what he is*. Thus, the first effect of existentialism is that it puts every man in possession of himself as he is, and places the entire

> responsibility for his existence squarely upon his own shoulders. And, when we say that man is responsible for himself, we do not mean that he is responsible only for his own individuality, but … *that he is responsible only for his own individuality, but that he is responsible for all men … If, moreover, existence precedes essence and we will to exist at the same time as we fashion our image … Our responsibility is thus much greater than we had supposed, for it concerns mankind as a whole.*

As a consequence, for Sartre:

> if indeed existence precedes essence, one will never be able to explain one's action by reference to a given and specific human nature … *there is no determinism—man is free, man is freedom.* Nor … if God does not exist, are we provided with any values or commands that could legitimize our behavior. Thus we have neither behind us, nor before us in a luminous realm of values, any means of justification or excuse.—*We are left alone, without excuse. That is what I mean when I say that man is condemned to be free. Condemned, because he did not create himself, yet is nevertheless at liberty, and from the moment that he is thrown into this world he is responsible for everything he does.*

This might sound like Nietzsche with a Gallic accent. But now we arrive at what Sartre called "the heart of existentialism," and the "creation and innovation" that form the common bond between moral choice and a work of art. He also famously, and controversially, claimed that each of us bears the responsibility for choices that commit not only ourselves, but "the whole of humanity":[132]

> What is at the *very heart and center of existentialism, is the absolute character of the free commitment, by which every man realizes himself in realizing a type of humanity*—a commitment always understandable, to no matter whom in no matter what epoch … I am obliged to choose my attitude to it, and in every respect I *bear the responsibility of the choice which, in committing myself, also commits the whole of humanity.*

Sartre concludes this work by declaring that the only universe is the godless "universe of human subjectivity." And furthermore, for him, existentialism is a humanistic philosophy of action, rather than a nihilistic and pessimistic philosophy of despair and hopelessness:

> There is no other universe except the human universe, *the universe of human subjectivity* … it is … *existential humanism. This is humanism, because we remind man that there is no legislator but himself; that he himself, thus abandoned, must decide for himself; also because we show that it is not by turning back upon himself, but always by seeking, beyond himself, an aim which is one of liberation … that man can realize*

> *himself as truly human … Existentialism is … an attempt to draw the full conclusions from a consistently atheistic position …* even if God existed that would make no difference … what man needs is to find himself again and to understand that *nothing can save him from himself,* not even a valid proof of the existence of God … *It is a doctrine of action.*

Sartre's existential "doctrine of action" would soon thereafter morph into a philosophy of political praxis, rather than a continuation of his recasting of the Heideggerean ontology of the world into which we are thrown.

During the 1950s and 1960s, Sartre became even more actively involved in leftist French political movements.[133] But unlike his other existential "fellow-travelers" (most notably Camus and Merleau-Ponty), Sartre became, for a while, a vocal admirer of the Soviet Union, although he never became a member of the French Communist Party. However, in 1956, when Soviet tanks rolled into Budapest in to suppress a short-lived revolt, Sartre's hopes for communism with a Russian accent were crushed. He then penned in his periodical *Les Temps Modernes* a long article called "Le Fantôme de Staline" ("The Ghost of Stalin") that condemned both the Soviet intervention and the submission of the French Communist Party to the dictates of Moscow. As a member of the non-Communist French left, Sartre spent his remaining years agitating against French and American imperialist ventures, especially in North Africa and Indochina, and promoting his own brand, sometimes Maoist-inspired, of "Sartrean socialism." Ontological similarities notwithstanding, a sharper contrast with Heidegger's life and deeds can scarcely be imagined than Sartre's political being-in-the-world …

Sartre's experiences during World War II and his on-again, off-again encounters with Merleau-Ponty, Camus, Raymond Aron, and other engaged intellectuals, contributed to awakening Sartre's increasing interest in the political dimension of human existence. *Sartre further developed his understanding of human beings and their world in a way in which he struggled to make his existential ontology compatible with Marxism.* This socially and politically critical attitude toward the existing global capitalist world "order"—i.e., Sartrean socialism—found its theoretical expression in the *Critique of Dialectical Reason* (*Critique de la raison dialectique*, 1960).

In this incomplete two-volume work, Sartre set out critically to examine Marxist dialectics and realized that its Soviet version was a distortion of what Marx had called for. Although Sartre still believed that Marxism was the only philosophy for his age, he conceded that it had become ossified and that, instead of adapting itself to particular situations, had forced the particular to fit a predetermined universal. Hence, according to Sartre, Marxist theory must learn to recognize the existential concrete circumstances that differ from one collectivity to another and also to respect the individual freedom of human beings.

A central idea in Sartre's *Critique* is the concept of *praxis*, an idea dating back to the ancient Greeks, especially to Aristotle, and extending beyond Sartre to the political theory of Hannah Arendt, the radical pedagogy of the Brazilian activist-intellectual Paolo Freire (1921–97), and to the "Praxis School" of philosophy associated with a diverse group of neo-Marxist thinkers in the former Yugoslavia from 1964 to 1974, among others.[134] For Aristotle, praxis referred to three basic human activities: *theoria* (conceptual thinking), *poiesis* (making useful or beautiful artifacts), and *praxis* (doing). For Sartre, praxis extends and transforms the notion of the "project" elaborated in *Being and Nothingness*.

According to Sartre, praxis both produces and is produced by individuals within existing political, social, economic, and historical structures. Individuals then define and actualize their own goals, thereby transcending and negating social constraints. However, the range of possibilities available for individual expressions of freedom is dependent on the existing social structures. Accordingly, the virtually "absolute" freedom Sartre ascribed to us in *Being and Nothingness* is circumscribed by the constraints of the political and historical situation. In his *Critique*, *the subjective world of the individual cogito has become for Sartre one in which personal freedom, must, via collective praxis, be defended and expanded, often in conflict with the existing social and political order.*

In the first volume of the *Critique*, Sartre outlines a theory of "practical ensembles," according to which praxis is no longer opposed to the in-itself, but rather conflicts with institutions that have become rigidified and constitute what Sartre calls the "practico-inert." *Human beings interiorize the universal features of the situation in which they are born, and they then develop their individual modes of praxis, which Sartre captures with the idea of the "singular universal."* In so doing, Sartre modifies his earlier conceptualization of Existentialism by locating individuals in specific social situations that strongly influence but do not determine them. This is because individuals surpass what is socially given and freely create their own aims and projects.

According to what Sartre calls the "regressive–progressive method," individuals develop both in a universal sense according in historical developments, and in a particular way as expressed in their individual projects. *By combining a Marxist theory of history with the existential ontology and psychoanalytic framework presented in Being and Nothingness, Sartre believes his "search for a method" to understand human being-in-the-world has come to fruition.* He would later apply this multidimensional method to his extensive analysis of the life and work of the great French writer Gustave Flaubert (1821–80). In the second volume of this *Critique*, Sartre attempted to explicate "the intelligibility of history." However, this project was never completed. The *Critique of Dialectical Reason,* although a flawed and not-infrequently obscurantist work that has suffered in comparison with *Being and Nothingness*, may nonetheless deserve more attention than it has gained so far.[135]

From 1960 until 1971, Sartre labored over his projected four-volume "total biography" of *The Family Idiot*: Gustave *Flaubert.* Sartre explored Flaubert's life and writing through the use of a double tool—Marx's historical-materialist methodology of class analysis on the one hand, and, on the other hand, Freud's psychoanalytic exploration of childhood and family relations. But like Sartre's *Critique, The Family Idiot* remained unfinished following the publication of the third volume of the French edition in 1972 (five volumes of this work have since appeared in English translation). From then until his death in 1980, as his eyesight and health deteriorated, Sartre did comparatively little theoretical writing. Instead, following his own motto that "commitment is an act, not a word," Sartre actively participated in activities that in his opinion were the way to promote "the revolution." In this way, he linked his theory of revolutionary change with the praxis he thought necessary to facilitate it.

Overall, Sartre's words and deeds after *Being and Nothingness* and *Existentialism is a Humanism* are distinguished by their volume and by their commitment to understanding the social–political world in order to change it for what Sartre considered the better—i.e., postcolonialist, anti-imperialist, and postcapitalist. However, as far as I can detect from the materials available to me—principally in the incomplete *Critique of Dialectical Reason*—apart from attempting somehow to render compatible his appropriation of Marxist theory with his lifelong pursuit of individual and collective freedom—Sartre did not essentially jettison the ontological framework on which the scaffolding of his, and our, world is constructed.

The World Without Sartre?

For Sartre, *the world is "the plentitude of being" encountered, "nihilated," and confronted by the consciousness that is the "for-itself." We are free, within situational limits, to make of it, and of ourselves, the "human reality" into which we are thrown and through which we choose live out our finite, incarnate existences. Our action in this world is an incomplete project,* as was Sartre's own complex and convoluted theoretical effort to put Marxist clothing on his existential ontology.

Maurice Merleau-Ponty would strive to flesh out Sartre's ontology with a full-blooded phenomenology of the perceptual and political worlds into which the French existentialists were cast and which they struggled to revolutionize. Merleau-Ponty, however, would also point out the dangers in too uncritical an adaptation of Marxism without a human face, something to which some of Sartre's words and deeds from the early 1950s until his death may have succumbed.

Michel Foucault once called Sartre as a "man of the 19th century trying to think the 20th." In this dismissive manner, Foucault may well have meant that Sartre's "biographies," except for his book on Jean Genet and his own, were of 19th-century writers, and also that Sartre's early emphasis on consciousness, subjectivity, freedom, responsibility, and the *cogito*, and his later commitment to Marxist dialectical thinking and a quasi-Enlightenment humanism, made Sartre into more of a "traditional" than a "radical" thinker and actor, and hence "irrelevant" for the structuralist, poststructuralist, and postmodernist "subject-less" inquiries of the late 20th and early 21st centuries.[136] (Foucault's own work will be examined in this book's successor volume, *The Reality of the World*.)

However, both Sartre's epigones and his critics may have misconstrued perhaps the most striking personification by any major Western philosopher of the dialectical interplay between theory and practice—the words and world of Jean-Paul Sartre—which articulated and explored his, and our, engagement with the world. Accordingly, Sartre's vision of and encounter with our all-too-human-reality remains both forceful and relevant.

The World Made Flesh

Merleau-Ponty's Perceptual and Political Worlds

Maurice Jean Jacques Merleau-Ponty (1908–61), was, with Sartre, the leading philosophical proponent of existential phenomenology in post-war France, and, arguably, in the world as a whole. While Merleau-Ponty has seldom, if ever, been paid the attention of his one-time philosophical colleagues and political comrades, Sartre, Camus, and de Beauvoir, this is unfortunate, because few, if any, other major Western thinkers have delved with such originality and insight into the importance of the human body and its perceptual schemes.

Furthermore, Merleau-Ponty was also among the scant number of philosophers who integrated empirical social–scientific findings (principally those of the Gestalt and child psychology from the 1930s to the 1950s) into his theoretical framework. And, perhaps more than any other 20th-century intellectual, Merleau-Ponty's work ranged widely over seminal topics in linguistics and literature, art, natural science, psychoanalysis, and, perhaps most significantly, political theory and praxis, particularly as actualized and debased within the former Soviet Union and by anti-humanist actors of all political stripes.

While it is unsurprising that Merleau-Ponty's stress on embodiment, or the "body-subject," has significantly influenced succeeding generations of phenomenological philosophers and social scientists of all nationalities, it has also had a major impact on writers and political activists who otherwise tended to reject core elements of his ontology. These include such poststructuralist thinkers as Michel Foucault (1926–84), Jacques Derrida (1930–2004), Gilles Deleuze (1925–95), and Jean-Luc Nancy (1940–), as well as such contemporary feminist thinkers as on Luce Irigaray (1930–), Hélène Cixous (1937–), and Julia Kristeva (1941), who have also, of course, been deeply influenced by the life and work of Simone de Beauvoir.[137]

Merleau-Ponty's Life-World

Merleau-Ponty was born in 1908 Rochefort-sur-Mer, France. His father, an artillery captain, was killed in 1913 during World War I, after which Merleau-Ponty moved with his family to Paris. Merleau-Ponty later described his childhood as incomparably happy, and he remained very close to his mother until her death in 1953. His university studies were at the elite École Normale Supérieure (ENS), where he befriended Simone de Beauvoir and the distinguished structuralist anthropologist Claude Lévi-Strauss (1908–2009), and completed his philosophy degree in 1930. Merleau-Ponty's philosophy professors at ENS included Léon Brunschvicg (1869–1944) and Émile Bréhier (1876–1952), who supervised his research on Plotinus for the Diplôme d'études supérieures as well his doctoral dissertations.

While a university student, Merleau-Ponty attended Edmund Husserl's 1929 Sorbonne lectures and the Russian-French sociologist Georges Gurvitch's (1894–1965) courses on German philosophy. After a year of military service, Merleau-Ponty taught at *lycées* in Beauvais and Chartres, and also had a year-long research scholarship to study the psychology of perception. From 1935 to 1940, he was a tutor (*agégé-répétiteur*) at the ENS, and attended the Russian-French philosopher Alexandre Kojève's (1892–1968) legendary lectures on Hegel, as well as the Lithuanian-American Aron Gurwitsch's (1901–73) lectures on phenomenology and Gestalt psychology. His first major book was originally titled *Conscience et comportement* (*Consciousness and Behavior*) and was published in 1942 as *La structure du comportement* (*The Structure of Behavior*, abbreviation: SB). In 1939, as the first outside visitor to the newly established Husserl Archives in Louvain, Belgium, he met the German philosopher Eugen Fink (1905–75) and perused Husserl's unpublished manuscripts, including *Ideen II* (*Ideas* II) as well as sections of what would be published as *Die Krisis* (*The Crisis*).

During World War II, Merleau-Ponty served as an infantry lieutenant, but was wounded in battle in 1940, just before the signing of the armistice between France and Germany. He was awarded the *Croix de guerre* ("cross of war") for his bravery in combat. After convalescence, Merleau-Ponty taught at a *lycée* in Paris, got married, had a daughter, and renewed his acquaintance with Jean-Paul Sartre, whom he first met while both were students at the ENS and who became involved with the resistance group *Socialisme et Liberté* ("Socialism and Liberty"). Just after the war, Merleau-Ponty collaborated with Sartre and Beauvoir in the founding of *Les Temps Modernes*, for which he served as political editor until 1952.

Also during the war, Merleau-Ponty completed the book for which he is best known, the *Phénoménologie de la perception* (*Phenomenology of Perception*, abbreviation PP), and in 1945, he replaced Sartre as an instructor at the *Lycée Condorcet* in Paris. Later that year, he began teaching psychology at the University of Lyon, where a few years afterwards he was awarded the position of Chair of Psychology. From 1947 to 1949, he also taught at the ENS, where one of his young students was Michel Foucault, and in 1947, Merleau-Ponty also published *Humanisme et terreur, essai sur le problème communiste* (*Humanism and Terror: An Essay on the Communist Problem*, abbreviation: HT), which is in part a major critique of Stalinism as well as some of its "free world" apologists and anti-Communist critics. The next year, he published a collection of essays on the arts, philosophy, and politics, *Sens et non-sense* (*Sense and Non-Sense*, abbreviation: SNS). In 1949, Merleau-Ponty was appointed Professor of Child Psychology and Pedagogy at the University of Paris. Three years later, he was, at the youthful age of 44, named to the Chair of Philosophy at the *Collège de France*, the most prestigious philosophical position in France, which he held until his death in 1961. Merleau-Ponty's inaugural lecture there was published as *Éloge de la Philosophie* (*In Praise of Philosophy*, abbreviation: IPP).

As a result of ongoing political disagreements with Sartre, Merleau-Ponty resigned from *Les Temps Modernes* in 1953. Two years later, he published *Les Aventures de la dialectique* (*Adventures of the Dialectic*, abbreviation: AD), in which he provided a nuanced critique of revolutionary Marxism as well as of what he called "ultrabolshevism," to which Sartre and de Beauvoir took offense and rebutted. During his years at the *Collège de France*, Merleau-Ponty's intellectual circle included Lévi-Strauss and the extremely influential if idiosyncratic psychoanalyst Jacques Lacan (1901–81). After spending time in Francophonic North Africa, he delivered a series of lectures on the concepts of race, colonialism, and development. And in 1957, Merleau-Ponty declined induction into France's Order of the Legion of Honor, possibly in objection to what he considered the inhumane actions by the French government, including the use of torture, forced displacements, and summary executions by the French army in response to the urban guerilla struggle for independence from 1956 to 1957 by the FLN (National Liberation Front), which has come to be called "The Battle of Algiers."[138] The last book Merleau-Ponty published during his lifetime, *Signes* (*Signs*, 1960, abbreviation: S), is a collection of essays on art, language, the history of philosophy, and politics. Merleau-Ponty died of a heart attack in Paris in 1961, at the age of 53, with a book by Descartes open on his desk.

Following Merleau-Ponty's death, his friend and former student, the French philosopher Claude Lefort (1924–2010) published two of his teacher's unfinished manuscripts *La prose du monde* (*The Prose of the World*, abbreviation: PW), and *Le visible et l'invisible* (*The Visible and the Invisible*, abbreviation: VI), a manuscript that was apparently part of a larger project, *Être et Monde* (*Being and World*).

Given the range and scope of Merleau-Ponty's oeuvre as well as the main focus of my book, I will concentrate on the world as conceived and perceived by Merleau-Ponty in his major work, the *Phenomenology of Perception* (PP), as well as in his political theory, mainly as enunciated in *Humanism and Terror* (HT).

The Incarnate Subject in the Perceptual World

Perhaps more than any other noteworthy philosopher, Merleau-Ponty was fascinated, perhaps obsessed, with our being-in-the-world as lived and constituted by our perceptual experience. Every aspect of our lives—our language, culture, history, nature, people, as well as other living creatures—may form the horizon, or, in a metaphor drawn from the Gestalt psychology he highly prized—the ground—of what is possible for us to be subjects in the world for ourselves, and objects in others' perceptual fields. *The world from this perspective is the "open and indefinite" unity that is the cradle, and grave, of our individual existence, and the shifting boundary for the consciousness and life-world of all sentient creatures that ever were and ever will be on our planetary home.*

The *Phenomenology of Perception* is Merleau-Ponty's paean to this world and to the incarnate subjectivity that emerges from and returns to it. It is also the third and final *chef d'oeuvre* of the trio of 20th-century existential–phenomenological masterpieces that comprise the core of "continental" philosophy bequeathed to us by Heidegger, Sartre, and Merleau-Ponty. With the premature termination of Merleau-Ponty's being-in-the-world in the early 1960s, and the death of Sartre about two decades, existential-phenomenology, and its political metamorphosis into existential Marxism, were also cut short, leaving the perpetuation of this tradition to epigones, commentators, and critics.

One of the enduring lacunae in the recent intellectual history of the West is a creative development of this most existential of all stories about the dynamics and significance of the human project—structuralism, poststructuralism, postmodernism, analytic philosophy, Germanic hermeneutics, and critical theory, notwithstanding. It is to the body of Merleau-Ponty's signature text that I now turn.

Toward the beginning of the *Phenomenology of Perception*, following some introductory remarks on phenomenology and its importance, Merleau-Ponty assesses the problematic nature of such traditional philosophical dichotomies as rationalist, Cartesian accounts of human existence (which Merleau-Ponty calls "intellectualist"), as well as more empirical and behavioristic attempts to objectify our being-in-the-world. He also critically analyzes the apparent dualism involving the mind and the body. According to Merleau-Ponty, *the constitution of the body as an "object" is a crucial moment in the construction of the idea of an objective world that exists "out there," as well as of the "empirical science" that describes it. Once this reifying conception of the body is called into question, so, too, is the notion of an outside world* entirely distinguishable from the thinking subject, a common but mistaken view, according to Merleau-Ponty, who *grounds the seemingly external world in the perceptual experiences of the lived body.*

Regarding the world, and our body, Merleau-Ponty declares early in the *Phenomenology of Perception* that "*The problem of the world, and … that of one's own body, consists in … that it is all there …*" and, "*Our own body is in the world as the heart is in the organism: it keeps the visible spectacle constantly alive, it breathes life into it and sustains it inwardly, and with it forms a system.*"[139] While our bodies, and hence what and who we are as existing subjects, are ontological, as well as historical, social, and political beings-in-the-world, our individual existence is firmly grounded in the field, or horizon, framing our particular phenomenology of perception. Accordingly, as Merleau-Ponty states,

> *the perception of the world is simply an expansion of my field of presence … and the body remains in it but at no time becomes an object in it. The world is an open and indefinite unity in which I have my place.*[140]

To understand, frame, and hence to construct and be constructed by the world, is to find, or lose, my place in the world, and that, at least for non-physically or mentally unimpaired people, is possible only through the sensory lenses, especially vision, of one's body. The *Phenomenology of Perception* provides its readers with Merleau-Ponty's particular view of his world in his words, and of our world as readers of that text.

The Body, My Body, My Place in the World

For Merleau-Ponty, *the body, my body, is not merely "my point of view on the world," but also "one of the objects" in that world, and the "constitution of our body as one of the objects of that world" is furthermore "a crucial moment in the genesis of the objective world," even* of science itself. Conversely, "the body, by withdrawing from the objective world, will carry with it the intentional threads linking it to its surrounding and finally reveal to us the perceiving subject as the perceived world."[141] Accordingly, it follows that "… *if it is true that I am conscious of my body via the world, that it is the unperceived term in the center of the world … it is true … that my body is the pivot of the world …*" leading to what Merleau-Ponty deems,

> *This paradox is that of all being in the world: when I move towards a world I bury my perceptual and practical intentions in objects which ultimately appear prior to and external to those intentions, and which nevertheless exist for me only in so far as they arouse in me thoughts or volitions.*[142]

That "move towards a world," is the thrust of my embodied subjectivity into a "physical world" I "inhabit," and:[143]

> *not merely the historical world in which situations are never exactly comparable—my life is made up of rhythm …* there appears round our personal existence a margin of *almost* impersonal existence … which I rely on to keep me alive; *round the human world which each of us has*

> *made for himself is a world ... to which one must first of all belong ... to be able to enclose oneself in the particular context of a love or an ambition.*

For Merleau-Ponty, the center of our worldly existence is simultaneously what enables both our "freedom and servitude," namely our incarnate, or bodily, being-in-the temporal world:[144] "*the ambiguity of being-in-the-world is translated by that of the body ... understood through that of time.*"

"Ambiguity" is a key ingredient of how Merleau-Ponty and his existential contemporaries Sartre, Camus, and Simone de Beauvoir not only "see" "the world," and all the ethical dilemmas arising from our being-in-the-world-with-others, but also how they write about the world, thus leading to not-infrequent puzzlement by their readers as to what they, and their texts, "really mean."[145]

According to Merleau-Ponty, ambiguity is inherent both in my perception of things and in my self-knowledge, primarily because of our inescapably ambiguous temporal situation. For him, "there is the absolute certitude of the world in general, but not of anything in particular"; hence, there is ambiguity because we are not capable of disembodied reflection upon our activities, but rather, as body-subjects, are intentionally directed toward the world as perceived. For Merleau-Ponty, both intellectualism and empiricism presuppose "a universe perfectly explicit in itself." But since he stresses the limitations of these two perspectives, the body-subject, as Merleau-Ponty sees it, engenders ambiguity and a kind of indeterminacy, reflecting both our contingency and our freedom.[146]

Merleau-Ponty *contrasts the world for humans with the subject-less environment in which animals enter and depart, and he characterizes our world as:*[147] *"... the common reason for all settings and the theatre of all patterns of behavior ..." The human world is*, in addition, *"a pure spectacle into which I am not absorbed, but which I contemplate and point out.*" And, toward the end of his *Phenomenology of Perception*, in one of his less ambiguous pronouncements, Merleau-Ponty describes the world in which we find our ontological roots in the following way:[148]

> The subject is a being-in- the-world and the world remains "subjective" *... the world as cradle of meanings ... standing on the horizon of our life as the primordial unity of all our experiences, and one goal of all our projects, is ...* the native abode of all rationality.

As body-subjects *whose lived, sensory being-in-the "primary" world is prior to our conception of the "self-evident" world and is the ground of all meaning:*[149]

> *the ... subject must ... have a world or be in the world ... sustain round about it a system of meaning ... These acquired worlds ... are ... carved out of a primary world which is the basis of the primary meaning.* In the same way there is a "world of thoughts."

As a phenomenologist, Merleau-Ponty views human consciousness as intentional, partially depersonalized, and projected into both the physical and cultural worlds in which its body-subject is immersed and, via sight and movement, constitutes the world as an "intersensory unity,"[150] since "*consciousness projects itself into a physical world and has a body, as it projects itself into a cultural world* ... towards the intersensory unity of a 'world.'"[151]

Embodied consciousness, according to Merleau-Ponty, "has" or "understands" its world without, or prior to, our using "symbolizing" or "objectifying functions" to impute a meaning to our bodily experience, which[152] *"provides us with a way of access to the world and the object ... My body has its world, or understands its world ..."* Hence, for Mearleau-Ponty:[153] "*To be a body, is to be tied to a certain world ... because the body is solidified ... existence, and existence a perpetual incarnation."*

But we are not just incarnate subjects; we are also what Merleau-Ponty calls a "historical idea," beings who are situated and whose freedom is never unconditional.[154] Furthermore, according to Merleau-Ponty, we are enmeshed in a "linguistic, intersubjective, and cultural world," from which our words, speech, and thought come into being:[155]

> *Thought is no "internal" thing, and does not exist independently of the world and ... words ... The linguistic and intersubjective world no longer ... distinguish it from the world itself, and it is within a world already spoken and speaking that we think.*

But, above all, for Merleau-Ponty, we are our bodies—our "natural self, the subject of perception, the opening on to our de facto world," and experience is the "beginning of knowledge":[156]

> I am my body ... *Our own body is in the world as the heart is in the organism ..., and ... we perceive the world with our body ... by remaking contact with the body and with the world, we shall also rediscover ourself* ... against the background of this world ... experience ... the opening on to our *de facto* world—is ... the beginning of knowledge, there is no ... way of distinguishing ... *what the world must necessarily be and what it actually is ... we are in the world—the senses ... as necessary to this world ... the only world which we can think of ... each sense should constitute a small world within the larger one.*

Merleau-Ponty then links our self-consciousness phenomenologically with "the constituted world" as "an integrated world" and the "perceptual ground" as the "general setting in which my body can co-exist with the world":[157]

> *the subject is absolutely nothing and ... part of the constituted world ... my body is geared onto the world ... This ... perceptual ground, a basis of my life, a general setting in which my body can co-exist with the world.*

Spatiality, especially depth perception, as the "most 'existential of all dimensions,' according to Merleau-Ponty, provides us with access to 'an intersubjective world,' as well as to the phenomenon of the world," "the primordial experience from which it springs":[158]

> the *equivalence of depth and breadth ... is part ... of ... an intersubjective world ... the phenomenon of the world ... its birth for us in that field into ... each perception ... where we are ... still alone ... other people will appear only later ...* we are destined to graduate to a world, and [spatial depth] is ... the most "existential" of all dimensions.

According to Merleau-Ponty, the natural, cultural, and intersubjective world(s) in which we exist may be both an object of wonder, following Husserl, and, in quintessentially existential terms, may also seem both "absolutely self-evident" and "absurd":[159]

> *existence projects round itself worlds ... against the background of one single natural world ... The world appears absurd ...* Absolute self-evidence and the absurd are equivalent, not merely as philosophical affirmations, but also as experiences.

Merleau-Ponty then links our consciousness of the world—"an open and indefinite unity in which I have my place"—not to our self-consciousness, *but to our "pre-conscious possession of the world" via the "pre-reflective cogito"* (a claim he would reexamine later in his life):[160]

> *Consciousness is removed from being, and from its own being, and ... united with them, by the thickness of the world ... The consciousness of the world is not based on self-consciousness ...* [but on] *pre-conscious possession of the world ...* perception of the world is ... an expansion of my field of presence ... *The world is an open and indefinite unity in which I have my place.*

In a series of passages that alternate between poetic expressiveness and philosophical pronouncements, Merleau-Ponty then explores our perceptual being-in-the-world, the inexhaustibility of "the real," and the ground of our living in and knowing the "logic of the world":[161]

> every perception is a communication or a communion ... *there is a logic of the world to which my body ... conforms, and through which things of intersensory significance become possible for us ... the human body ... has running through it a movement towards the world itself ... Human behavior opens upon a world (Welt) ... The natural world ... is the schema of intersensory relations ... The world has its unity.*

What immediately follows are declarations proclaiming that "all our logical operations" are based upon "our experience of the world," and that my "experiences of the world" are bound into "one single world" in "its entirety":[162]

> *The world remains the same world throughout my life, because it is that permanent being within which I make all corrections to my knowledge …* impossible to conceive a subject with no world … *any definition of the world would be merely a … schematic outline, conveying nothing to us … if we did not … know it by virtue of the mere fact that we are. It is upon our experience of the world that all our logical operations … must be based, and the world itself … is not … some idea which breathes life into the matter of knowledge … My experiences of the world are integrated into one single world.* My point of view is … a way I have of infiltrating into the world in its entirety.

Merleau-Ponty then proceeds to describe the "mystery of the world" and the "ambiguity" of our consciousness or existence. He focuses on "the thing and the world" within my subjective and spatiotemporal, finite, and death-constricted landscape of perception:[163]

> if the thing and the world could be defined … if the world [is] conceived from no point of view … I should hover above the world, so that all times and places … would become unreal, because I should live in none of them and would be involved nowhere. *Things and instants can link up … to form a world only through … that ambiguous being known as a subjectivity, and can become present … only from a certain point of view and in intention … This is what is … expressed by … the thing and the world are mysterious … The world … is not an object …* The *thing and the world exist only in so far as they are experienced by me or by subjects like me …* the world itself lives outside me, just as absent landscapes live on beyond my visual field.

In concluding this section of his *Phenomenology of Perception*, Merleau-Ponty appropriately, and perhaps in intentionally ambiguous language, portrays the "perceived world" as the "vague theater of all experiences," which, prior to "any science and verification," frames both our subjectivity and the things forming the "real" and "phantasmic" experiences constituting our ambiguous self-knowing being-in-the-world, since[164] "*I know myself only in my inherence in time and in the world, that is, I know myself only in ambiguity.*"

The next section of the *Phenomenology of Perception* is called "Other Selves and the Human World." It focuses on my incarnate location within the natural and intersubjective worlds, and how my body is thrown into and perceived within the human world, the venue of our contingency and freedom.

Regarding the natural world—seldom if ever discussed within the Western intellectual canon of "great works" in philosophy and related fields—Merleau-Ponty declares that:[165] "I am thrown into a nature, and that nature appears not only as outside me ... but it is also discernible at the center of subjectivity." And, regarding the cultural world, which I share with others, who are objects for me while subjects for themselves, Merleau-Ponty elucidates the difficulty in considering each individual self as "constituting the world."[166]

Merleau-Ponty then returns to the dual theme of my body "as the potentiality of the world" and of "objective thought" (i.e., science etc.) as grounded in my perceptual world, as my "phenomenal body." And, the consciousness it actualizes lies situated within the "sensory field" of the "primordial world." In so doing, he also alludes to the "problem of other minds" (consciousnesses in bodies other than mine), as well as to the grounding of such "objective thought" as "physicomathematical correlations" in "my perception of the world,"[167] *"because my body is a movement towards the world, and the world my body's point of support."*

Merleau-Ponty is keenly aware of the existence of others' bodies, and hence of the "interworld" in which we "coexist through a common world." At the same time, he takes it for granted that *"I am given to myself" in my body, through which I am also "inserted" into a natural, cultural, and social world in which my individual freedom is simultaneously "fated" to be "thrown into the world"* and also capable of "withdrawal" from the social world and thus to "strip" it, and others, of "significance":[168]

The physical, social, and natural worlds constitute the "external world," whose relation to our "internal world" has, in Merleau-Ponty's opinion, been "falsified" by both idealism and realism. And, in language at times eerily akin to that of Wittgenstein, Merleau-Ponty "resolves" this classical philosophical problem by "dissolving" it within "the truly transcendental" and "transparent world," and the "paradox of time beneath the subject, the world, the thing, others, and the body." In so doing, "we shall understand that beyond these there is nothing to understand." In this quasi-mystical way, Merleau-Ponty concludes the second part of his *Phenomenology of Perception*.[169]

> *The physical and social world always functions as a stimulus to my reactions ... positive or negative ... We must ... rediscover, after the natural world, the social world, not as an object or sum of objects, but as a permanent field or dimension of existence ... Our relationship to the social is, like our relationship to the world, deeper than any ... perception or judgement. But if ... we relate to the paradox of time those of the body, the world, the thing, and others, we shall understand that beyond these there is nothing to understand.*

In the third and final part of his *chef d'oeuvre*, entitled "Being-for-Itself and Being-in-the-World," and recollecting the ontologies of Hegel, Heidegger, and Sartre, Merleau-Ponty further elaborates how we, as subjects to ourselves and objects to others, and how, by virtue of or body's motion in our perception of the world, constitute the "being" of the world. This world is not only (our) idea, in contrast with Schopenhauer, or merely "a synthesis of objects," as in Kantian epistemology—but is the open "totality of things toward which we project ourselves":[170]

> *I must see the existing world appear at the end of the constituting process, and not only the world as an idea, otherwise I shall have no more than an abstract construction, and not a concrete consciousness, of the world ... The world around us must be, not a system of objects which we synthesize, but a totality of things, open to us, towards which we project ourselves.*

The existence and "shape" of the world, as well as the status of "other possible worlds," and especially "why this particular world has come into being" are, as Kant had previously argued, questions we raise due to "our psychophysiological make-up," but which, however, are "out of our reach" satisfactorily to answer. Furthermore, in Heideggerian terms, Merleau-Ponty proclaims that the "ontological contingency of the world" is not some "deficiency in being," but rather is the basis "for all of our ideas of truth."

For Merleau-Ponty, "the world is that reality of which the necessary and the possible are merely provinces," and the original ontological questions—"why is there something rather than nothing" is "impossible" to pose:[171]

> *the shape of this world and the very existence of a world are ... consequences of necessary being ... bound up with my psycho-physiological constitution and the existence of this world ... other possible worlds, can be conceived as having the same claim to reality as this one ... the question why there is something rather than nothing seems apposite, and why this particular world has come into being, but the reply is ... out of our reach, since we are imprisoned in our psycho-physiological make-up ... There is no other world possible ... in which mine is ... because any "other world" ... would set limits to this one ... and would ... fuse with it.*

In one of the most striking parts of this remarkable book, Merleau-Ponty then articulates his *"vision," both literally and metaphorically, of the world, both as a "field of experience" and whose "being" is an "unfinished work," "inseparable" from our own subjectivity and "view of the world."* He also suggests an incipient philosophy of language regarding speech, words, discourse, meaning, truth, and the "spoken" and "tacit" *cogito* ("the "I think" or "thinking thing," pace Descartes and Husserl).[172]

In the concluding passages of this section of the book, Merleau-Ponty further articulates, alluding to Cartesianism, how "my existence as subjectivity is merely one with my existence as a body and with the existence of the world." Hence, "the world is wholly inside and I am wholly outside myself." And, he then restates and somewhat amplifies this bedrock existential proposition that "we are in the world," which, for Merleau-Ponty, entails that "the belief in an absolute mind, or in a world in itself detached from us is no more than a rationalization of this primordial faith." This is so, because, citing the 19th-century French poet Arthur Rimbaud:[173]

> "we are not of the world," as the horizon of our particular commitments, as ... the world's phantom. *Inside and outside are inseparable. The world is wholly inside and I am wholly outside myself ... I understand the world because there are for me things near and far, foregrounds and horizons, and this ... is because I am situated in it ... because my existence as subjectivity is merely one with my existence as a body and with the existence of the world, and because the subject that I am ... is inseparable from this body and this world. The ontological world and body which we find at the core of the subject are not the world or body as idea, but ... the world ... contracted into a comprehensive grasp, and ... the body ... as a knowing-body.* [174]

The final sections of Merleau-Ponty's *Phenomenology of Perception*, are, like Heidegger's *Being and Time*, focused on our temporality and our meaning-giving (German: *Sinngebung*) existence in the world as historical beings. Unlike Heidegger, however, and more akin to Sartre, Merleau-Ponty also delineates our freedom and responsibility for being in and of the world.

Once again, he stresses "the presence of the world in the heart of the subject" and the meaning of "meaning" for us. But Merleau-Ponty, presaging his later works on art and literature, also alludes to "the Logos of the aesthetic world":[175]

> the ... question is ... understanding ... the relation between *meaning* and *absence of meaning* ... I come ... bringing ... *a universal setting in relation to the world. At the heart of the subject himself ... the presence of the world ... and ... every ... process of signification ... as derivative ... to that pregnancy of meaning within signs which could serve to define the world ... beneath the intentionality of acts* [is] *a "Logos of the aesthetic world," an "art hidden in the depths of the human soul ..."*

And then, in one of the most compact and concise passages in his *Phenomenology of Perception,* Merleau-Ponty summarizes his view of *the world as our "cradle of meanings," its possibility with or without human consciousness, and our "subjective" being-in-the-world. Importantly, he also stresses our "prescientific experience of the world" in which we live as the unspoken foundation of our scientific knowledge of anything whatsoever.*[176]

At times in tones strikingly familiar to those of Sartre in *Being and Nothingness*, Merleau-Ponty then discusses our "bodily being, our social being, and the preexistence of the world" as the "basis of our freedom." Furthermore, our freedom and our consciousness—which is the "nothingness," or the "non-being" that "constitutes us"—are "impinged on" by the "plenum" (for Stoic philosophers, this "plenum" denotes space filled with matter) of the "external world," the limit of our action and that to which we confer meaning and significance. He also interweaves our freedom with "our fate," and, recalling Kierkegaard, depicts our "choice of ourselves" as "conversion" involving our "whole existence":[177]

Merleau-Ponty then returns, as it were, to *the world, as the "collection of things which emerge from a background of formlessness by presenting themselves to our body"* in the *"natural world"* as the *"indissoluble" "place of all possible themes and styles for our being-in-the-world as well as for intersubjectivity."*[178]

Merleau-Ponty further elaborates on who we are—"an intersubjective field" and a manner of "refusal to be anything"—and on the nature of human freedom, which destroys the "world's 'worldliness'" and enables our consciousness to "make its life in the world" and to "take responsibility" for this "general refusal to be anything.[179]

Then to conclude this massive work, Merleau-Ponty poses and addresses the ultimate existential questions: what is the relationship between human consciousness and the world? What is freedom? And who and what am I? He concludes his *Phenomenology of Perception* with a citation from the 20th-century French writer Antoine de Saint-Exupery, in a phrase reminiscent of Karl Marx, that we are nothing but "a network of relationships":[180]

> *What then is freedom? To be born is both to be born of the world and ... into the world. The world is already constituted, but also never completely constituted ... We choose our world and the world chooses us ... I am a psychological and historical structure, and have received, with existence, a manner of existing, a style ... even a philosopher's thought is merely a way of making explicit his hold on the world, and what he is* ... Nothing determines me from outside ... *because I am from the start outside myself and open to the world ... by the mere fact of belonging to the world, and not merely being in the world in the way that things are ... "Man is but a network of relationships, and these alone matter to him."*[181]

Merleau-Ponty's "answers" to his existential questions may seem vague, ambiguous, even "meaningless" to logical positivists, or, alternatively, may appear poetic to readers not in need of concrete particulars. But they are consistent with his own, as well as his existential–phenomenological predecessors' and comrades', style of life and writing. To borrow a phrase from Merleau-Ponty himself—they are the "prose of the world," the literary expressions of Merleau-Ponty's personal way of being-in-the-world as an engaged writer–philosopher, one who, unlike Heidegger, but similar to Marx, Sartre, Camus, and de Beauvoir, *sought to transform the world, in all its social, historical, and political complexities and ambiguities, into a more humane, and humanistic, ground of our lived existence.*

From the Philosophical to the Political World

It has been said that Merleau-Ponty described his philosophical career as falling into two distinct phases, the first of which, up to and including the *Phenomenology of Perception*, involved his attempt to restore the world of perception to philosophical prominence and to affirm the primacy of the prereflective cogito.[182] During this period of his work, Merleau-Ponty emphasized the fundamental grounding of the world for our thinking and reflection. The second phase refers predominantly to Merleau-Ponty's late book *The Visible and the Invisible* and to his abandoned project *Prose of the World*, and is characterized as an attempt "to show how communication with others, and thought, take up and go beyond the realm of perception."[183] According to Merleau-Ponty, what he had in his earlier work had called the "tacit," or "prereflective cogito," had become problematic.[184] Had he lived, he might have also, as it were, fleshed out further the "visible" and "invisible" arcs of our perceptual, philosophical, and intersubjective being-in-the-world.[185]

In his later work, Merleau-Ponty also makes more explicit the "world of the philosopher" and the relations between philosophical "dialectical ambiguity" and the role played by "true philosophy" in "relearning to look at the world" and, as radical reflection on our individual "being-in-this world." As Merleau-Ponty says in his speech *In Praise of Philosophy*, amplifying remarks he made toward the beginning of his *Phenomenology of Perception* about the role of philosophy in returning to "return to the actual world of experience":[186]

> *It is in the world of the philosopher that one saves the gods and the laws by understanding them, and to make room on earth for the life of philosophy,* it is precisely philosophers like Socrates who are required ... the philosopher, in order to experience more fully the ties of truth which bind him to the world and history, finds ... a renewed image of the world and of himself placed within it among others. *His dialectic, or his ambiguity, is only a way of putting into words what every man knows well- the value of those moments when his life renews itself and continues on, when he gets hold of himself again, and understands himself by passing beyond, when his private world becomes the common world.*

Merleau-Ponty continues in the same vein about "true philosophy":[187]

> *True philosophy consists in relearning to look at the world, and in this sense a historical account can give meaning to the world quite as "deeply" as a philosophical treatise. We take our fate in our hands, we become responsible for our history through reflection, but equally by a decision on which we stake our life ... Philosophy, as radical reflection ... is in history, it too exploits the world and constituted reason ... The first philosophical act would appear to be to return to the world of actual*

> *experience which is prior to the objective world, since it is in it that we shall be able to grasp the theoretical basis no less than the limits of that objective world, restore things their concrete physiognomy, to organisms their individual ways of dealing with the world, and to subjectivity its inherence in history.*

In his political works and deeds, Merleau-Ponty exhibited his individual way of "dealing with the world," and, in so doing, in giving "meaning" to both that world and to his own life.

The Tragic Contingencies of the Political and Historical Worlds

Merleau-Ponty's political writings, particularly *The Adventures of the Dialectic* and *Humanism and Terror*, further situate our perceptual being-in-the-world—our organic, incarnate subjectivities—within the ever-shifting, historically unpredictable networks of power and class relations. In *Humanism and Terror*, published in 1947 in response to Arthur Koestler's novel *Darkness at Noon* and serving as a critique of both the adoption of state terrorism by Stalin and the growing anti-Communism climate in the West, Merleau-Ponty stresses how the historically "pure" logic of Hegel and Marx became objectified and misapplied, with terrifying results, in the Soviet Union. For Merleau-Ponty, communism with a Stalinist visage is a perversion, not an ineluctable consequence, of Marx's historical materialism.[188]

> Marxism had understood that it is inevitable that our understanding of history should be partial since every consciousness is itself historically situated. But instead of concluding that we are locked in our subjectivity and sworn to magic as soon as we try to act on the world, Marxism discovered ... a new foundation for historical truth ... on the profound idea that *human perspectives, however relative, are absolute because there is nothing else and no destiny. We grasp the absolute through our total praxis.*

For the Marxists of the early-to-mid-20th century, according to Merleau-Ponty, the subjective life of one's inner world has been jettisoned for the "intersubjective truth" of a world in which history, not individuality, is the center of existence:[189]

> The Marxist has recognized the mystification involved in the inner life; he lives in the world and in history ... it consists in questioning our situation in the world, inserting ourselves in the course of events, in properly understanding and expressing the movement of history ... Marxism is neither the negation of subjectivity and human action nor ... scientistic materialism ... It is much more a theory of concrete

subjectivity and concrete action—of subjectivity and action committed within a historical situation.

However, the Marxian historical dialectic has, alas, not led to a classless society and a pacified world (following violent proletarian revolutions in Russia, China, Cuba, Vietnam, and elsewhere), but has instead led to Stalinism in the former Soviet Union, following Lenin's death in the early 1920s, and to Soviet mass terror starting in the 1930s.

Against this, however, remains Marx's *vision of a world without classes, of a "true" communism without exploitation and state tyranny, and that is a dream-world we should not abandon, in favor of the "liberal world order" that "triumphed" over fascisms in World War II,* and the Soviet perversion of communism, leading to socialism without a human face in much of the post-World-War II "communist world":[190]

> The development of communism ... expresses the ... the decomposition of the liberal world ... and ... violence can only be transcended in the violent creation of a new order ... *Everything depends on a fundamental decision not just to understand the world but to change it, and to join up with those who are changing the world as a spontaneous development in their own lives.*

Unfortunately, according to Merleau-Ponty, the "fate of history" has a way of subverting the best intentions of individuals and classes who would try to bend it to their wills, especially in the infamous Stalinist "show trials," where:[191]

> a *drama transpires that is rooted in the ... structures of human action, the real tragedy of historical contingency ... A dialectic whose course is not entirely foreseeable can transform a man's intentions into their opposite ... as Napoleon said, and ... Bukharin ..., "fate is politics,"—destiny here not being a fatum already written down unbeknown to us, but the collision in the very heart of history between contingency and the event* ... Man can neither suppress his nature as freedom ... nor question ... history's tribunal, *since in acting he has ... the fate of humanity.*

"The fate of humanity" is "history's tribunal" for Merleau-Ponty (as well as for me!). Following Hegel, the ultimate judgment that "world history if the final verdict," is, not merely our "political fate," according to Napoleon, Bukharin, and Merleau-Ponty, but is also the revelation that "the real tragedy of historical contingency" is another name for "the fate of humanity."[192] And, for Merleau-Ponty:[193]

> The Terror of History culminates in Revolution and History is Terror because there is contingency ... There might be an absolute truth ...

> if ever there was an end to history and the world … a finished world … But … *we are actors in an open history, our praxis introduces the element of construction rather than knowledge as an ingredient of the world, making the world not simply an object of contemplation but something to be transformed.* What we cannot imagine is a consciousness without a future and a history with an end. Thus, as long as there are men, the future will be open.

But perhaps the very "contingency" of history does not ineluctably lead either to (violent) Revolution or to Terror, since by its very nature history is undetermined and revolutions may be (mainly) non-violent, as in the relatively peaceful disintegration of the former Soviet Union and the emergence of, at least nominally, liberal constitutional democracies in its former Eastern European satellite nations and in many other once fully autocratic countries as well.

The World Without and Beyond Merleau-Ponty

Yes, Monsieur Merleau-Ponty, the human world, with all its "order and disorder"—and whose world market, the dynamics of which, as you yourself noted, were most perceptively described a century-and-a-half ago by Karl Marx—must not merely be contemplated but also has to be transformed.[194] Although the "proletarian socialist" revolutions in Russia about a century ago and in China more than 70 years ago have engendered more terror than humanism, for Merleau-Ponty:[195]

> The decline of proletarian humanism … still validates a critique of the present and alternative humanisms … it *cannot be surpassed. Even if it is incapable of shaping world history, it remains powerful enough to discredit other solutions* … Marxism … *is the simple statement of those conditions without which there would be neither any humanism, in the sense of a mutual relation between men, nor any rationality in history … Marxism … is* <u>the</u> *philosophy of history and to renounce it is to dig the grave of Reason in history.*

Well-stated, however, *any contemporary or future version of post-proletarian revolutionary praxis should be as nonviolent as possible since, in a nuclear world, our history might indeed be ended by any agent's use of weapons of mass destruction, and the world as it has existed since the dawn of civilization might be destroyed. Accordingly, the non-violent creation of a new postliberal, and post-actually-existing-communist, world remains a task for the 21st century.*

Merleau-Ponty pointed us in the right direction; it is up to succeeding generations to bring to fruition his vision of a perceivable world without unnecessary violence and exploitation.[196] Will the rulers of the world within which we now live do so?

As Merleau-Ponty, following Trotsky, said regarding this world's possible future, either toward a Marxist classless society or to a neo-Hegelian authoritarian state: *"the world might evolve in the direction of a monopolistic and authoritarian capitalism …"* This rings true as the arc of history at the moment seems to be bending more toward authoritarianism than to global harmony and equity. Will this continue? As Merleau-Ponty, in a quintessentially existential formulation declares, "*The world in which we live is ambiguous on this question.*"[197]

Merleau-Ponty's vision of the latter possibility of a world-future has remained incomplete, as his incarnate subjectivity prematurely ceased to exist in the world he had so acutely perceived and dissected. His intellectual projects both to further elaborate a phenomenology of the actual world we perceive and also to create a viable alternative to the political world in which we work and struggle remain illuminating but inchoate indicators of what might have been had he lived longer.

Many philosophers and political activists have carried on versions of these theoretical and practical projects, especially in continental Europe, but usually without the phenomenological humanism Merleau-Ponty deemed essential for completing them.[198] But "across the pond" in Great Britain, there arose a competing, vision of the world, one largely rooted in linguistic analysis and the philosophy of language—to which Merleau-Ponty himself made significant but largely ignored contributions during his own "linguistic turn" in such late works as *The Visible and the Invisible*[199]—and whose impact has far exceeded that of existential phenomenology, at least among English-speaking philosophers. It is to the world and words of Ludwig Wittgenstein, the most influential of all "analytic" philosophers, quite possibly *malgré lu*i, that I now turn.

Notes

1 For overviews of Husserl's life, major works, and phenomenology more generally, see the IEP article by Marianne Sawicki, https://iep.utm.edu/husserl/; Christian Beyer, "Edmund Husserl," *The Stanford Encyclopedia of Philosophy* (Summer 2018 Edition), Edward N. Zalta, ed., https://plato.stanford.edu/archives/sum2018/entries/husserl/; *Internet Encyclopedia of Philosophy* https://iep.utm.edu/phenom/; and David Woodruff Smith, "Phenomenology," The *Stanford Encyclopedia of Philosophy* (Summer 2018 Edition), Edward N. Zalta, ed., https://plato.stanford.edu/archives/sum2018/entries/phenomenology/.

2 Due to limited access to research materials during the pandemic, my main source of Husserl's works has been Donn Welton, ed., *The Essential Husserl Basic Writings in Transcendental Phenomenology* (Bloomington and Indianapolis, IN: Indiana University Press, 1999), abbreviated as EH. Major works by Husserl in English translations include: "Philosophy as Rigorous Science," trans. in Q. Lauer (ed.), *Phenomenology and the Crisis of Philosophy* (New York: Harper, 1965); *Formal and Transcendental Logic*, trans. D. Cairns (The Hague: Nijhoff, 1969); *The Crisis of European Sciences and Transcendental Philosophy*, trans. D. Carr (Evanston, IL: Northwestern University Press, 1970); Logical Investigations,

2nd revised edition, trans. J.N. Findlay (London: Routledge, 1973); *Experience and Judgement*, trans. J.S. Churchill and K. Ameriks (London: Routledge, 1973); *Ideas Pertaining to a Pure Phenomenology and to a Phenomenological Philosophy—Third Book: Phenomenology and the Foundations of the Sciences*, trans. T.E. Klein and W.E. Pohl (Dordrecht: Kluwer, 1980); *Ideas Pertaining to a Pure Phenomenology and to a Phenomenological Philosophy—First Book: General Introduction to a Pure Phenomenology*, trans. F. Kersten (The Hague: Nijhoff, 1982. *Abbreviated Ideas 1);* Cartesian Meditations, trans. D. Cairns (Dordrecht: Kluwer, 1988); *Ideas Pertaining to a Pure Phenomenology and to a Phenomenological Philosophy—Second Book: Studies in the Phenomenology of Constitution*, trans. R. Rojcewicz and A. Schuwer (Dordrecht: Kluwer, 1989. *Abbreviated Ideas 2); On the Phenomenology of the Consciousness of Internal Time*, trans. J.B. Brough (Dordrecht: Kluwer, 1990); *Early Writings in the Philosophy of Logic and Mathematics*, trans. D. Willard (Dordrecht: Kluwer, 1994); and *The Crisis of European Sciences and Transcendental Phenomenology: An Introduction to Phenomenological Philosophy*, trans. David Carr (Evanston, IL: Northwestern University Press, 1970; abbreviated The Crisis).

Secondary sources on Husserl include: David Bell, *Husserl* (London: Routledge, 1990); Rudolf Bernet, Iso Kern, and Eduard Marbach, *Introduction to Husserlian Phenomenology* (Evanston, IL: Northwestern University Press, 1993); David Carr, *Interpreting Husserl* (Dordrecht: Nijhoff, 1987); Jacques Derrida, Edmund Husserl's 'Origin of Geometry,' trans. J.P. Leavy (New York: Harvester Press, 1978); Hubert Dreyfus (ed.), Husserl, Intentionality, and Cognitive Science (Cambridge, MA: MIT Press, 1982); Emmanuel Levinas, The Theory of Intuition in Husserl's Phenomenology (Evanston, IL: Northwestern University Press, 1973); J.N. Mohanty and William McKenna (eds.), Husserl's Phenomenology: A Textbook (Lanham, MD: University Press of America, 1989); Barry Smith and David Woodruff Smith (eds.), The Cambridge Companion to Husserl (Cambridge: Cambridge University Press, 1995); Robert Sokolowski, (ed.) Edmund Husserl and the Phenomenological Tradition (Washington, DC: Catholic University of America Press, 1988); and Dan Zahavi, *Husserl's Phenomenology* (Stanford, CA: Stanford University Press, 2003).

3 See https://en.wiktionary.org/wiki/phenomenon.

4 In the opening pages of Volume II of his *Logical Investigations*, Husserl gives his famous phenomenological battle-cry, back to the things themselves:

> Our great task is now to bring the Ideas of logic, the logical concepts and laws, to epistemological clarity and definiteness. Here phenomenological analysis must begin. Logical concepts, as valid thought-unities, must have their origin in intuition: they must arise out of an ideational intuition founded on certain experiences, and must admit of indefinite reconfirmation, and of recognition of their self-identity, on the re-performance of such abstraction … we can absolutely not rest content with "mere words," i.e. with a merely symbolic understanding of words, meanings inspired only by remote, confused, inauthentic intuitions- if by any intuitions at all- are not enough. We must go back to the "things themselves." We desire to render self-evident in fully-fledged intuitions that what is here given in actually performed abstractions is what the word-meanings in our expression of the law really and truly stand for.
>
> (Husserl as cited in: https://en.wikipedia.org/wiki/Early_phenomenology)

5 See David Woodruff Smith, "Phenomenology," *op. cit.*

6 Phenomenology is to be distinguished from phenomenalism, an epistemological stance claiming that all statements about physical objects are synonymous with statements about persons having certain sensations, or sense-data. The philosopher Bishop George Berkeley, for example, was a phenomenalist but not a phenomenologist.

7 Edmund Husserl, "The world of the natural attitude and my surrounding world" Par. 27–8, in *Ideas Pertaining to a Pure Phenomenological Philosophy, First Book*, trans. F. Kersten (The Hague: Martinus Nijhoff, 1982), Vol. 2. *Ideas* 1. ID1. In EH, 60–2.

8 According to Sawicki,

> psychologism for Husserl is the error of collapsing the normative or regulative discipline of logic down onto the merely descriptive discipline of psychology. It would make mental operations (such as combination) the source of their own regulation. The "should" of logic, that utter necessity inhering in logical inference, would become no more than the "is" or facticity of our customary thinking processes, empirically described … Psychologism, Husserl charges, would place logical inferences on the same plane with mental operations, and this would make even mathematics into a branch of psychology. Indeed, math and logic do have structures that are isomorphic to those of mental operations, such as combination and distinction. But given that similarity, how then would one distinguish the regulation of any of these processes from the description of it? Under psychologism, there's no way. But Husserl makes the distinction in a way that also shows how regulation (i.e., the laws of logic) comes from elsewhere than the plane of mental activity.
>
> (IEP, *op. cit.*)

See also: Martin Kusch, "Psychologism," *The Stanford Encyclopedia of Philosophy* (Spring 2020 Edition), Edward N. Zalta, ed., https://plato.stanford.edu/archives/spr2020/entries/psychologism.

9 *Ibid.*

10 Husserl, *Ideas I*, 32. "The phenomenological Epoché," in, EH, *op. cit.*, 65.

11 Husserl, "The Structure of Intentionality," *Ideas* 1, par. 88, in EH, *op. cit.*, 88.

12 Husserl, *Ideas*, *op. cit.*, *passim*, and Gaston, *op. cit.*, 50.

13 See Gaston, *op. cit.*, 62 and 65, where these passages from *The Crisis* and *Ideas* are cited. As Gaston himself says: "For Husserl, the path to the one true world is transformed by the affirmation of a new range of constituted and possible worlds." This raises "the possibility of coexisting virtual worlds we see today … Phenomenology never questions the concept of world itself; for Heidegger, the world will always be a problem." *Ibid.*

14 Husserl, *Ideas II*, in EH, "Objective Reality, Spatial Orientation, and the Body," par. 1., 163–4.

15 See Husserl, *Ideas II, passim,* as well as Gaston, *op. cit.*, 53–4.

16 For the concept of the "Lifeworld," see https://en.wikipedia.org/wiki/Lifeworld. Also see Beyer, "Edmund Husserl," *Stanford Encyclopedia of Philosophy*, *op. cit.*

17 Husserl, The Crisis, "Elements of a Science of the Life-World," par. 33, in EH, *op. cit.*, 363.

18 *Ibid.*, 368–9.

19 *Ibid.*, 375–6.

20 *Res* is Latin for (the) thing or object. Husserl, *Ideas* 1, "The Basic Approach of Phenomenology," in EH, 82.

21 Accessible overviews of Heidegger's life and thought include articles in the *Internet* and *Stanford Encyclopedias of Philosophy*, respectively: https://iep.utm.edu/heidegge/, and Michael Wheeler, "Martin Heidegger," *The Stanford Encyclopedia of Philosophy* (Fall 2020 Edition), Edward N. Zalta, ed., https://plato.stanford.edu/archives/fall2020/entries/heidegger/.

Heidegger's collected works in German, the *Gesamtausgabe,* are published by Vittorio Klostermann in Frankfurt am Main, Germany. It was started by Heidegger himself and is not completed yet. There are four series, (I) Publications, (II) Lectures, and (III) Unpublished material, lectures, and notes, and (IV), *Hinweise und Aufzeichnungen* ("Notes and Sketches").

I've used the following translations into English of Heidegger's works: *Being and Time* (abbreviation BT), trans. John Macquarrie and Edward Robinson (New York: Harper & Row, 1962) (first published in 1927). [NB: page numbers refer to the Macquarrie and Robinson translation]; David F. Krell (ed.), *Martin Heidegger: Basic Writings (BW)*, revised and expanded edition (London: Routledge, 1993); Heidegger, *Contributions to Philosophy (Of the Event) (CP)* trans. Richard Rojcewicz and Daniela Vallega-Neu (Bloomington, IN: Indiana University Press, 2012); *History of the Concept of Time (HCT)*, trans. T. Kisiel, (Bloomington, IN: Indiana University Press, 1985); *An Introduction to Metaphysics*, trans R. Manheim (New York: Doubleday, 1961); *Kant and the Problem of Metaphysics (KPM)*, trans. R. Taft (Bloomington, IN: Indiana University Press, 1997); "Letter to the Rector of Freiburg University, November 4, 1945," in K.A. Moehling, *Martin Heidegger and the Nazi Party: An Examination*, Ph.D. Dissertation, Northern Illinois University, 1972, translated by and reprinted in Richard. Wolin (ed.), *The Heidegger Controversy: a Critical Reader* (Cambridge, MA: MIT Press, 1993); "Only a God can Save Us," *Der Spiegel's* Interview with Martin Heidegger, *Der Spiegel*, May 31, 1976, trans. M.O. Alter and J.D. Caputo and published in *Philosophy Today* XX (4/4): 267–85, translation reprinted in R. Wolin, *op. cit.*; *The Basic Problems of Phenomenology*, trans. A. Hofstadter (Bloomington, IN: Indiana University Press, 1982); and "The Self-Assertion of the German University," trans. W.S. Lewis, in R. Wolin, *op. cit*.

From the immense and growing secondary literature on Heidegger, the following texts are useful: T. Carman, *Heidegger's Analytic: Interpretation, Discourse, and Authenticity in 'Being and Time,'* (Cambridge: Cambridge University Press, 2003); Rudolf. Carnap, "The Elimination of Metaphysics Through Logical Analysis of Language," in A.J. Ayer (ed.), *Logical Positivism* (Glencoe, Scotland: Free Press, 1959); T. Clark, *Routledge Critical Thinkers: Martin Heidegger* (London: Routledge, 2001); S. Critchley, *Continental Philosophy: a Very Short Introduction* (Oxford: Oxford University Press, 2001); H.L. Dreyfus, *Being-in-the-World: A Commentary on Heidegger's Being and Time, Division I* (Dreyfus) (Cambridge, MA: MIT Press, 1991); H.L. Dreyfus, H.L. and H. Hall, (eds.), *Heidegger: a Critical Reader* (Oxford: Blackwell, 1992); H.L. Dreyfus, and M. Wrathall, M. (eds.) *Heidegger Reexamined* (4 Volumes) (London: Routledge, 2002); P. Gorner, *Heidegger's* Being and Time: *an Introduction* (Cambridge: Cambridge University Press, 2007); C. Guignon, *Heidegger and the Problem of Knowledge* (Indiana, IN: Hackett, 1983); C. Guignon (ed.), *The Cambridge Companion to Heidegger* (Cambridge:

Cambridge University Press, 1993); T. Kisiel, *The Genesis of Heidegger's Being and Time* (Berkeley, CA: University of California Press, 1993); C. Macann, C. (ed.), *Heidegger: Critical Assessments* (4 Volumes), (London: Routledge, 1992); W. Marx, *Heidegger and the Tradition*, translated by T. Kisiel and M. Greene (Evanston, IL: Northwestern University Press, 1970); S. Mulhall, *Routledge Philosophy Guidebook to Heidegger and 'Being and Time.'* (second edition), London: Routledge, 2005); F. Olafson, *Heidegger and the Philosophy of Mind* (New Haven, CT: Yale University Press, 1987); G. Pattison, *The Later Heidegger* (London: Routledge, 2000); O. Pöggeler, *Martin Heidegger's Path of Thinking*, trans. D. Magurshak and S. Barber (Atlantic Highlands, NJ: Humanities Press International, 1993); R. Polt, *Heidegger: an Introduction* (London: Routledge, 1999); R. Rorty, *Essays on Heidegger and Others (Philosophical Papers, Volume 2)*, Cambridge: Cambridge University Press, 1991); M. Wrathall, *How to Read Heidegger* (London: Granta, 2003); J. Young, *Heidegger's Later Philosophy* (Cambridge: Cambridge University Press, 2002).

Regarding Heidegger's Nazism and his possible philosophical affinity with their ideology, see T. Adorno, *The Jargon of Authenticity*, trans. Knut Tarnowski and Frederic Will (Evanston, IL: Northwestern University Press, 1973); V. Farias, V., *Heidegger and Nazism* (Philadelphia, PA: Temple University Press, 1989); G. Neske and E. Kettering, *Martin Heidegger and National Socialism: Questions and Answers*, trans. Lisa Harries (New York: Paragon House, 1990); H. Ott, *Martin Heidegger: a Political Life* (London: Harper Collins, 1993); T. Rockmore, *On Heidegger's Nazism and Philosophy* (London: Wheatsheaf, 1992); T. Rockmore and J. Margolis (eds.) *The Heidegger Case: on Philosophy and Politics* (Philadelphia, PA: Temple University Press, 1992); H. Sluga, *Heidegger's Crisis: Philosophy and Politics in Nazi Germany* (Cambridge, MA: Harvard University Press, 1993); R. Wolin, *The Politics of Being: The Political Thought of Martin Heidegger* (Cambridge, MA: MIT Press, 1990); J. Young, *Heidegger, Philosophy, Nazism* (Cambridge: Cambridge University Press, 1997); and https://en.wikipedia.org/wiki/Martin_Heidegger_and_Nazism.

22 Robert Pippin, Andrew Feenberg, and Charles Webel (eds.), *Herbert Marcuse: Critical Theory and the Promise of Utopia* (Greenwood CT, New York, & London: Greenwood Press and Macmillan Publishers, 1987).

23 In his *Ponderings*, the *Black Notebooks*, Heidegger, in a characteristically ambiguous manner, declared

> How National Socialism can never be the principle of a philosophy but must always be placed under philosophy as the principle. How, nevertheless, National Socialism can take up quite definite positions and thus can coeffectuate a new basic posture toward being! But even this only under the presupposition that National Socialism knows itself in its limits—i.e., realizes that it is true only if it is able—only if it is in condition—to prepare and set free an original truth.
>
> (BN, *op. cit.*, 139)

Alfred Rosenberg was the head of the NSDAP (Nazi Party) Office of Foreign Affairs during their rule of Germany (1933–45), and led the *Amt Rosenberg* ("Rosenberg's bureau"), an official Nazi body for cultural policy and surveillance. During World War II, Rosenberg was the head of the Reich Ministry for the Occupied Eastern Territories (1941–45). After the war, he was convicted of crimes against peace; as well as of planning, initiating, and waging wars of aggression; war crimes; and crimes against humanity, at the Nuremberg trials in 1946. He was sentenced to death and executed later that year. Rosenberg was the author of a seminal work of Nazi ideology, *The Myth of the Twentieth Century* (1930), and is considered one of the main fabricators of such key Nazi ideological creeds as its racial theory, the persecution of the Jews, *Lebensraum* (increased "Living Space" for Germans via conquest of other lands, principally in Eastern Europe), the abrogation of the Treaty of Versailles, opposition to what was the Nazis deemed "degenerate" modern art, and their rejection of traditional Christianity. See https://en.wikipedia.org/wiki/Alfred_Rosenberg.

24 For the vow of allegiance to Hitler, see https://en.wikipedia.org/wiki/Vow_of_allegiance_of_the_Professors_of_the_German_Universities_and_High-Schools_to_Adolf_Hitler_and_the_National_Socialistic_State.

The "Bekenntnis der Professoren an den Universitäten und Hochschulen zu Adolf Hitler und dem nationalsozialistischen Staat," translated into English as the "Vow of allegiance of the Professors of the German Universities and High-Schools to Adolf Hitler and the National Socialistic State" was a document issued in November 1933 in Leipzig, Germany. This was part of the Nazi purge to remove from the civil service academics and other employees with Jewish ancestry, starting with the law passed in April 1933. This document was signed by those in support of the Third Reich. Heidegger, in his inaugural lecture in May 1933 as "Rektor," and who was later in October appointed "Führer of the university" in Freiburg, proclaimed (translated):

> The much celebrated "academic freedom" is being banished from the German university; for this freedom was not genuine, since it was only negative. It meant primarily freedom from concern, arbitrariness of intentions and inclinations, lack of restraint in what was done and left undone. The concept of freedom of the German is now brought back to its truth.

25 See https://en.wikipedia.org/wiki/Black_Notebooks. The *Black Notebooks* (German: *Schwarze Hefte*) are a set of notebooks written by Heidegger between 1931 and the late 1940's. Originally, a set of small notebooks with black covers in which Heidegger jotted his "notes and sketches," they were subsequently organized into a 1,000-page transcript. Peter Trawny is editing the transcripts and publishing them in the Heidegger *Gesamtausgabe* (*GA*). The first transcript was published in 2014. As of 2021, 14 notebooks have been published, covering the years 1931–41 (*GA*, 94–6). The notebooks from 1942 to 1945 are in private possession, but have been prepared for publication. For a commentary on *The Black Notebooks*, see Eric S. Nelson, "Heidegger's Black Notebooks: National Socialism, Antisemitism, and the History of Being," December 2017, page proofs available at www.researchgate.net/publication/321822510.

26 Donatella De Cesare, *Heidegger and the Jews: The Black Notebooks* (Medford, MA: Polity Press, 2018). Also, see Adam Kirsch's review in the New York Times of Emmanuel Faye's book *Heidegger: The Introduction of Nazism Into Philosophy* (New Haven, CT: Yale University Press, 2009): www.nytimes.com/2010/05/09/books/review/Kirsch-t.html.

27 Adorno, *Jargon of Authenticity, op. cit.* and Jiirgen Habermas, *The Philosophical Discourse of Modernity. Twelve Lectures* (Cambridge: Polity Press, 1987). Also see Rudi Visker, "Habermas on Heidegger and Foucault Meaning and Validity in The Philosophical Discourse of Modernity," in *Radical Philosophy* 61, Summer 1992, 15–22, available at: www.radicalphilosophyarchive.com/article/habermas-on-heidegger-and-foucault/.

28 See https://en.wikipedia.org/wiki/Martin_Heidegger_and_Nazism. For Arendt's view of Heidegger, whom she called "the fox," see Michael T. Jones: "Heidegger the Fox: Hannah Arendt's Hidden Dialogue," *New German Critique,* No. 73, (Winter, 1998), 164–92.
29 For Kant, see my discussion earlier in this book.
30 Heidegger, BN 1, *op. cit.*, 5.
31 See Heidegger, *Ponderings*, aka BN, *op. cit.*, 188.
32 Along with Sartre's *Being and Nothingness*, which many find derivative of Heidegger, and Merleau-Ponty's *The Phenomenology of Perception,* both of which will be discussed later in this chapter.
33 Especially in Rorty's books *Essays on Heidegger and Others: Philosophical Papers* II (Cambridge: Cambridge University Press, 1991), and *Truth and Progress: Philosophical Papers* III (Cambridge: Cambridge University Press, 1998).
34 In the first division of *Being and Time*, 122–3, Heidegger provides a lengthy exposition and critique of Descartes' epistemology and ontology, focusing on the latter's definition of the world, which Heidegger viewed as ontologically *res extensa* (spatially "extended thing"), and grounded in the "cogito sum," and which Heidegger attempted phenomenologically to "destroy," by overcoming the Cartesian split between *res cogitans* (the "thinking thing" called the mind) and *res extensa* including our own bodies).
35 A somewhat abbreviated glossary of some key terms in Heidegger's *Being and Time* is available at: https://en.wikipedia.org/wiki/Heideggerian_terminology. More detailed explanations are dispersed through much of the secondary literature on Heidegger, especially Hubert Dreyfus's *Being-in-the-World*, *op. cit.*, *passim.*
36 Heidegger, *Being and Time* (BT), *op. cit.*, 19.
37 *Ibid.*, 32.
38 *Ibid.*, 33.
39 BN, *op. cit.*, 173.
40 BT, *op. cit.*, 78.
41 *Ibid.*, 91–4.
42 *Ibid.*, 94.
43 *Ibid.*, 106–7.
44 *Ibid.*, 133–5.
45 *Ibid.*, 149–68.
46 *Ibid.*, 364.
47 *Ibid.*, 220.
48 *Ibid.*, 230.
49 *Ibid.*, 390.
50 *Ibid.*, 250.
51 *Ibid.*, 277, 281, 310, and 346.
52 *Ibid.*, 329, 331, and 342–4.
53 *Ibid.*, 346–7. See Jean-Paul Sartre, *Situations*, trans. Benita Eisler (New York: George Braziller, 1965).
54 *Ibid.*, 368, 374, and 381.
55 *Ibid.*, 414–17.
56 *Ibid.*, 417.
57 *Ibid.*
58 *Ibid.*, 440.
59 See *Ibid.*, 93. Also see: Ravindra Raj Singh. *Heidegger, World, and Death* (New York: Lexington Books, 2013), and https://en.wikipedia.org/wiki/Heideggerian_terminology/.

60 *Ponderings*. BN, *op. cit.*, 135.
61 See Martin Heidegger, *Contributions to Philosophy (Of the Event)*, trans. Richard Rojcewicz and Daniela Vallega-Neu (Bloomington, IN: Indiana University Press, 2012).
62 BT, *op. cit.*, 204.
63 Heidegger, "Letter on Humanism," in *Heidegger Basic Writings*, (BW) *op. cit.*, 237.
64 Heidegger, "What Calls for Thinking?" *Ibid.*, 424–5.
65 Heidegger is referring to Hitler's designation as German Chancellor … *Der Spiegel* Interview with Martin Heidegger," *op. cit.*
66 *Ibid.*
67 *Ibid.*
68 Heidegger, "Letter on Humanism," *op. cit.*, 170–4.
69 *Ibid.*, 174–5.
70 *Ponderings, BN*, *op. cit.*, 153–60.
71 *Ibid.*
72 *Ibid.*
73 *Ibid.*
74 "Letter on Humanism," BW, *op. cit.*, 229 and 243. For an analysis of feeling "at home," of "dwelling close to the earth," and of existential homelessness, see Tony Fry, "Homelessness: A Philosophical Architecture," *Design Philosophy Papers*, 3:3, 2005, 191–203, http://dx.doi.org/10.2752/144871305X13966254124798/.
75 "Letter on Humanism," *BW*, *op. cit.*, 249 and 252.
76 *Ibid.*
77 "Only a God can Save Us," *Der Spiegel's* Interview with Martin Heidegger, *Der Spiegel* (*Spiegel*), May 31, 1976, trans. M.O. Alter and J.D. Caputo and published in *Philosophy Today* XX (4/4): 267–85.
78 *Ibid.*
79 *Ibid.*
80 *Ibid.*
81 Martin Heidegger, *Ponderings II–VI, Black Notebooks 1931–1938*, (*BN*), trans. Richard Rojcewicz (Bloomington and Indianapolis, IN: Indiana University Press, 2016).
82 See Rudolf Carnap, "The Elimination of Metaphysics Through Logical Analysis of Language," *op. cit.*, original: Carnap, "Uberwindung der Metaphysik durch logische Analyse der Sprache." *Erkenntnis* 1931, 2, 219–41, and "Heidegger and Carnap on the Overcoming of Metaphysics," https://people.ucsc.edu/~abestone/papers/uberwindung.pdf.
83 For the influence of Existentialism and Hegel on Martin Luther King Jr., see www.critical-theory.com/martin-luther-king-jr-hegel/.
84 Primary sources by Sartre are: *Being and Nothingness*, abbreviated BN, trans. and Introduction. Hazel E. Barnes (New York: Washington Square Press, 1984), print; online: http://yunus.hacettepe.edu.tr/~cin/Being%20And%20Nothingness/Being%20and%20Nothingness%20-%20Sartre.pdf. NB: even though the print and digital texts are identical, the pagination differs significantly. So I will refer to the pagination in the following way, with the online text (which is overall significantly shorter) going first, and then the print edition, e.g., 5/10 (online/print, respectively). *Psychology of the Imagination*, trans. Bernard Frechtman (London: Methuen, 1972); *Sketch for a Theory of the Emotions*, trans. Philip Mairet (London: Methuen, 1972); *The Transcendence*

of the Ego: An Existentialist Theory of Consciousness, trans. and ed. Forrest Williams and Robert Kirkpatrick (New York: Noonday, 1957); *Nausea*, Lloyd Alexander and H.P. van den Aardweg trans., Hayden Carruth, Introd., (New York: New Directions Press, 1958); *Existentialism and Humanism*, abbreviated EH, trans. Philip Mairet (London: Methuen, 1973), shorter version available at http://homepages.wmich.edu/~baldner/existentialism.pdf; *Critique of Dialectical Reason 1: Theory of Practical Ensembles*, abbreviation CDR 1, transl. Alan Sheridan-Smith, ed. Jonathan Rée (London: Verso, 2004); *Critique of Dialectical Reason*, vol. 2, *The Intelligibility of History*, abbreviation CDR 2, tr. Quintin Hoare, foreword by Frederic Jameson (London: Verso, 2006), both of which are available online at: https://libcom.org/library/critique-dialectical-reason-vol-1-2-jean-paul-sartre; *The Problem of Method*, trans. Hazel E. Barnes (London: Methuen, 1964); *The Words*, trans. Bernard Frechtman (New York: Braziller, 1964), *Anti-Semite and Jew: An Exploration of the Etiology of Hate,* George J. Becker, trans., Michael Walzer, Preface (New York: Schocken, 1995); and *The Family Idiot*, trans. Carol Cosman. 5 vols. (Chicago, IL: University of Chicago Press, 1971–94).

85 For concise overviews of Sartre's life and works, see Thomas Flynn, "Jean-Paul Sartre," *The Stanford Encyclopedia of Philosophy* (Fall 2013 Edition), Edward N. Zalta, ed., https://plato.stanford.edu/archives/fall2013/entries/sartre/; Wilfrid Desan, www.britannica.com/biography/Jean-Paul-Sartre; and IEP, https://iep.utm.edu/sartre-ex/.

Other secondary sources and related books by Simone de Beauvoir include: Joseph Catalano, *A Commentary on Jean-Paul Sartre's Being and Nothingness* (Chicago, IL: University of Chicago Press, 1980) and *A Commentary on Jean-Paul Sartre's Critique of Dialectical Reason,* vol. 1 (Chicago, IL: University of Chicago Press, 1986); Simone de Beauvoir, *The Force of Circumstances*, tr. Richard Howard (New York: G. P. Putnam's Sons, 1964-65), *Adieux: A Farewell to Sartre*, tr. P. O'Brian (New York: Pantheon, 1984), and *Letters to Sartre* tr. and ed. Quentin Hoare (New York: Arcade, 1991); Andrew Dobson, *Jean-Paul Sartre and the Politics of Reason* (Cambridge: Cambridge University Press, 1993); Joseph P. Fell, *Heidegger and Sartre: An Essay on Being and Place* (New York: Columbia University Press, 1979); Thomas R. Flynn, *Sartre and Marxist Existentialism: The Test Case of Collective Responsibility* (Chicago, IL: University of Chicago Press, 1984) and *Sartre, Foucault and Historical Reason*, vol. 1 *Toward an Existentialist Theory of History* (Chicago, IL: University of Chicago Press, 1997); Christina Howells, ed. *Cambridge Companion to Sartre* (Cambridge: Cambridge University Press, 1992); Francis Jeanson, *Sartre and the Problem of Morality*, tr. Robert Stone (Bloomington, IN: Indiana University Press, 1981); William Leon McBride, *Sartre's Political Theory* (Bloomington, IN: Indiana University Press, 1991) and ed., *Sartre and Existentialism*, 8 vols. (New York: Garland, 1997); Paul Arthur Schilpp, ed., *The Philosophy of Jean-Paul Sartre* (La Salle, IL: Open Court, 1981); William Schroeder, *Sartre and His Predecessors* (Boston, MA: Routledge & Kegan Paul, 1984): Charles Taylor, *The Ethics of Authenticity* (Cambridge, MA: Harvard University Press, 1991); Adrian Van den Hoven, and Andrew Leak, eds., *Sartre Today. A Centenary Celebration* (New York: Berghahn Books, 2005); Jonathan Webber, ed., *Reading Sartre: On Phenomenology and Existentialism* (London: Routledge, 2011).

86 Sartre, *Being and Nothingness* (BN), "Introduction: The Pursuit of Being," *op. cit.*, xlv–lvii/3–20.

87 The English word "nihilate" was first used by Helmut Kuhn in his *Encounter with Nothingness.* To "nihilate" (*neantir*), for Sartre, is to encase with a shell of non-being. BN, *op. cit.*, 632/804.
88 Toward the end of *Being and Nothingness*, Sartre depicts our "human reality." *Ibid.*, 615/784.
89 *Ibid.*, 17–18/50–2.
90 For Sartre, "Consciousness is prior to nothingness and … 'derived' from being." BN, *op. cit.*, lvi/16.
91 *Ibid.*, li–lii/10–1.
92 *Ibid.*, 639/807.
93 *Ibid.*, 553/707.
94 *Ibid.*, 4/34.
95 *Ibid.*, 104–5.
96 *Ibid.*, 157.
97 *Ibid.*, 115–16/169–70.
98 *Ibid.*, 138/199 and 146/207.
99 *Ibid.*, 181–3/251–3.
100 *Ibid.*, 195/268.
101 *Ibid.*, 198–200/272–5.
102 *Ibid.*, 200/275.
103 *Ibid.*, 202–4/277–9.
104 *Ibid.*, 233/316 and 245/330.
105 *Ibid.*, 267–70/358–61.
106 *Ibid.*, 291–2/387–8.
107 *Ibid.*, 303–7/402–6.
108 *Ibid.*, 306–7/405–6.
109 *Ibid.*, 314/415 and 317–18/419.
110 *Ibid.*
111 *Ibid.*, 325–6 and 328–330/429 and 433–5.
112 *Ibid.*, 352, 359, 369/462–3, 470, and 481.
113 *Ibid.*, 391–2/508–9.
114 *Ibid.*, 423/547–48.
115 *Ibid.*, 433, 439, 444/559, 566–7, and 572.
116 *Ibid.*, 457, 460–1, 463/588–9, 593, and 596.
117 *Ibid.*, 477/614.
118 *Ibid.*
119 Sartre, BN, *op. cit.*, 486–7/625–6.
120 *Ibid.*, 509–10/653–4.
121 *Ibid.*, 530/679, 532/681–3, and 546–7/697–9.
122 *Ibid.*, 549/702.
123 *Ibid.*, 575–8/735–6.
124 *Ibid.*, 615/784.
125 *Ibid.*, 580/741.
126 *Ibid.*, 617–18, 785–6.
127 *Ibid.*, 625–6/795–6.
128 *Ibid.*, 627–8/798.
129 Sartre, *Existentialism is a Humanism*, EH, trans., Philip Mairet, reproduced under "Fair Use" provisions. Available at: http://homepages.wmich.edu/~baldner/existentialism.pdf. Not paginated. In this work, Sartre famously declared: "I believe I am becoming an existentialist."

130 *Ibid.*
131 *Ibid.*
132 *Ibid.*
133 See Thomas Flynn, "Jean-Paul Sartre," *The Stanford Encyclopedia of Philosophy, op. cit.*; Wilfrid Desan, *Jean-Paul Sartre, op. cit.*, and IEP, *Sartre, op. cit.*
134 On praxis, see *https://en.wikipedia.org/wiki/Praxis*(process).
135 See Fredric Jameson, "Foreword," CDR 1, *op. cit.*, xiii–xxxiii.
136 See Thomas Flynn, *ibid.*
137 Good, concise online overviews of Merleau-Ponty's life, works, and influence are: Ted Toadvine, "Maurice Merleau-Ponty," *Stanford Encyclopedia of Philosophy* (SEP), Spring 2019 Edition, Edward N. Zalta, ed., https://plato.stanford.edu/archives/spr2019/entries/merleau-ponty/; and Jack Reynolds, *Internet Encyclopedia of Philosophy* (IEP), https://iep.utm.edu/merleau/.

Primary works of Merleau-Ponty in English translations also include: *Phenomenology of Perception*, Donald Landes, trans., (London: Routledge, 2012); *Adventures of the Dialectic*, Joseph Bien (trans.), (Evanston, IL: Northwestern University Press, 1973, AD); *Humanism and Terror: An Essay on the Communist Problem*, John O'Neill (trans.) (Boston, MA: Beacon Press, 1969, HT); *In Praise of Philosophy*, John Wild and James Edie (trans.), (Evanston, IL: Northwestern University Press, 1963); *Signs*, Richard McCleary (trans.), (Evanston, IL: Northwestern University Press, 1964); *The Primacy of Perception*, James Edie (ed.), (Evanston, IL: Northwestern University Press); *The Prose of the World*, John O'Neill (trans.), (Evanston, IL: Northwestern University Press, 1973); *Consciousness and the Acquisition of Language*, Hugh J. Silverman (trans.), (Evanston, IL: Northwestern University Press); *The Structure of Behavior*, trans. Fischer, (London: Methuen, 1965, SB in text); *The Visible and the Invisible*, trans. Alfonso Lingis, (Evanston, IL: Northwestern University Press, 1968, VI); *The Essential Writings of Merleau-Ponty* (EW), edited by Alden L. Fisher (New York: Harcourt, Brace & World, Inc., 1969), available at: www.bard.edu/library/arendt/pdfs/Merleau-Ponty-EssentialWritings.pdf; and *Maurice Merleau-Ponty, Texts and Dialogues, On Philosophy, Politics and Culture*, Hugh Silverman and James Barry Jr. (eds), (New Jersey, NJ: Humanities Press, 1992).

Selected secondary works on Merleau-Ponty include: David Abram, *The Spell of the Sensuous* (New York: Vintage Books, 1996); Diana Coole, *Merleau-Ponty and Modern Politics After Anti-Humanism*, (Lanham, MD: Rowman and Littlefield, 2007): M.C. Dillon, *Merleau-Ponty's Ontology*, 2nd edition, (Evanston, IL: Northwestern University Press, 1997); Rosalyn Diprose and Jack Reynolds (eds), *Merleau-Ponty: Key Concepts*, (Stocksfield: Acumen, 2008); Galen Johnson, *The Retrieval of the Beautiful: Thinking Through Merleau-Ponty's Aesthetics*, (Evanston, IL: Northwestern University Press, 2010); Donald Landes, *Merleau-Ponty and the Paradoxes of Expression*, (London: Bloomsbury, 2013); Scott Marratto, *The Intercorporeal Self: Merleau-Ponty on Subjectivity*, (Albany, NY: SUNY Press, 2012); Dorothea Olkowski and Gail Weiss, 2006, *Feminist Interpretations of Maurice Merleau-Ponty*, (University Park, PA: The Pennsylvania State University Press, 2006); John Sallis, *Phenomenology and the Return to Beginnings*, 2nd edition, (Pittsburgh, PA: Duquesne University Press, 2003); Jean-Paul Sartre,

Jean-Paul, "Merleau-Ponty," in *Situations*, Benita Eisler (trans.), (New York: Braziller, 1965), 227–326; Jon Stewart, Jon (ed.), *The Debate Between Sartre and Merleau-Ponty*, (Evanston, IL: Northwestern University Press, 1998); Evan Thompson, *Mind in Life: Biology, Phenomenology, and the Sciences of Mind*, (Cambridge, MA: Harvard University Press, 2007); Ted Toadvine, ed., *Merleau-Ponty: Critical Assessments of Leading Philosophers*, vols. 1–4, (London: Routledge, 2006); Louse Westling, Louise, *The Logos of the Living World: Merleau-Ponty, Animals, and Language*, (New York: Fordham University Press, 2014); and Kerry Whiteside, *Merleau-Ponty and the Foundation of an Existential Politics* (Princeton, NJ: Princeton University Press, 1998).

138 "The Battle of Algiers" is also the title of a legendary 1966 film by the Italian director Gillo Pontecorvo (1919–2006) with a rousing musical score by the late, great Italian composer Enrico Morricone (1928–2020). See https://en.wikipedia.org/wiki/Battle_of_Algiers_(1956%E2%80%931957); and https://en.wikipedia.org/wiki/The_Battle_of_Algiers.

139 Merleau-Ponty, *Phenomenology of Perception* (PP), *op. cit.*, 230 and 235.

140 *Ibid.*, 354.

141 *Ibid.*, 83.

142 *Ibid.*, 94–5.

143 *Ibid.*, 96–7.

144 *Ibid.*, 98.

145 See, for example, Simone de Beauvoir's *The Ethics of Ambiguity,* trans. Bernard Frechtman (Philosophical Library/Open Road Media Reissue Edition, 2015), one of the very few book-length texts in the entire existentialist canon that deals systematically with ethics.

146 See PP, *op. cit.*, 41, 136, and 344, and Reynolds, IEP, *op. cit.*

147 *Ibid.*, 121 and 500.

148 *Ibid.*, 500.

149 *Ibid.*, 149–50.

150 *Ibid.*, 158–62.

151 *Ibid.*

152 *Ibid.*, 169–71.

153 *Ibid.*, 170.

154 *Ibid.*, 198.

155 *Ibid.*, 213–14 and 229.

156 *Ibid.*, 231, 235, and 256–7.

157 *Ibid.*, 276 and 292.

158 *Ibid.*, 298.

159 *Ibid.*, 343–5.

160 *Ibid.*, 347 and 354.

161 *Ibid.*, 373–4 and 378.

162 *Ibid.*, 382–4.

163 *Ibid.*, 387–9.

164 *Ibid.*, 394 and 400–2.

165 *Ibid.*, 403.

166 *Ibid.*, 406–7.

167 *Ibid.*, 408–9.

168 *Ibid.*, 412–19.

169 *Ibid.*, 420–5.

170 *Ibid.*, 435, 437, and 450.
171 *Ibid.*, 460 and 462–4.
172 *Ibid.*, 469–72. In Merleau-Ponty's own words:

> language presupposes nothingness ... a silence of consciousness embracing the world of speech in which words first receive a form and meaning ... Behind the spoken *cogito*, the one which is converted into discourse ... into ... truth, *there lies a tacit cogito, myself experienced by myself.* But this subjectivity ... *does not constitute the world, it divines the world's presence round about it as a field not provided by itself ... the project towards the world that we are ... the world's being inseparable from our views of the world ... subjectivity conceived as inherence in the world ... the world is the field of our experience, and ... we are nothing but a view of the world ... the most intimate vibration of psycho-physical being already announces the world* ... which ... never gets beyond being an "unfinished work," ... does not require ... a constituting subject.

173 *Ibid.*, 474–5. For Merleau-Ponty,

> *The subject is a being-in- the-world and the world remains 'subjective ... the world as cradle of meaning ..., and ground of all thinking ... The world ... as standing on the horizon of our life as the primordial unity of all our experiences, and one goal of all our projects ...* the native abode of all rationality ... *But every one of these words, like every equation in physics, presupposes our pre-scientific experience of the world ... What ... do we mean ... that there is no world without a being in the world? Not indeed that the world is constituted by consciousness, but ... that consciousness always finds itself already at work in the world ...* the world will outlast me, and other men will perceive it when I am no longer here ... indeed my presence in the world is the condition of the world's possibility?"

He does not give a clear answer to this ontological/epistemological question.
174 For "the problem of rationality," see Jan Van der Veken, *Merleau-Ponty on the Ultimate Problems of Rationality*, available at: www.utpjournals.press/doi/pdf/10.3138/uram.12.3.202, who states:

> His [Merleau-Ponty's] position ... can be summarized as a strict phenomenological position, according to which man brings meaning to a world that is meaningless without him. The question—is man really the origin of meaning?—forced Merleau-Ponty to have a greater sensitivity to the fundamental and equally undeniable givenness of both the subject and the world. Man and world are sustained through and through by a more encompassing mystery, which itself is the possibility of their mutual dialogue: man interprets what is waiting to be expressed in vision, in language, in thought. There is something to see and to say ... He clearly reacted against "the Absurd Emperor of the world," "the necessary being" ("a God who is force)." Metaphysics, as he understood it, cannot be "reconciled with the manifest content of religion positioning an absolute thinker of the world."

God becomes a term of reference for a human reflection which, when it considers the world such as it is, condenses in this idea what it would like the world to be. For a more comprehensive historical, political, and philosophical analysis of rationality and reason, see Charles Webel, *The Politics of Rationality*, *op. cit.*

175 *Ibid.*, 477, 493, and 498.

176 *Ibid.*, 500–1.

177 *Ibid.*, 503 and 507–8.

178 *Ibid.*, 512, 515, 521, and 523.

179 *Ibid.*, 525–7.

180 *Ibid.*, 525–30.

181 The last sentence in the *Phenomenology of Perception* is a citation from Antoine de Saint-Exupéry's 1942 novel *Pilote de Guerre* (*War Pilot*).

182 See Reynolds, *IEP*, *op. cit.*

183 See EW, *op. cit.*, 367–8.

184 Reynolds, *ibid.*, and Merleau-Ponty, VI, *op. cit.*, 179.

185 According to Toadvine, SEP, *op. cit.*:

> Sensible flesh—what Merleau-Ponty calls the "visible"—is not all there is to flesh, since flesh also "sublimates" itself into an "invisible" dimension: the "rarified" or "glorified" flesh of ideas. Ultimately we find a relation of reversibility within language like that holding within sensibility: just as, in order to see, my body must be part of the visible and capable of being seen, so, by speaking, I make myself one who can be spoken to (allocutary) and one who can be spoken about (delocutary). While all of the possibilities of language are already outlined or promised within the sensible world, reciprocally the sensible world itself is unavoidably inscribed with language.

186 See EW, *op. cit.*, 18, 21, and 26, and PP, *op. cit.*, xxiii and 66.

187 *Ibid.*

188 Merleau-Ponty, *Humanism and Terror* (HT), *op. cit.*, 18.

189 *Ibid.*, 21 and 23.

190 *Ibid.*, 34–6.

191 *Ibid.*, 64. Nikolai Ivanovich Bukharin (1888–1938) was a Bolshevik revolutionary, Marxist theorist, and politician, who, together with Joseph Stalin, in 1927, ousted Leon Trotsky, Grigory Zinoviev, and Lev Kamenev from the Communist Party of the Soviet Union. Two years later, Bukharin himself was expelled from the Politburo. See https://en.wikipedia.org/wiki/Nikolai_Bukharin. And when the Great Purge of his former comrades, rivals, and alleged traitors began in 1936, Stalin had Bukharin charged with conspiring to overthrow the Soviet state. After a show trial that alienated many Western leftist sympathizers with communist ideals (including Merleau-Ponty), Bukharin was executed in 1938. In this context, Merleau-Ponty reflects on what Trotsky called "The Revolution Betrayed," and then goes on to cite Hegel. See HT, *op. cit.*, 67–8.

192 See "*Hegel: The History of the World is the World's Court of Judgment*" in Chapter Two of this book.

193 Merleau-Ponty, HT, *op. cit.*, 91–2. In commenting on Marx(ism), Machiavelli, and the "rationality" of revolutionary violence in giving birth to a "new world" from the embers of the "old world," Merleau-Ponty remarks: "As social life … affects each individual … the … manner of his being in the world, the Revolution in the Marxist sense … takes a long time … against harmful reversals to the old world." *Ibid.*, 104.

194 See HT, 113–14, where Merleau-Ponty says:

> The development of production, says Marx, created a *world market,* i.e., an economy in which every man depends for his life on what happens everywhere else in the world. Most … experience this relation to the rest of the world as a fate and draw from it only resignation.

This capitalist "world market" has been recently recast under the guise of "globalization" and "liberal democracy." And, as Merleau-Ponty elucidates the Marxist revolutionary dialectic: "A dialectical world is a world on the move … where values can be reversed … not a bewitched world …" *Ibid.*, 120. And in his conclusion to *Humanism and Terror*, Merleau-Ponty declares:

> *The human world is an open or unfinished system* … cannot tell us *that* humanity will be realized … a *philosophy which arouses in us a love for our times … a view which … embraces … all the order and … disorder of the world.*
>
> (*Ibid.*, 188–9)

195 *Ibid.*, 153.

196 For glimpses of how to bring "peace" to Earth and what "peace" in the contemporary world denotes, see Charles Webel and Marcel Kaba, "Definitions of Peace," Chapter 1 of *A Cultural History of Peace in the Modern Era* (Volume 6) (London: Bloomsbury 2020, 21–40; and Charles Webel and Sofia Khaydari, "Toward a Global Ethics of Nonviolence," in *Solidarity beyond Borders: Ethics in a Globalizing World*, Janusz Salamon, editor (London: Bloomsbury Publishing, 2015), 153–69. The positions presented here are in stark contrast to Merleau-Ponty's rendering of pacifism as "unrealistic," at least from what he characterizes as a Marxist perspective: "He who condemns all violence puts himself outside the domain to which justice and injustice belong. He puts a curse upon the world and humanity …" HT, *op. cit.*, 108.

But, for better or worse, the "proletariat" (the working class, locally and globally) has not on the whole emerged as a progressive, anti-capitalist revolutionary agent but, in many countries, as a regressive, procapitalist "populist" promoter of ethno-nationalism. Marx and (Western) Marxism remain largely unread and negatively stereotyped outside a relatively few intellectual elites and political fringes. And "justice and injustice" are mere abstractions if divorced from the concrete legal, political, economic, racial, and gender-based norms and practices of actually existing social institutions.

197 Merleau-Ponty, HT, *op. cit.*, 150–1.

198 Merleau-Ponty's "humanism in extension" is cosmopolitan and inclusive, in contrast with the restrictive and parochial "Western humanism" that cloaks its ethnocentric and militaristic imperialism with flowery phrases, since, according to Merleau-Ponty: "… *Western Humanism has nothing in common with a humanism in extension, which acknowledges in every man a power more precious than his productive capacity … as a being capable of self-determination and of situating himself in the world.*" HT, *op. cit.*, 176.

199 For Merleau-Ponty's "linguistic turn" and nascent philosophy of language, which impacted generations of French, but few Anglophonic, philosophers, see Reynolds, IEP, *op. cit.*, who states:

> According to Merleau-Ponty, the tacit cogito is therefore a product of language, and the language of the philosopher, in particular. *He continues to*

> *speak of a world of silence, but the concept of the pre-reflective cogito imports the language of the philosophy of consciousness* ... and hence misrepresents the relationship between vision and speech. The famous phenomenological reduction to the things themselves, which tries to bracket out the outside world, is hence envisaged as a misplaced nostalgia rather than as a real possibility ... *his abandonment of the idea of a pre-reflective cogito, or consciousness before linguistic significance ... serves to radicalize phenomenology. It also means that language comes to play a far more important role in his philosophy than it previously had. Indeed, Merleau-Ponty used both linguistics, and the language-based emphasis of structuralism to critique Sartre ... who only accorded language a minimal role.*

Reynolds also emphasizes that Merleau-Ponty

> was also friends with, and used the work of people like Jacques Lacan (a psychoanalyst who suggested that the unconscious is structured like a language), Claude Levi-Strauss (a structuralist anthropologist who dedicated his major work *The Savage Mind* to the memory of Merleau-Ponty), and also Ferdinand De Saussure (a linguist who showed what a pivotal role differences play in language, and whose work has inspired many recent philosophers including Derrida). Merleau-Ponty was hence very much involved in what is termed the linguistic turn, and *one curious aspect of Merleau-Ponty's place within the philosophical tradition is that despite the enduring attention he accords to the problem of language, the work of thinkers such as those cited above, and others who have been inspired by them (Derrida and Foucault for example), has been used to criticize him* ... he paradoxically laid the groundwork for his own denigration and unfashionability in French intellectual circles, and it is only in the last 15 years that it has been realized that his phenomenology took very seriously the claims of such thinkers, and even pre-empted some aspects of what has come to be termed "postmodern" thought. Levi-Strauss actually finds *The Visible and the Invisible* to be a synthesis of structuralism with phenomenology.

4 Talking About the World

Ludwig Wittgenstein's Language of the World

Language has, of course, from the "ancients to the postmoderns" been used to talk about, analyze, and describe the world. But, it is only in the past century or so that language itself has been the intense focus of conceptual analysis. And, no one since 1920 has been more influential in analyzing the language used to "picture" and give "meaning" to the world than Ludwig Wittgenstein.

Ludwig Wittgenstein (1889–1951) was one of the most influential philosophers of the 20th century—many academic philosophers would say *the* most important—and is regarded by some contemporary thinkers as the most important philosopher since Immanuel Kant.[1] While Wittgenstein is widely regarded as a founder of "analytic" philosophy, in large part due to his early work in logic and the philosophy of language—most notably in the only book he published during his lifetime, the *Tractatus Logico-Philosophicus* (abbreviation: TLP)—it is his posthumously published *Philosophical Investigations* (abbreviation: PI) that has mainly enshrined him in the small pantheon of modern philosophical titans. And, although Wittgenstein did not write extensively about "the world" (German: *die Welt*), what he did have to about it is as succinct and suggestive, as it is, like much of his writing, elusive and often enigmatic.

Wittgenstein's philosophical "world" (German: *Welt*—and it should be noted that his most influential works were written in German and then translated into English) can, of course, be interpreted solely within the contexts within which this term is used in his texts. However, I believe this a necessary but insufficient condition for understanding Wittgenstein's *Welt*. Accordingly, in addition to analyzing the inner-textual "language-game" (Wittgenstein's memorable term *Sprachspiel* from his PI) of *die Welt* in Wittgenstein's works, I will also refer to what I will call his overall worldview (*Weltanschauung*) and to the world as a whole in which he lived and wrote, insofar as these can be detected from the works to which I have had recent access.[2]

A thinker of Wittgenstein's magnitude does not exist in a context-free zone of pure abstraction. So, without falling prey to the intellectual "sin" of psychologism—so acutely dissected by Gottlob Frege (1848–1925),

DOI: 10.4324/9781315795171-5

and Edmund Husserl—I will attempt to interweave the textual "world" of Wittgenstein's words with his lived world (*Lebenswelt*).[3]

Wittgenstein's Life (*Leben*) and World (*Welt*)

It is curious and possibly remarkable that the person widely considered to have been one of the most influential figures in modern Occidental intellectual history in general, and especially in Western philosophy since the end of World War II, seems to have exhibited little knowledge of or interest in the history of philosophy, or indeed the history of anything, at least in his published works. This, however, is hardly unique to Wittgenstein. That is because most recent and contemporary publications in analytic and "ordinary language" philosophy, stemming, perhaps unintentionally, from Wittgenstein, as well as his precursors and mentors Bertrand Russell (1892–1970) and Frege, have been primarily focused not on "metaphysical" theories of the world, or on anything "speculative" for that matter—since they have tended to view both sweeping theories and metaphysics as "meaningless" and/or "ungrammatical"—but instead have probed central issues in logic, the philosophy of language, epistemology, and (the philosophy of) mathematics. While Russell did write a lengthy and idiosyncratic *History of Western Philosophy*,[4] was committed to pacifist and libertarian–socialist political struggles for most of his adult life, and was anything but ascetic, his "strictly" philosophical work was done mostly when he was relatively young (he lived until the age of 97) and was largely confined to those quite technical fields. But Wittgenstein himself seems to have lived a largely "apolitical," even monastic life, in and out of the cloisters of Cambridge University, where he was tutored by Russell and with which he was affiliated for decades. His roots in the moneyed Austrian aristocracy of *fin de siècle* and early 20th-century Viennese culture may have been the catalyst for his turning away from the decadent materialism in which he grew up and his turn toward the "purity" of a philosophy shorn of its historical and metaphysical trappings.[5]

Ludwig Wittgenstein was born in 1889 in Vienna to a wealthy and large family, a prized meeting place for intellectual and cultural Viennese circles. Ludwig's father, Karl Wittgenstein, was one of the most successful businessmen in the Austro-Hungarian Empire's steel and iron industries. His father's parents were born Jewish but converted to Protestantism and his mother was Catholic, but her father was Jewish. Wittgenstein himself was baptized in a Catholic church and was given a Catholic burial, although between baptism and burial he did not appear to have been a practicing Catholic. Wittgenstein's home attracted many prominent artists and musicians. And music was a *Leitmotif* throughout Wittgenstein's life, as were loneliness, alienation, suffering, and death. Ludwig was the youngest of eight children, and three of his four brothers committed suicide.

After a rigorous Austrian secondary-school scientific education in Linz, where he became acquainted with the works of the physicists Heinrich

Hertz and Ludwig Bolzmann, Wittgenstein enrolled in the *Technische Hochschule* (Technical University) in Berlin to study mechanical engineering. In 1908, at his father's suggestion, Wittgenstein transferred to Manchester University in England as an aeronautical engineering research student. There he devised and patented a novel aero-engine employing an airscrew propeller driven by blade tip-jets.[6] Throughout his life, Wittgenstein retained an interest in the logic of machines, and engineering. Accordingly, mechanical examples and metaphors were present his later philosophical writings on the philosophy of language, which might in part explain his indifference to history, aversion to theoretical speculation, and search for "purity" "perfection," "rigor," and "simplicity" in style, architecture, morality, philosophy, and his own life.

While in Manchester, Wittgenstein became increasingly interested in the philosophy of physics as well as in logic and pure mathematics, which led him to communicate with Frege, who recommended that he study with Russell, the most noted logician and philosopher of mathematics in England at the time, and a major proponent of "realism" in contrast with the "idealism" that was widespread among his peers and predecessors at Cambridge University and elsewhere in Great Britain. After having written Russell to inquire if he should study philosophy and having sent him a short essay, at Russell's encouragement, Wittgenstein transferred in 1911 to Cambridge. Wittgenstein greatly impressed Russell and his colleague, the distinguished philosopher of language and moral philosophy. G.E. Moore (1873–1958). Russell wrote, upon meeting Wittgenstein: "An unknown German appeared … obstinate and perverse, but I think not stupid." After working with Wittgenstein for about a year, Russell commented: "I shall certainly encourage him. Perhaps he will do great things … I love him and feel he will solve the problems [in logic and the philosophies of mathematics and language] I am too old to solve."[7] Russell's hunch proved to be on the mark.

At Cambridge between 1911 and 1913, Wittgenstein conversed not only with Moore and Russell, with whom he had a complex relationship, but also with the great economist John Maynard Keynes (1883–1946). For long periods, Wittgenstein left Cambridge to think and live in Norway in relative isolation. In 1913, he returned to Austria, and the following year, at the start of World War I (1914–1918), Wittgenstein joined the Austrian army, and for a while served with distinction on the front. He was taken captive in 1918 and spent the remaining months of the war in a prisoner-of-war camp.

It was during that war that Wittgenstein drafted the notes for his first important philosophical work, *Tractatus Logico-Philosophicus* (TLP). He also wrote biographical jottings about his wartime experiences, in which, inter alia, he pondered his life's possible meanings. After the war with Russell's support, the TLP was published in German and translated into English. It caused an immediate stir in the philosophical world and was for a time seen as a major argument for logical positivism. The *Tractatus*, like

The World as Will and Representation by Schopenhauer (a very different thinker and person who may or may not have influenced Wittgenstein's thinking about "the world"), is from beginning to end replete with references to *die Welt*, and, in a few significant places toward its end, to "the will" as well.[8]

In 1920, Wittgenstein, having, in his own opinion, "solved" all philosophical problems in the *Tractatus* (largely by "dissolving" them), gave away his part of his family's fortune and pursued several non-academic activities, including working as a gardener, elementary school teacher, and architect of a house for his sister Gretl, in the Vienna region. Nine years later, he returned to Cambridge to resume his philosophical career after having been engaged in discussions on the philosophy of mathematics, logic, and the philosophy of science with members of the Vienna Circle, whose conception of logical empiricism was indebted to Wittgenstein's accounts in the *Tractatus* of the nature and importance of logic, and of philosophy reconceived as logical syntax.[9] During this time in Cambridge, Wittgenstein's conception of philosophy and its problems underwent dramatic changes, which are recorded in several volumes of conversations, lecture notes, and letters (e.g., in *The Blue and Brown Books*, and in his *Philosophical Grammar*). This "middle period" of Wittgenstein's philosophical development, according to many of his interpreters, led to his rejection of what he deemed "dogmatic" philosophy, which included not only previous "metaphysical" dogmas, but also his own *Tractatus*.

In the 1930s and 1940s, Wittgenstein conducted seminars at Cambridge, where he became Professor of Philosophy. During World War II, he worked as a hospital porter in London and as a research technician in Newcastle. In 1945, he prepared the final manuscript of the *Philosophical Investigations*, but, at the last minute, withdrew it from publication (and authorized only its posthumous publication). After the war, he returned to university teaching but resigned his Cambridge professorship in 1947 to concentrate on writing. Much of this he did in Ireland, preferring isolated rural places for his work. He spent the last two years of his life in Vienna, Oxford, and Cambridge, where he died of prostate cancer in 1951. His work from these last years has been published as *On Certainty*. It is said that, in 1951, on his death-bed, Wittgenstein's last words were "Tell them I've had a wonderful life."[10]

The World (*Die Welt*) in Wittgenstein's Works and Worldview (*Weltanschauung*)

Wittgenstein's "world," in the "ordinary meaning" of the term involving the geographical contexts in which he lived, thought, and died—while providing him with what the Austrian philosopher of science Karl Popper (1902–94), among others, called a "context of discovery" for the generation of his overall worldview (*Weltanschauung*) on life in general and his life in particular—is by no means exhaustive of the quite technical ways in which Wittgenstein used the word *Welt* in his philosophical writings.[11] In turn, while Wittgenstein studiously attempted to refrain from, and even reject, "speculative" expressions and instead saw philosophy as "linguistic therapy" rather than "theory-formation," his uses of *Welt* and *die Welt* (the world)—which, like much of his writing, are far from perspicuous or univocal—in the works published under his name may plausibly be interpreted from within the framework of his more general view of "the world," or his *Weltanschauung*, especially as he experienced it during wartime and outside philosophical circles.

The only systematic exploration of Wittgenstein's usage of *die Welt* with which I am familiar is a very recent piece by Hans Sluga. As Sluga has pointed out in that article and in email correspondence with me, "the term [*die Welt*] is the first one in the *Tractatus* and occurs [over 40] times in that work and no one has ever really taken notice of it."[12] Sluga appropriately focuses on the statements in the TLP where Wittgenstein uses *die Welt*, as well as on Wittgenstein's "Secret Notebooks," mostly written during his service in World War I, and his much later work "*On Certainty*," and so shall I.[13]

Wittgenstein wrote the notes for the *Tractatus* during World War I and completed the book during a military leave in the summer of 1918. In his wartime diaries and those notes, Wittgenstein struggled with what the French existentialists, as well as Kierkegaard, Nietzsche, and Heidegger, might have called fundamental "existential" issues, although Wittgenstein, coming from a very different tradition, abstained, to the best of my knowledge, from using that term.

Among the ultimate concerns addressed by Wittgenstein are the meaning and the "right direction" of life (*Sinn und Ziel des Lebens*) in general and of his life in particular, often in very personal, religious, and ethical terms.[14] For example, Michael Mauer claims:[15]

> there is sufficient autobiographical and biographical evidence to show, that also for the rest of his life, Wittgenstein's inner moral struggle *with* his own imperfection and continuously growing feelings of meaninglessness never came to a standstill. Throughout his life, Wittgenstein continues to search for the *right* form of life. Tormented with inner strain and restlessness, in permanently new drafts he will try to make vanish the *problems of philosophy* and the *problem of life,* which stand in a complex relationship to each other.

Mauer continues:[16]

> In the *Tractatus* we come across the attempt to give the *problems* a logically strict form in order to make them dissolve ... At a decisive point in the *Tractatus* the question of the *meaning of life* gains paradigmatic importance for Wittgenstein's philosophical method.

According to Wittgenstein:

> One notices the solution to the problem of life in the disappearance of the problem. (Is this not the reason why people, to whom after long doubting the meaning/sense ((*Sinn*)) of life became clear, could not then say wherein this sense/meaning consisted?)[17]

Maurer continues:

> As a reaction to the logic of the *problem of life*—which is the *prime example* for the paradoxical nature of philosophical problems—Wittgenstein carries out the radically therapeutic project of a final distinction between sense and nonsense. The philosophical questions turn out to be unsolvable, i.e. meaningless in the medium in which they are posed to us—*language.* The author of the *Tractatus* wants to make the problems vanish by leading the reader to the understanding that their existence is based on a meaningless way of asking, which is inherent to the limited logic of our language.

Maurer also quotes a "War Diary" by Wittgenstein and comments:[18]

> In a state of deep disconcertment about the painful disruption of human existence, the young Wittgenstein asks himself the metaphysical question *par excellence*: "Kann man aber so leben daß das Leben aufhört problematisch zu sein? Daß man im Ewigen lebt und nicht in der Zeit?" "But can one live in such a way that life ceases to be problematic? That one lives in eternity and not in time?" In asking this question Wittgenstein recapitulates a traditional philosophical thought: the idea of an approach *sub specie aeternitatis. The finite nature of human understanding is to be abolished in favor of a deeper insight into the hidden rules of the world.* This is the expression of a passionate desire to be *redeemed* by the ordering of metaphysical knowledge, wherein the untiring longing for the *right* form of life *comes to an end.*

From what might be called a "logical existentialist" point of view, in his "War Diaries," "Secret Notes," "Letters" to such friends and colleagues as Norman Malcolm, and even in the *Tractatus*, Wittgenstein put in writing his lifelong struggle to understand the world, and himself (possibly including his own sexuality).[19] In parts of the *Tractatus*, however, he seems to confine any "metaphysical" questions about life's possible sense or meaning, as well as the ultra-important domains of ethics or aesthetics, to what would seem to be an "ungrammatical," and hence "nonsensical" domain beyond logical and linguistic analysis, and thus beyond "the limit of the world." But later in his life, he would make more explicit his "ultimate concerns," which transcended "abstruse questions of logic," etc. As he wrote to Malcolm in 1939:[20]

> what is the use of studying philosophy if all that it does for you is to enable you to talk with some plausibility about some abstruse questions of logic, etc., and if it does not improve your thinking about the important questions of everyday life, if it does not make you more conscientious than any ... journalist in the use of the DANGEROUS phrases such people use for their own ends? ... I know that it's difficult to think *well* about "certainty," "probability," "perception," etc. But it is, if possible, still more difficult to think, or *try* to think, really honestly about your life and other people lives. And the trouble is that thinking about these things is *not thrilling*, but often downright nasty. And when it's nasty then it's *most* important.

For the Western metaphysical, and by implication, ethical, aesthetic, and religious traditions, Wittgenstein's *Tractatus* may have been both "dangerous" and "nasty," since it called into question many of their most basic assumptions about what might be meaningfully known and said.

The *Tractatus* was first published in German in 1921 as the *Logisch-Philosophische Abhandlung* (*Logical-Philosophical Treatise*). G.E. Moore originally suggested that the work's Latin title is a kind of tribute to the *Tractatus Theologico-Politicus* (*Theological-Political Tractatus*) by Spinoza. The book was translated into English the following year by C.K. Ogden with the help of the tragically short-lived Cambridge mathematician and philosopher Frank P. Ramsey (1903–30). The *Tractatus* was influential chiefly among the logical positivist philosophers of the Vienna Circle, especially Rudolf Carnap and Friedrich Waismann, and later became recognized as one of the most influential philosophical books during the interwar period, in part due to the "popularization" of logical positivism and its proponents by the Oxford philosopher A.J. Ayer (1910–89) in his 1935 book *Language, Truth and Logic*. In addition, Bertrand Russell's article "The Philosophy of Logical Atomism" is, in part, a working out of ideas he had gleaned from Wittgenstein.[21]

The *Tractatus* begins with a number of the most extraordinary—and most-often interpreted—sentences in the history of Western philosophy:[22]

> 1.0 The world is everything that is the case.[23]
> 1.1 The world is the totality of facts, not of things.
> 1.11 The world is determined by the facts ...
> 1.13 The facts in logical space are the world.
> 1.2 The world divides into facts.

For Wittgenstein, an atomic fact, namely, a combination of objects (entities, things), determines what is the case, and hence the world.

Wittgenstein's aforementioned sentences seem to be propositional, i.e., statements in everyday language that can be either true or false, or, more technically, in philosophy, that may refer to the non-linguistic meaning ostensibly behind the statement. *Hence, the English word "world" is simply a noun (substantive) whose semantic senses are the ways people use that term in ordinary, everyday language* (apparently the "later" Wittgenstein's view) *or whose "meanings" are its synonyms in a dictionary. Alternatively, "world/ the world" may refer to the factual existence ("is") of some extra-linguistic entity, or to "the totality of facts"* (arguably the position of Wittgenstein in his *Tractatus*). As Wittgenstein says, *"the proposition constructs a world with the help of a logical scaffolding"* (4.014).[24]

These statements, or propositions, are presented in declarative, almost deductive, form, without warrant or immediate justification, as though they are axioms, or self-evident truths, whose validity will be demonstrated by succeeding statements, and the numerical sequence of these sentences seems chosen by Wittgenstein to approximate the deductive logic with which he was very familiar. "The world" (*die Welt*) is a *Leitmotiv* running through the movements of this text (in sharp contrast with the paucity of its usage in Wittgenstein's later works, especially his *Philosophical Investigations*). But to unravel the "meaning(s) of 'the world'" for Wittgenstein, one must also address the existential, semantic question infamously posed during former U.S. President Bill Clinton's impeachment debacle in 1998, when he disingenuously and mendaciously claimed that his assertion that "there's nothing going on between us" (his sexual relationship with Monica Lewinsky) had been truthful because he had no ongoing relationship with Lewinsky at the time he was questioned, and notoriously said "It depends upon what the meaning of the word 'is' is."[25]

According to Hans Sluga, when one poses the question "What is the 'is?' if one asks what 'is' *die Welt* for Wittgenstein":[26]

> the expression "the world." … was for the Wittgenstein of the *Tractatus* an "essentially philosophical word." But what kind of word is it? Is it a name? We certainly cannot determine the meaning of the expression "the world" with a pointing gesture. We cannot place ourselves outside the world, point to it and say: this is the world. One is tempted to explain the word with an embracing gesture, in which one draws a circle with outspread arms and says: All this is the world.

Sluga continues:[27]

> But the later Wittgenstein pointed out the helplessness of such gestures in philosophical discourse … the world is not an object in the sense of the *Tractatus*. Because objects are simple. But the world as everything that is the case, or … a totality of facts, and as such cannot be simple and therefore cannot be a nameable object.

Wittgenstein himself writes in the *Tractatus*, *"The total reality is the world"* (2.063),[28] in addition to propositions 1.1 ff. So, one might conclude that Wittgenstein's description (not explanation, which he claims should not be attempted by philosophy) of "the world" is, in line with logical positivism, a neutral, impersonal "picture" (*Bild*) of the scaffolding, or skeleton, of what there is, namely, the "totality of facts," and by implication, the exclusion of what is not the world (or not "in" the world?), namely any thing, or combination of entities, that do not exist, because they are not factual, and therefore "not the case" (*nicht der Fall*).

Sluga also argues that Wittgenstein does *not* use the term "the world" to denote *just the present* state of affairs or totality of facts. Rather, according to Sluga, Wittgenstein:[29]

> means that the world is always, immortal, or timelessly identical with everything that is the case ... The *Tractatus* knows of two kinds of existence ... But does the world ever exist? Shouldn't we instead speak of a world course in which only some but not all is the case ... If the first sentence [of the *Tractatus*] is an identity sentence, then the expression "the world" means something like ... "everything that is the case" or ... "the totality of the facts." ... Wittgenstein tells us that such a totality can only be seen in the fact that we have variables that span all facts. *But that also means that the term "the world" actually has no meaning and with that ... the first sentence of the Tractatus also loses its meaning.*

Wittgenstein writes at the end of the *Tractatus*:

> The correct method of philosophy would be ... to say nothing other than what can be said- ... and then ... when someone ... wanted to say something metaphysical ... he gave no meaning to certain signs in his sentences.
>
> (TLP: 6.53)

So, at the very beginning of the Tractatus, is Wittgenstein undermining his own anti-metaphysical project to eliminate such "nonsensical, ungrammatical" terms as "the world" from the "meaningful propositions" of natural science and the "tautologies" (self-evident truths) of mathematics or logic? This, like many things in Wittgenstein's *oeuvre*, is open to multiple responses, readings, or interpretations (many of which have been provided by others) and does not seem to admit of a clear, factual answer.[30] But before consigning ourselves to resignation or mere puzzlement, or proceeding further to explore the "flesh and bones," Wittgenstein often fitfully offers to provide some content to his bare-bones "logical/linguistic" factual scaffolding of the world, it is important to keep "in mind," that according to him, "All philosophy is 'Critique of language'" (4.0031).[31] And further, "*The object of philosophy is the logical clarification of thoughts. Philosophy is not a theory but an activity ... essentially of elucidations ...* to make propositions clear."(4.111–12)[32]

But are Wittgenstein's own philosophical works, including the Tractatus, and how he uses the term "the world," in strict keeping with the "clarification" of propositions, rather than in their positing?

But he also declares that: "*we should have to be able to put ourselves with the propositions outside logic … outside the world …*" and "*What can be shown cannot be said*" (4.1212). *But how does one know that nothing can be said, logically demonstrated, or factually stated about what is "outside the world" (including religious, ethical, and aesthetic claims), or even that there is anything at all outside "the totality of things that are the case*?" Are such statements "clear activities" of "elucidation" and linguistic critique, or do they instead verge on "opaque and blurred thoughts?" While I'm inclined toward the latter reading of these and many other passages in Wittgenstein's texts, my readers, like those of Wittgenstein, will provide their own interpretations. And that, following Nietzsche, is as it should be, since, as Wittgenstein argued that philosophy is not a natural science.[33]

After firing the opening declarative salvos, Wittgenstein proceeds to refer to "the real world," "an imagined world," the "substance of the world." *The "total reality" of the "configuration of the objects" constitutes" the total reality of the world*" (2.022–3 and 2.063).[34]

Wittgenstein then goes on to speculate about "a god" who "creates a world in which certain propositions are true …" (5.123), but also to deny the existence of "the soul, the subject, as … conceived in contemporary superficial psychology" (5.5421).[35] And then, in possibly some of the most memorable, and puzzling or even "mystical" proclamations in modern thought, Wittgenstein announces that:[36]

> *The limits of my language mean the limits of my world.*
>
> (5.6)

> *Logic fills the world: the limits of the world are also its limits … we cannot therefore say what we cannot think.* That the world is my world … the limits of the language … mean the limits of my world.
>
> (5.61)

> *The world and life are one.*
>
> (5.621)

> *I am my world.*
>
> (5.63)

> *The subject does not belong to the world but it is a limit of the world.*
>
> (5.632)

> the "world is my world …" The philosophical I is not the man, not the human body or the human soul … but *the metaphysical subject, the limit—not a part of the world.*
>
> (5.641)

Many trees have been uprooted to provide the paper needed for the thousands of pages of printed text devoted to commenting on, interpreting, and "explaining" what the aforementioned statements "mean." I will not add to those ... except to say that these passages sometimes sound like Schopenhauer ("my will") and at other times like philosophical "eliminative materialists" who wish to deny the ("metaphysical") "subject's" and "philosophical I's" worldly (and possible "otherworldly") existence.[37] That is to say, they are quintessential Wittgenstein, the thoughts of a person who, while seemingly bracketing the history of philosophy has reinvigorated it! And he has done so, in part, by rethinking "the world" as the boundary or "limit" between the "meaningful" and the "metaphysical" in such a way that the "mystical" is somehow preserved. For, as he declares later in the *Tractatus*:[38]

> *Not how the world is, is the mystical, but that it is.*
>
> (6.44)

> The feeling of the world ... is the mystical feeling.
>
> (6.55)

But to move from the ineffability of what is mystical, and the "feeling of the world ... the mystical feeling," back to the logical "scaffolding of the world," implies, according to Wittgenstein, that:[39]

> *logic has nothing to do with ... whether our world is really of this kind or not ... logical propositions describe the scaffolding of the world ... they present it ...* They presuppose that names have meaning, and ... elementary propositions have sense ... this is their connection with the world.
>
> (6.1233–6.124)

Without providing a reference—Kantian, Husserlean, or otherwise—Wittgenstein then claims, in a decidedly "transcendental" moment, that[40] "Logic is not a theory but a reflection of the world. Logic is transcendental." (6.13). Furthermore, he proceeds to make the stunning pronouncement that at the basis of "the whole modern view of the world" and natural laws lies the "illusion" that they explain natural phenomena.[41] And then, in words eerily reminiscent of, but without citing, Schopenhauer, Wittgenstein declares that:[42]

> The world is independent of my will ... Even if everything we wished were to happen, this would only be, *a favor of fate, for there is no logical connection between will and world.*
>
> (6.373–4)

And in one of his more memorable pronouncements, he announces that:

> *The sense of the world must lie outside the world. In the world everything is as it is and happens as it does happen. In it there is no value.*
>
> (6.41)

Wittgenstein attempts to situate ethics (and aesthetics) "outside the world," ostensibly because they are "transcendental" (perhaps a synonym for "outside the" factual "world," but if so, where and what is this "transcendental" x?):[43]

> If there is a value … *It must lie outside the world* … Hence … there can be no ethical propositions … Ethics are transcendental. Ethics and aesthetics are one. Of the will as the bearer of the ethical we cannot speak.
>
> (6.41–6.423)

Wittgenstein seems then to reinforce the boundary between "the world" of facts that can "be expressed in language," but also, and stunningly, also makes a series of "personal," even prophetic, statements about "the world of the happy" (man), death, life, the immortality of the soul, "the riddle of life in time and space" (pace Nietzsche?), "the problems of life," and God—topics central to the history of philosophy and theology, but that also might seem to be "off-limits" for such a "hard-headed" thinker as the "early" Wittgenstein:[44]

> *If good or bad willing changes the world, it can only change the limits of the world,* not the facts … *the world … must … wax or wane as a whole. The world of the happy is quite another than that of the unhappy.*
>
> (6.43)

> *As in death … the world does not change, but ceases.* (6.431). Death is not an event of life … Our life is endless in the way.
>
> (6.4311)

> The temporal immortality of the soul of man … its eternal survival … after death, is … in no way guaranteed … *The solution of the riddle of life in space and time lies outside space and time.*
>
> (6.4312)

> *How the world is, is completely indifferent for what is higher. God does not reveal himself in the world.*
>
> (6.432)

> *even if all possible scientific questions be answered, the problems of life have still not been touched … there is then no question left, and just this is the answer.*
>
> (6.52)

So, the "answer" or "solution" to "the problems of life" is to realize that even if "all possible scientific questions" have been "answered," there will be "no question left?" That sounds confounding, even questionable, or "unphilosophical." Regarding this, Wittgenstein proposes "the right method of philosophy" (having nothing to do with "the propositions of natural science") would be "to say nothing except what can be said," and that to "see the world rightly," one must "surmount these" (Wittgenstein's own!) "propositions" (6.54).[45]

And, in one of the most memorable last sentences one can imagine, Wittgenstein concludes his sole completed book by stating:[46] "whereof one cannot speak, thereof one must be silent" (7).

Words and Worlds in Wittgenstein's Post-*Tractatus* Works

But Wittgenstein, of course, did not remain silent after the publication of his *Tractatus*, even if what has since appeared under his name was not completed by Wittgenstein himself. Furthermore, in the works now assigned so Wittgenstein's "middle" and "later" periods, there are scant references to "the world." Accordingly, my treatment of those works will be correspondingly skeletal, especially since his *Philosophical Investigations* has drawn more commentaries than perhaps any other book of philosophy since the end of World War II and has also been enormously influential in further steering academic philosophy away from the "big questions" that had occupied it for millennia and toward ahistorical and generally apolitical linguistic analyses of the meanings of words.

So, what appear to be the meanings of "the world" in Wittgenstein's post-*Tractatus* books? First, I will consider his *Blue and Brown Books*, then the few relevant references in the *Philosophical Investigations*, and will conclude with some observations on his *On Certainty*.

In *The Blue Book* of 1933–34, Wittgenstein presages his "later" philosophy by focusing on the "grammar of our expressions," the description of one of his most famous concepts, language-games (*Sprachspiele*), and the meaning of words as based on how they are used in ordinary language.[47] These themes would also be a major focus of his *Philosophical Investigations*. Among the most relevant passes in *The Blue Book* are those in which he discusses these themes to examine "thinking," "mental," and "physical worlds" in the contact of ordinary language:[48]

> [it] is liable to mislead us if we say "thinking is a mental activity." … And it is extremely important to realize how, by misunderstanding the grammar of our expressions, we are led to think of one in particular of these statements as giving the *real* seat of the activity of thinking.

He then introduces the key notion of "language games":

> *Language games are the forms of language with which a child begins to make use of words. The study of language games is the study of primitive forms of language or primitive languages* … A word has the meaning someone has given it … It is wrong to say that in philosophy

> we consider an ideal language as opposed to an ordinary one. *For that would make it appear as though we thought we could improve on ordinary language. But ordinary language is all right.* Whenever we make up "ideal languages," it is not in order to replace our ordinary language by them, but just to remove some trouble caused in someone's mind by thinking that he has got hold of the exact use of a common word.

Wittgenstein then goes on to discuss "mental" and "physical worlds," the former being a kind of "ethereal subterfuge," and "dissolves" the mind/matter problem by looking at how these "gaseous" words are used in ordinary language:[49]

> *At first sight it may appear ... that here we have two kinds of worlds, worlds built of different materials; a mental world and a physical world. The mental world in fact is liable to be imagined as gaseous, or rather, ethereal ... we already know the idea of "ethereal objects" as a subterfuge, when we are embarrassed about the grammar of certain words, and when all we know is that they are not uses as names for material objects. This is a hint as to how the problem of the two materials, mind and matter, is going to dissolve ... The thing to do in such cases is always to look how the words in question are actually used in our [ordinary] language.*

So in those preceding sentences, the "problem" of "the two materials, mind and matter," is going to "dissolve," as is the "gaseous" "mental world" and the substance dualism that had bedeviled philosophy for centuries, especially Cartesian idealism.[50] *But does simply looking at how the words in question are actually used in ordinary language "dissolve," or just defer, in depth consideration of "a mental and a physical world"* Wittgenstein's subsequent works would indicate the former; however, the more-recent debates in contemporary philosophy of mind, consciousness studies, and neurophilosophy about how mental and physical states are related would indicate a deferral of rather than a (dis)solution of this problem.

In his *Philosophical Investigations*, Wittgenstein further develops these ideas and many others—notably including "language games," "forms of life," "rule-following," "private language," and "meaning as use," etc.—which have been explored by many other commentators on this text and, in any event, are tangential for "the world." While there is a paucity of explicit references to *Welt* in his *Philosophical Investigations*—unlike the *Tractatus*, which is replete with them—Wittgenstein's late masterpiece does include relevant passages that demand attention, the first of which is:[51]

> Other illusions come from various quarters to attach themselves to the special one spoken of here. *Thought, language, now appear to us as the unique correlate, picture, of the world* [German: *Weltbild]. These concepts: proposition, language, thought, world, stand in line one behind*

> *the other, each equivalent to each.* (But what are these words to be used for now? The language-game in which they are to be applied is missing.)

Then, in a particularly striking passage, *Wittgenstein relegates words like "world" and "experience" to the same linguistic status of "doors" and "tables," although "the language game in which they are applied is missing."* Hence, while Wittgenstein mentions (without clarification) "the a prior order of the world ... the order of possibilities, which must be utterly simple" and, in a Kantian phrase, "is prior to all experience" and "must run through all experience," *Wittgenstein seems to dismiss the "illusion" of a ("super") "order," or "super-concepts" in favor of a "humble use" of the words "language," "experience," and "world*":[52]

> Thought is surrounded by a halo—*Its essence, logic, presents an order, in fact the a priori order of the world: that is, the order of possibilities, which must be common to both world and thought. But this order, it seems, must be utterly simple. It is prior to all experience, must* run through all experience; no empirical cloudiness or uncertainty can be allowed to affect it—It must rather be of the purest crystal. But this crystal does not appear as an abstraction; but as something concrete, indeed, as the most concrete, as it were the *hardest* thing there is ... We are under the illusion that what is peculiar, profound, essential, in our investigation, resides in its trying to grasp the incomparable essence of language. That is, the order existing between the concepts of proposition, word, proof, truth, experience, and so on. This order is a *super-order* between—so to speak—*super-concepts. Whereas, of course, if the words "language," "experience," "world," have a use, it must be as humble a one as that of the words "table," "lamp," "door."*

So, all words, or at least all nouns, according to Wittgenstein, have the same grammatical status? Ok, *but that would also seem to imply that "special" words, or "super-concepts," like "language, experience, and world," have no "profound" or "essential" status. If so, then the entire history of invoking and using those nouns to denote extra-linguistic states of mind ("experience") and spatiotemporal–historical entitles like the world is at best irrelevant, and more likely has committed the category error of mistaking words for things.* If that is the case, then it's no accident that Wittgenstein (and most of his "ordinary-language" philosophical followers) pay scant attention to anything except how words are used (primarily in colloquial English). Or, alternatively, perhaps as Wittgenstein also says in the PI:[53]

> *The essence is hidden from us:* this is the form our problem now assumes. We ask: "*What is* language?", "*What is* a proposition?" And the answer to these questions is to be given once for all; and independently of any future experience ... One person might say "A proposition is the most ordinary thing in the world' and another: "A proposition—that's

> something very queer"—And the latter is unable simply to look and see how propositions really work. The forms that we use in expressing ourselves about propositions and thought stand in his way.

But to claim that "the essence is hidden from us" is to assume the existence of the/an "essence," even if we cannot perceive it. And not to argue for or against the existence of such an "essence" is to beg, or at least, to defer the question.

Wittgenstein, in the posthumously published *On Certainty*, seems to call into question our very ability to know for sure "how things are," even though he also seems to describe the "world-picture" forming "the starting point of belief for me," as well as "what is reasonable to believe."

For example, regarding "the external world," he states that:[54] "Doubting the existence of the external world does not mean, for example, doubting the existence of a planet, which later observations proved to exist." And, regarding "certainty," he says,[55] "Certainty is as it were a tone of voice in which one declares how things are, but one does not infer from the tone of voice that one is justified." In tones recalling his *Philosophical Investigations*, Wittgenstein also advocates that[56] "The propositions which one comes back to again and again as if bewitched—these I should like to expunge from philosophical language." But he also says, implying possible epistemological relativism:[57] "*when language-games change, then there is a change in concepts, and with the concepts the meanings of words change.*"

Regarding his "picture of the world," Wittgenstein declares that:[58]

> But I did not get my picture of the world by satisfying myself of its correctness; nor do I have it because I am satisfied of its correctness. No: it is the inherited background against which I distinguish between true and false … *The propositions describing this world-picture might be part of a kind of mythology. And their role is like that of rules of a game; and the game can be learned purely practically, without learning any explicit rules.*

Furthermore, according to Wittgenstein,[59] "The existence of the earth is rather part of the whole picture which forms the starting-point of belief for me." Here and elsewhere, Wittgenstein seems to exemplify what he calls "the reasonable man" ("person"), especially in a period of radical, "post-modern" doubts about scientific evidence and the subjectivity of belief:[60]

> There cannot be any doubt about it for me as a reasonable person … The reasonable man does not have certain doubts … *Thus we should not call anybody reasonable who believed something in despite of scientific evidence* … When we say that we know that such and such …, we mean that any reasonable person in our position would also know it, that it would be a piece of unreason to doubt it … But who says what it is reasonable to believe in this situation? … So it might be said:

> "The reasonable man believes: that the earth has been there since long before his birth, that his life has been spent on the surface of the earth, or near it, that he has never, for example, been on the moon" ... *But what men consider reasonable or unreasonable alters. At certain periods men find reasonable what at other periods they found unreasonable. And vice-versa.*

Again, this implies a kind of relativism, but not a radical skepticism, about what is true and reasonable to believe. But what is it reasonable to believe about Wittgenstein? Well, at certain periods during his life and just after his death, his students and disciples may have promoted an image of Wittgenstein's "mental and material worlds" that differs somewhat from what one might now believe Wittgenstein's *Weltanschauung* to have been. What, if anything, might one therefore conclude?

Wittgenstein: Logical Existentialist? Liberator? Mystic? Or as Enigmatic as the World?

As the most influential philosopher—at least for Anglophonic academic philosophers—of the past century, and as a person whose life has inspired almost as many interpretations as his written works, Wittgenstein has been read primarily as a revolutionary thinker who has engendered the most important philosophical movement of the post-World-War II period—ordinary language, or analytic, philosophy. This movement has been so dominant, especially in elite philosophy departments in the United States, United Kingdom, Canada, Australia, New Zealand, and much of Scandinavia, that prospective students and faculty primarily interested in the history of philosophy and/or in "continental" philosophy are frequently marginalized or even excluded from the power-and-rewards-structure of mainstream academic philosophy. And professional philosophical publications and conferences in the Anglophonic sphere, while drawing upon writers and participants worldwide, usually conduct their business ignorant of or indifferent to the plentitude of past and present important thinkers and works outside the analytic mainstream, especially if they're not Anglophonic. (Importantly, the reverse usually does not hold, since many, perhaps most, philosophers and other theorists in "continental" Europe, Latin America, and elsewhere cite Anglophonic writers and invite them as guests.)

Until very recently, when we may finally be entering a "post-analytic," or even (lamentably) a "post-philosophical" era, much Anglophonic philosophy has not incorrectly been perceived by many students and the general educated public to be dry, overly technical, arid, irrelevant and, perhaps worst of all, boring, even to many people who are otherwise quite interested in, or even within, the field.[61] It might be unfair to lay blame on Wittgenstein for this situation. But the general absence of references in his works to the "grand European and historical traditions," and the often laser-like focus on the philosophy of language and, in his early works, logic and semantics, together with the "strictly philosophical" works of the younger Bertrand Russell and the generally apolitical and ahistorical culture in which they lived and worked, have laid the foundation for these developments.

On the other hand, very recently there have appeared readings of Wittgenstein's works that seem to interpret them as, at least in part, as actually or potentially "existential," or even as "liberatory." According to Hans Sluga, for example, Wittgenstein's *Weltanschauung* might be called "existential." But this seems to be a peculiar kind of existentialism, one akin more to Schopenhauer and Nietzsche than to the Existentialists per se. As Sluga argues, Wittgenstein's view of the world is as "a totality of accidental, senseless circumstances; a picture ... Wittgenstein ... saw coming from Schopenhauer and confirmed by his war experience ... an existential view of the world."[62] And Sluga also believes Wittgenstein's *Weltanschauung*, especially in the *Tractatus*, to have been a "separation" between the world of personal experience and the "philosophical world."[63]

In addition to contemplating Wittgenstein as a "logical existentialist," Rupert Read adds the designation of "liberatory philosophy" to his work:[64]

> The idea of "logical existentialism" ... is not quite so far away from Wittgenstein's intentions as has been thought ... for Wittgenstein, one is/we are condemned to be free [pace Sartre], even ... in matters of logic ... *we* are the ones who are responsible for that which we do. There is no higher (metaphysical) motivator. So the paradoxical phrase "logical existentialism" does seem at times—*almost*—to catch some of the complexity and surprisingness of Wittgenstein's *treatment of rule following, of his realistic vision of the full fluxing nature of language-in-action in the world.* A vision in which human agency is not subordinate to rule- tyrannies ... but rather, rules are understood ... *through* their practical presence in actual and possible human and social life. A presence which, as Sartre might have said, is most present in their seeming absence: for it is when rules are acted *from*, and thus in effect *no longer* stand there like signposts.

Reed also says that he is seeking in reading the PI:[65]

> to read Wittgenstein in such a way that the book's major key (liberation) and its minor key (ethics) come to be seen as two ways of unlocking the same door. The two come to be seen as internally related, in honesty in particular and the intellectual virtues in general ... they come to be seen as internally related in the intimate connection that emerges between autonomy in Wittgenstein's sense (as opposed, not to others, but to captivity by delusion) and relationality to other beings ... *Seeing* the other (and not as: an Other*). It is life, as opposed to the machine*

> *which our world is being remade in the image of* ... Philosophy is the love of true freedom, in the intellectual realm; that *is* wisdom. Liberatory philosophy ... is a freeing—and *simultaneously,* a thoroughgoingly uniting-with-others—way of seeing Wittgenstein's philosophical activity. The final and crucial liberation we need ... is from banal, widespread fantasies of "liberty." We need to be freed from the crude delusion of freedom "itself" by which we are possessed. We need to be freed from our obsession in politics and economics, across most of the contemporary world, with (such) "freedom ..."

While it is unclear to me the extent to which such admirable ethical and "liberatory" sentiments are Wittgenstein's own, it is obvious *that commentators on such influential works as those penned by Wittgenstein are, inevitably, to some extent projecting their own interests and desires onto the texts they interpret.* And I, of course, am no exception. Accordingly, perhaps it is most prudent in reading Wittgenstein to return to what Wittgenstein himself says at the very end of his *Tractatus*, when he uses the words "inexpressible" and "mystical," and, hence, that thereof "one cannot speak ..." one "must be silent."

Wittgenstein also famously declared toward the end of the *Tractatus* that "The solution of the problem of life is seen in the vanishing of the problem" (6.521).[66] And in his *Philosophical Investigations*, he proclaimed:[67]

> *Philosophy is a battle against the bewitchment of our intelligence by means of language ... What we want to do is to bring words back from their metaphysical to their everyday use ...* When philosophers use a word—"knowledge," "being," "object" "I,"—and try to grasp the *essence* of the thing, one must always ask ...: is the word ever actually used in this way in the language-game which is its original home?

In the same work, Wittgenstein announces:[68] "The philosopher's treatment of a question is like the treatment of an illness."

And in the *Cambridge Lectures*, Wittgenstein declares:

> This idea of *one* calculus is connected with [the] consequence that certain words are on a different level from others, e.g., "proposition", "world", "word", "grammar," "Logic." I had the idea ... that certain words were *essentially* philosophical words ... But it's not the case that those words have a different position from the others. e.g., *I could now just as well start [the] Tractatus with a sentence in which "lamp" occurs, instead of "world."*[69]

But perhaps the linguistic–analytic (as opposed to psychoanalytic ☺) "treatment" Wittgenstein may have proposed is killing, rather than curing, the "patient"—Western philosophy in particular and the Occidental intellectual tradition in general. And perhaps the best "solution" to the "problem" of understanding and interpreting Wittgenstein's words is to abandon the attempt to do so ...

Still, I have not been silent in writing about Wittgenstein. But he cannot reply, at least not in this world, or possibly not in any other as well. And it is briefly to other, "possible worlds" that I now turn.

The Possible Worlds, and Words, of David K. Lewis

Concepts of the world in the post-Wittgensteinian "analytic tradition" of Western philosophy are generally notable for their focus on "possible worlds" rather on this, the "actual," world, and for logical leaps of imagination in constructing the world we inhabit. This is a development that gathered steam in the 1960s and, by the 1980s, had become quite mainstream in Anglophonic philosophical circles, especially those with an American accent.

One of the notable aspects of this dimension of analytic philosophy is its avoidance of or indifference to not only "continental" philosophical discussions of "the world"—with the possible exception of infrequent references to some of the more technical aspects of Kantian epistemology or to Leibniz's logic and metaphysics—but also to the findings of the empirical social and natural sciences—the American philosopher Willard Van Orman Quine (1908–2000) being an exception regarding the latter. This is quite curious given the prominence of the philosophy of science (and its intellectually disreputable offshoot, scientism) in mainstream philosophical circles since the turn of the 20th-century.

Another remarkable feature of this mode of philosophizing is its ostensible abstention from theory-formation and metaphysical posturing, in the spirit of Wittgenstein, while at the same time demonstrating a striking utilization of suppositions and speculations about "other worlds," "Martians," the "common man," and "transworld individuals." There is also within much modal logic and analytic philosophy, more generally, a significant reliance on thought experiments that might make even the most fanciful works of science fiction seem unimaginative in contrast. And, even the style of many of the technical and rigorous professional articles and monographs in this tradition, while seemingly sparse and impersonal, is often strewn with first-person pronouncements but typically devoid of references to philosophers, as well as to other thinkers, scholars, and even scientists, outside a relatively narrow network of like-minded peers. The result is the development of an often insular and a self-referential professional discourse seemingly rigorous and clear, but usually shorn of interest in anything outside contemporary professional fads and strikingly devoid of "real-world" concerns.

Although any number of logicians, semanticists, and philosophers of language, including Saul Kripke (1940–) and Quine,[70] have contributed to "modal logic" and have occasionally dabbled in "possible worlds theories," perhaps the thinker most closely associated with this topic was

David K. Lewis, who, according to a previously mentioned "ranking" of influence among Anglophonic philosophers placed third among 20th-century philosophers, behind only Wittgenstein and Russell but ahead of Heidegger and everyone else in the "continental tradition."[71] While Lewis is virtually unknown outside academic philosophical circles—in sharp contrast with such as "public intellectuals" as Sartre, Camus, and de Beauvoir (the latter two French celebrity-writers having virtually no standing within Anglo-American academic philosophy)—the ahistorical methodology of logical and linguistic analysis utilized by Lewis and his peers is representative of the analytic philosophical tradition's not-infrequently narrow and culturally provincial approach to conceiving the world.

The encomiums paid to Lewis are exemplified by the influential article on Lewis by Brian Weatherson in *The Stanford Encyclopedia of Philosophy*, who states:[72]

> David Lewis was one of the most important philosophers of 20th Century. He made significant contributions to philosophy of language, philosophy of mathematics, philosophy of science, decision theory, epistemology, meta-ethics and aesthetics. In most of these fields he is essential reading; in many of them he is among the most important figures of recent decades. And this list leaves out his two most significant contributions. In philosophy of mind … But his largest contributions were in metaphysics … his modal realism. Lewis held that the best theory of modality posited concrete possible worlds. A proposition is possible if and only if it is true at one of these worlds … It is hard to think of a philosopher since Hume who has contributed so much to so many fields. And in all of these cases, Lewis's contributions involved … articulating, a big picture theory of the subject matter … Because of all his work on the details of various subjects, his writings were a font of ideas even for those who didn't agree with the bigger picture.

And in his *Internet Encyclopedia of Philosophy* online article, Scott Dixon has written a similar paean to Lewis's philosophical work, albeit in a somewhat less rhapsodic and prolix vein.[73]

What are missing from these accounts, professionally speaking, are any indications of Lewis's (or his peers') interests in the history of philosophy, political and social philosophy (except for John Rawls, Robert Nozick, and their "liberal" and "libertarian" acolytes and commentators), and, of course, "continental" thought. That is unsurprising given the cultural context of Anglophonic philosophy, especially in such elite philosophy departments as those at Harvard, Oxford, Princeton, Cambridge, Pittsburgh, Rutgers, Berkeley, and New York University, et al, where many of the most influential analytic philosophers studied, were trained, and have taught. However, what is even more striking is the relative paucity of details regarding Lewis's life—in sharp contract with the abundance of information in the same *Encyclopedias* about the lives of Heidegger, Sartre, and even Wittgenstein. Perhaps this reflects a tendency in the analytic tradition to exclude history, culture and society, politics, and psychology and to focus like a laser beam on the technical issues published in professional journals and books. In keeping with this tradition, my comments will be similarly sparse and chiefly focused on the more accessible and relevant aspects of Lewis's main monograph, *On the Plurality of Worlds*.

What is known is about Lewis's life is that he was born in 1941 in Ohio to two academics. While an undergraduate at Swarthmore College near Philadelphia, he spent a year at Oxford, where he heard the eminent philosopher of language J.L. Austin's last lectures and had Iris Murdoch as a tutor. He received his Ph.D. in philosophy at Harvard under the supervision of Quine, and his dissertation became his first book, *Convention*. He published more than 100 professional papers and a few book-length monographs. He taught at Princeton University for most his life, and he died in 2001. Unlike many philosophers, he was married. That's it.

Of greatest relevance for my purposes are Lewis's contributions to "modal logic," or "concrete modal realism," and, above all, his view of "possible worlds." Modal logic (or ML) investigates the usage of such terms as "possibly," "probably," and "necessarily":[74]

> Modal logic is ... the study of the deductive behavior of the expressions "it is necessary that" and "it possible that." However, the term "modal logic" may be used more broadly for a family of related systems. These include logics for belief, for ... temporal expressions, for the deontic (moral) expressions such as "it is obligatory that" and "it is permitted that ..." Modal logic also has important applications in computer science.

Saul Kripke was such a major contributor to ML that the most familiar logics in the modal family are constructed from a weak logic called K, after his last name.[75]

Lewis's most famous contribution to ML is what is often called "model realism" (MR), or sometimes "concrete modal realism." In *On the Plurality of Worlds* (OPW), he defines MR as:[76] *"the thesis that the world we are part of is but one of a plurality of worlds, and that we who inhabit this world are only a few out of all the inhabitants of all the worlds."* And in his early book *Counterfactuals*, Lewis wrote about "possible worlds" in the following passage:[77]

> I believe, and so do you, that things could have been different in countless ways. But what does this mean? Ordinary language permits the paraphrase: there are many ways things could have been besides the way they actually are. I believe that things could have been different in countless ways; I believe permissible paraphrases of what I believe ... I *therefore believe in the existence of entities that might be called "ways things could have been." I prefer to call them "possible worlds."*

I believe Lewis says "I believe" quite a bit …

For Lewis:[78]

> *the world is a big physical object; or maybe some parts of it are entelechies* [79] or spirits or auras or deities … nothing is so alien … as not to be part of our world …

So, it seems that, for Lewis *"the world" may be physical, or meta-physical (as in "spirits or auras or deities*"), or, more generally, *"this entire world is … the way things are." But they might have been otherwise, or "different, in ever so many ways."*

Even a non-philosopher might be tempted to say, "duh, really?" This sounds tautologous, since saying that "the world is either physical or meta (non) physical," and "things are the way they are," but "things might have been otherwise" are propositions that *prima facie* seem self-evidently true by definition and devoid of substance (predicates). But let's proceed to see what Lewis has to say about "possible worlds."

Of the huge literature on "worlds," *especially as actual,* Lewis has virtually nothing to say, which is not surprising given the absence of references to his philosophical predecessors from non-analytic traditions. However, the absence of a serious discussion of Leibniz—probably the most important progenitor of "possible worlds" theories (see the section on Leibniz earlier in this book)—is stunning. Lewis mentions this omission and excuses himself from commenting on Leibniz with the demurral that: "Anything I might say about Leibniz would be amateurish, undeserving of others' attention, and better left unsaid."[80]

This apologia (as in a Socratic "defense") may or may not be disingenuous. However, the omission from a book about "possible worlds" of such distinguished, if sometimes ambiguous, views as those of Leibniz (not to mention Kant and the entire Scholastic tradition) is in keeping with the analytic tradition's general ahistorical and anti-exegetical, anti-hermeneutical approach.

In an opening section of OPW called "A Philosophers' Paradise," Lewis discusses our world and other worlds, from the framework of his thesis of "a plurality of worlds," or modal realism:[81]

> There are ever so many ways that a world might be*; and one of these … is the way that this world is.* Are there other worlds that are other ways? I say there are. I advocate a thesis of plurality of worlds, or "*modal realism," which holds that our world is but one world among many.* There are countless other worlds … *Our world consists of us and all our surroundings … it is one big thing having lesser things as parts … likewise … other worlds have lesser otherworldly things as parts. The worlds are something like remote planets … There are so many other worlds … every way … a world could possibly be is a way … some world is …* The other worlds are of a kind with this world of ours.

A lot of "I say," "I advocate," and "there are" proclamations for a renowned logician and philosopher of language ... Lewis continues in the same vein:[82]

> The worlds are not of our ... making. It may happen that one part of a world makes other parts, as we do; and as other-worldly gods and demiurges do on a grander scale ... *none of these things we make are the worlds themselves. Why believe in a plurality of worlds?—Because the hypothesis is serviceable, and that is a reason to think ... it is true.*

A "serviceable hypothesis" is a (good?) reason "to think that is true," really? Physicists, chemists, cosmologists, millions of ordinary people, and, yes, even some philosophers (including Aristotle!) have had "serviceable" but ultimately false, hypotheses about the "geocentric" cosmos, the "eternity of the world," "spontaneous generation," "the flatness of the earth," the "aether," "phlogiston," etc. And, yes, indeed, many analytic philosophers have referred to possible worlds, but usually less frequently than Lewis refers to them; actual individuals in this actual world are usually outside this analytic universe of discourse. But what, for Lewis, is "the actual world," but, pace Wittgenstein's *Tractatus*, "whatever actually is the case":[83]

> the *world we are part of is the actual world ... That is one possible way for a world to be. Other worlds are ... unactualized, possibilities.* If there are many worlds, and every way that a world could possibly be is a way that some world is, then ... there is some world where such-and-such is the case.

Yup ... "Other worlds" are "unactualized possibilities." Presumably, *our* world is a *realized* "possibility," and therefore "the *actual* world," or "what goes on here ..." But there are alternatives to our "actual world." And Lewis then proceeds to discuss "worldmates" and the world in "mereological" (wholes and parts) terms:[84]

> *A possible world has parts, namely possible individuals. If two things are parts of the same world, I call them worldmates. A world is the mereological sum of all the possible individuals that are parts of it, and so are worldmates.*

Lewis continues by describing worlds as "individuals" and as "particulars," as well as "spatiotemporally and causally isolated from one another." He also "says" a lot about "alien worlds" and "alien natural properties," and "says that":[85]

> *worlds are individuals, not sets ... worlds are particulars, not universals. So ... worlds are concrete ... Worlds are spatiotemporally and causally isolated from one another; else they would be not whole worlds, but parts of a greater world.*

Lewis then offers a preliminary summary of his position regarding "ours ... the actual world, or 'this-worldly,' as well as" "my world" and "my worldmates":[86]

> *ours is one of many worlds. Ours is the actual world; the rest are not actual. Why so?... "actual" mean[s}... "this-worldly."... applies to my world and my worldmates; to this world we are part of, and to all parts of this world.*

In one of the most interesting and unexpected passages in OPW, Lewis then restates his previous articulation of the "ways of 'being in a world,'" without, of course, referring to Heidegger or any continental philosophical discussions of this most "existential" of all topics:[87]

> I distinguished three ways of "being in a world": (1) being *wholly* in it ... being part of it; (2) being *partly* in it ... having a part that is wholly in it; and (3) existing *from the standpoint of* it ... something exists according to a world—for instance, Humphrey both exists and wins the presidency according to certain worlds other than ours.[88]

I assume that the "Humphrey" to whom Lewis refers is Hubert Humphrey, who lost the 1968 U.S. presidential election to Richard Nixon. But substitute "Trump" for "Humphrey," and to millions of Trump's supporters and enablers, he "won" the 2020 U.S. presidential election and not Joe Biden. Do, therefore, Trump and the Trumpistas exist "according to world other than ours?" Or are they simply wrong and don't know it, or know they are wrong and lie about and hence deny the truth, and therefore continue to exist as parts of *this* world, but as cognitively and/or ethically impaired individuals? Or maybe, as humans we are all like this author and you, my readers, since, while we have reason, we are not gods and are "only so smart," as Lewis says?[89]

Departing from "an impossible god," Lewis poses (seemingly as a straw man to his own MR), the following alternative, called "ersatz modal realism," but also considers "common sense" and the common opinion that there is "one concrete world":[90]

> a popular ... alternative to my own view [is] *ersatz modal realism* ... instead of an incredible plurality of concrete worlds, *we can have one world only, and countless abstract entities representing ways that this world might have been ... common sense opinion about what there is must be respected. There is one concrete world, and one only. It includes all the concrete beings there are. There are no other worlds, and no other-worldly possible individuals ... the one concrete world is not any more extensive ... not any less of a spatiotemporal unity, than common sense opinion supposes it to be.*

In this case at least, common sense makes good sense, at least unless and until astrophysicists and cosmologists can scientifically demonstrate the existence of other worlds with actual, not "ersatz," individuals.

Lewis then goes on further to discuss "ersatz worlds" and "world-making language." For one of the few times in his OPW, he refers to Wittgenstein and, surprisingly, to Leibniz as well:[91]

> if the worldmaking language is plain English … *the state-descriptions represent Leibniz's possible worlds or Wittgenstein's possible states of affairs.*

Then, he concludes his OPW with statements about what must be "true of it at a world" (?), the (im?)possible existence of "transworld individuals," "possible worlds," and, finally, "the world as a totality of things":[92]

> *What is true of it at a world will … be … that it has its constant essential intrinsic nature; and … it has various relationships … to other things of that world.*

Interesting logic for a logician is entwined in this declaration:[93]

> *I do not deny the existence of trans-world individuals, and yet there is a sense in which … they cannot possibly exist.*

And to conclude the book, Lewis summarizes his main claims:[94]

> *Possibilities are not always possible worlds …* not all possible individuals are possible worlds. Only the biggest ones are … *The world is the totality of things. It is the actual individual that includes every actual individual as a part … a possible world is a possible individual big enough to include every possible individual …* a way … an entire world might possibly be.

It might also be possible, of course, *that the only way to answer scientifically and, therefore, empirically, whether any other possible world (besides our own) is actual, is not to engage entirely in logic games and imaginative, but fanciful, suppositions, but to combine the evidence provided by scientific investigation with the procedures of inferential and deductive reasoning characteristic of modern logic and linguistic analysis*. But Lewis, like most of his peers, is content with the latter. A serious inquiry into the possibility and actuality of worlds demands more than that.

More generally, Lewis and other analytic philosophers, many of whom, like Russell, Wittgenstein, and their successors, were originally mathematicians, logicians, and/or engineers or scientists prior to moving into philosophy, carry with them the assumptions of those disciplines, as do continental philosophers from less technical backgrounds bring to their work their humanistic and/or social–scientific or artistic backgrounds. Furthermore, analytic thinkers implicitly or explicitly tend to look askance at what are often deemed the "unscientific," "unphilosophical," and/or "ungrammatical" methods and interests of philosophers with more humanistic and/or social scientific inclinations, and, as gatekeepers of mainstream philosophy departments and publishing vehicles, routinely exclude from their power structures would-be university-based philosophers who don't share their substantive and methodological orientations.

When attention is paid by mainstream analytic philosophers to non-Anglophonic thinkers and more "synthetic" philosophers, those cited are mostly German-speaking theorists who have made a significant contribution to modern logic, the philosophy of science, the philosophy of language, and/or contemporary philosophy of mind—especially from logical positivist and/or physicalist/materialist frameworks. This is clearly so with Frege and Leibniz—who were frequently cited by Russell and Wittgenstein—and, to a somewhat lesser degree, is also the case with the German philosopher/logician Rudolf Carnap (1891–1970, whose 1928 book *The Logical Structure of the World* is a central text of logical positivism and logical empiricism). The Austrian–British philosopher of science Karl Popper's (1902–94) book *Logic of Scientific Discovery* is a classic in that field and Popper also published *The Open Society and Its Enemies*, a critique of totalitarianism, after the annexation of Austria by Nazi Germany in 1938. The latter work is one of the few books in social and political theory by a philosopher with principally scientific interests—although it might be criticized on many grounds, including its labeling of such thinkers as Plato and Hegel as "authoritarian" and "proto-totalitarian" in part because of what Popper calls their "holist" and "historicist" presuppositions.[95] In any event, works by Anglophonic philosophers on "grand" themes from the history of Western philosophy are few and far between …

David Chalmers on Constructing the World

The recent work of David Chalmers (1966–) the very influential Australian philosopher of mind and language, especially in his book *Constructing the World*, is illustrative of the central tendency of analytic thinkers mainly or exclusively to cite only those non-Anglophonic philosophers deemed essential for their purposes. In Chalmers' case, it is Rudolf Carnap who figures prominently in *Constructing the World*, a lengthy and ambitious work.

Chalmers is most noted for his arguments regarding "the hard problem" of consciousness, one version of which is why do the feelings [subjective experiences called "qualia"[96]] that accompany awareness of sensory information as generated by the brain exist at all? He is also noted for his views regarding the possibility of philosophical "zombies."

In his *Constructing the World* (abbreviation: CW), Chalmers departs from Carnap's attempt in *The Logical Structure of the World* to derive all concepts, including logical ones, from a "single primitive concept," from which all other concepts are used to build up our concepts—spatial, temporal, bodily, behavioral, and mental—of the "external world." He proposes "definitional scrutability," as "a compact class of truths from which all truths are definitionally scrutable," as exemplified by Carnap's "single world-sentence D that definitionally entails all truths," as the epistemological foundation for the "knowledge of the base truths about the world [that] might serve as a basis for knowledge of all truths about the world." Although Carnap's project was widely considered an ambitious failure, Chalmers argues that a version of Carnap's project might succeed, so that *with the right basic elements and derivation relation, we can indeed "construct the world" and with the right framework in "metaphysical epistemology," a "global picture of the world" might be generated.*[97]

Typically, in Chalmers' CW, there are no citations from, or references to, any thinkers in the European "continental tradition," except for logicians and philosophers of science central to analytic frames of reference. Nor is there any sustained discussion of "the world" that is to be logically "constructed." This is in sharp contrast to such contemporary "continental" philosophers and social theorists as Jürgen Habermas (1929–).[98]–).

Philosophical parochialism, even while its practitioners are analyzing and constructing "the world," is, however, not limited to its English-speaking practitioners.[99] Recent and contemporary views of the world by French-speaking thinkers are not entirely free of intellectual ethnocentrism, either. However, if Francophonic philosophers "err," it is in the direction of at times being perhaps excessively historical, social, and political in their orientations—in sharp contrast with the often dessicated and deracinated discussions of actual and possible worlds by those in the analytic philosophical tradition. One of the most influential French philosophers of the past century was Jacques Derrida.

Derrida's World Deconstruction/Destruction

A phonocentric tradition casts language not in the world (the exterior, mundane, empirical, and contingent) but as the "origin" of the world. Language is the transcendental possibility of the world. It is itself entirely free of the world: intelligible, necessary, and universal ... Without the ideality of the *phōnḗ* and the *logos,* there would be no "idea of the world."[100]

Jacques Derrida, *Voice and Phenomenon*

From this tradition, without speaking and hearing, no language, no concept of the world ... "the game of the world" ("le jeu due monde") is to think of writing which is neither in the world nor in "another world" of writing that marks "the absence of the here-and-now of another transcendent present, of another origin of the world."[101]

Derrida, *Of Grammatology*

Jacques (aka "Jackie") Derrida was born in 1930 into a Sephardic Jewish family near Algiers, Algeria. Renowned as the founder of "deconstruction," a sometimes contentious and controversial way of criticizing both texts and political institutions he introduced in 1967, Derrida himself at times seemed to have regretted the fate of his linguistic/philosophical creation. Deconstruction subsequently became quite influential in continental (but not analytic) philosophy, literary criticism, critical theory with a French accent, art and architectural theory, and, to a somewhat lesser degree, in political theory with a leftist orientation. For a time during the late 20th-century, especially in, perhaps ironically, Francophilic circles in the humanities (but, importantly, not in mainstream philosophy) departments of many Anglophonic universities (notably at Yale and the Universities of California at Irvine and Berkeley), Derrida's fame nearly reached the status of a rock star, with people thronging to lecture halls to hear him speak, films and televisions programs devoted to him, and numerous books and articles on his thought. The closest contemporary "public-intellectual" equivalent is probably the Slovenian philosopher/critic Slavoj Žižek (1949–), whose performance pieces and myriad publications generate an almost cultish-following.[102]

After a childhood during which Derrida often felt himself to have been a victim of anti-Semitic-inspired discrimination, in 1949, Derrida moved to Paris.[103] Three years later, he was admitted to the elite École Normale Supérieure (the ENS, where Sartre, Beauvoir, and many other distinguished French intellectuals had studied). At the ENS, Derrida came into direct or indirect contact with the who's who of post-war French intellectual culture, mostly notably with the influential leftist philosopher Gilles Deleuze (1925–95), whom Derrida befriended, as well as with the Marxist theorist Louis Althusser (1918–90), the ostensible progenitor of postmodernist "philosophy" Jean-Francois Lyotard (1924–98), and the literary critic and semiotician Ronald Barthes (1915–80).[104] Furthermore, Merleau-Ponty, Sartre, de Beauvoir, the structuralist anthropologist Claude Lévi-Strauss (1908–2009), the unorthodox psychoanalyst Jacque Lacan (1901–89), and the hermeneuticians and religious philosophers Emmanuel Levinas (1905–95) and Paul Ricoeur (1913–2005) were also active during Derrida's coming-of-age in Paris in the 1950s—as was Michel Foucault (1926–84), with whom Derrida famously feuded.[105] This was also a period in France in which philosophy was dominated by phenomenology and existentialism. Accordingly, Derrida wrote his master's thesis on *The Problem of Genesis in Husserl's Philosophy*, and he studied Hegel's works with Jean Hyppolite (1907–68), who supervised Derrida's unfinished doctoral dissertation, "The Ideality of the Literary Object," which focused on the problem of language in Husserl's theory of history.[106]

During the 1960s, Derrida perused seminal works by Heidegger, Nietzsche, Saussure, Freud, Levinas, et al, and in 1967 he published three books—*Writing and Difference*, *Voice and Phenomenon*, and, most importantly, *Of Grammatology* (abbreviation: OG). In those works, especially in OG, Derrida used the word "deconstruction" *en passant* to describe his project, which also highlighted such other signature concepts as *différance*, crucially distinguished from but related to *difference*, "logocentrism," and the "metaphysics of presence," inter alia. These terms are notoriously elusive and perhaps too many pages have been printed in interpreting them. In addition, Derrida's style of writing (his own "grammatology") is very often obscured or enhanced, encumbered or enlivened—depending on one's point of view—by word-plays, neologisms, etymological genealogies, and other linguistic devices that have contributed not only to his popularity within literary circles but also to not-inconsiderable derision and perplexity among many Anglophonic philosophers and readers outside faddish critical theory (with a French accent) elites.[107]

In the early 1980s, Derrida became the founder, and for a time director, of the Collège Internationale de Philosophie in Paris, and was also appointed "Director of Studies" in "Philosophical Institutions" at the École des Hautes Études en Sciences Sociales, a position he held until his death. He also held many appointments in American universities, in particular at Johns Hopkins University, Yale University, and the University of California at Irvine, where the Derrida archives was established. Derrida was also at times closely associated with a group of thinkers investigating how philosophy is taught in the high schools and universities in France.

During the 1990s, Derrida's published works focused mainly on politics, religion, and notably, animal rights. His 1993 book, *Specters of Marx, inter alia* argued that a "deconstructed Marxism" consisting in a "new messianism of a democracy to come" is highly relevant for this "age of globalization." And in *Rogues* (2003), Derrida, like Noam Chomsky, claimed that even the most democratic nations—the United States in particular—are perhaps the most "roguish" of all states. Based on lectures first presented during the summer of 1998, *The Animal that Therefore I am* appeared in 2006 as his first posthumous work. "Animality" and the concept of world as historicized by Heidegger are central themes of Derrida's last courses, also posthumously published as the multivolume *The Beast and the Sovereign*. While conducting these seminars (2002–03), Derrida was diagnosed with pancreatic cancer. He departed this world in 2004.

De(con)structing Deconstruction

Deconstructing Derrida's printed signs in detail is a wordy, worldly activity best left to those who have dedicated time, space, and careers to that unenviable task. Given my limited resources, expertise, and interest, I shall not attempt to write more than the barest statements necessary to focus on how the Derridean signs disclose "the world" as articulated—in all their intricacy and ambiguity—in the few texts by the absent "author" Jacques Derrida available to me during the pandemic.[108]

Among Derrida's signature terms, perhaps the most notable and notorious is "deconstruction."[109] Derrida's original use of the word "deconstruction" may have been his version of Martin Heidegger's concept of *Destruktion*. Heidegger's term referred to the exploration and disclosure of the concepts that tradition and history have imposed on a word. Here is how Derrida uses the word "deconstruction" in "Signature Event Context":[110]

> Deconstruction cannot limit itself or proceed immediately to a neutralization; it must, by means of a double gesture, a double science, a double writing, practice an *overturning* of the classical opposition *and* a general *displacement* of the system ... *Deconstruction does not consist in passing from one concept to another, but in overturning and displacing a conceptual order, as well as the nonconceptual order with which the conceptual order is articulated.*

And in his at times remarkably clear and pithy "Letter to a Japanese Friend," Derrida, while refraining from giving either a clear definition or examples of "deconstruction," does provide a somewhat minimal "context" for the term's usage:[111]

> especially in the U.S., the motif of deconstruction has been associated with "poststructuralism" [a word unknown in France until its "return" from the U.S.] ... deconstruction is neither an analysis nor a critique ... Deconstruction is not a method and cannot be transformed into one ... If deconstruction takes place everywhere it takes place, when there is something ... we still have to think through what is happening in our world, in modernity, at the time when deconstruction is becoming a motif ... The word "deconstruction," like all other words, acquires its value only from its inscription in a chain of possible substitutions, in what is too blithely called a "context."

Vive la Différance?

Another key term in Derrida's lexicon of neologisms is *différance*, differing from but related to *différence*. In his 1972 book *Margins of Philosophy*, Derrida describes the "movement of *différance*," and in the Introduction to selections from that work, Peggy Kamuf states:[112]

> Derrida's invented word ... welds together difference and deferral and thus refers to a configuration of spatial and temporal difference together ... Because it cannot be differentiated in speech, the mark of their difference [*difference* and *différance*] is only graphic; the a of *différance* marks the difference of writing within and before speech.

According to Derrida, there is a "general law of *différance*" namely that "the principle of language is "spoken language minus speech":[113]

> *différance* as temporization, *différance* as spacing. How are they to be joined? Let us start ... from the problematic of the sign and writing ... the subject [...its self-consciousness] is inscribed in language, is a "function" of language, becomes a speaking subject only be making its speech conform ... to the system of the rules of language as a system of differences ... by conforming to the general law of *différance*, or by adhering to the principle of language that Sausurre says is "spoken language minus speech."

In *Speech and Phenomena*, Derrida states that *différance* "moves" to "produce the subject," via its "general law," pace Saussure, that[114] "The movement of *différance* is not something that happens to a transcendental subject; it produces the subject."

And in *On Grammatology (*OG*)*, broadly understood as the study or science of writing and written signs, Derrida segues from *differance* as "differing/deferring" to his three other main conceptual terms, deconstruction, presence, and logocentrism:[115]

> with respect to what I shall call *differance,* an economic concept designating the production of differing/deferring ... Operating necessarily from the inside ... the enterprise of deconstruction always in a certain way falls prey to its own work ... This deconstruction of presence accomplishes itself through the deconstruction of consciousness, and therefore through the irreducible notion of the trace (German: *Spur*), as it appears in both Nietzschean and Freudian discourse ... the logos as the sublimation of the trace is theological. Infinist theologies are always logocentrisms, whether they are creationisms or not.

Of course ...

Signifying Logocentrism and Erasing the Privileged Metaphysics of Presence

In OG, by appropriating the theory of signs as articulated by Saussure, and by stressing the crucial distinction between signifier and signified, Derrida attempts to connect, deconstruct, and seemingly to "undo" the "logocentrism," or the "presence and consciousness" of the "metaphysics of the logos"—in its "absolute" form, "the face of God—and thereby to reveal "the history of writing" as "erected on the history of the *gramme*":[116]

> It is the book dreamed up by logocentrism ... This distinction [between signifier and signified] is always ultimately grounded in a pure intelligibility tied to an absolute logos: the face of God. *The concept of the*

> *sign, whose history is coextensive with the history of logocentrism, is essentially theological.* How then to understand writing differently? To write differently? … The formal essence of the signified is presence, and the privilege of its proximity to the logos as *phōnḗ* [Ancient Greek for "voice" or "sound"] is the privilege of presence … it is always already in the position of the signifier, is the apparently innocent proposition within which the metaphysics of the logos, of presence and consciousness, must reflect upon writing as its death and its resource … *On what conditions is a grammatology possible? Its fundamental condition is the undoing of logocentrism. But this condition of possibility turns into a condition of impossibility* … The history of writing is erected on the base of the history of the *gramme* [from the Ancient Greek *grámma,* a mark or sign, or what is written] as an adventure of relationships between the face and the hand.

And in *Speech and Phenomena*, Derrida attempts to "restore the original and non-derivative character of signs, in opposition to classical metaphysics" (presumably Aristotle) but which is also simultaneously and paradoxically a "concept whose whole history and meaning belong to the metaphysics of presence":[117]

> A phoneme or grapheme is necessarily always to some extent different each time that is presented in an operation or perception. But, it can function as a sign, and in general as language, only if a formal identity enables it to be issued again and to be recognized … Within the sign *the difference does not* take place between reality and representation … the gesture that confirms this difference is the very effacement of the sign … But because it is … the philosophy and history of the West—which has so constituted and established the very concept of signs, the sign is from its origin and to its core marked by this will to derivation or effacement. Consequently, *to restore the original and nonderivative character of signs, in opposition to classical metaphysics, is, by an apparent paradox, at the same time to effect a concept of signs whose whole history and meaning belong to the metaphysics of presence.*

How to read and interpret the elusive "meanings" of the preceding passage is a challenge I will leave to Derrideans as well as to my more intrepid readers …

That "said," Derrida, in *Margins of Philosophy*, in a kind of phenomenological regression, asks:[118]

> But what is consciousness?

His "answer," however, seems as far removed from Husserl (curiously, Derrida's initial "authorial" inspiration) as imaginable:

> Most often ... consciousness offers itself to thought only as self-presence, as the perception of the self in presence ... so the subject as consciousness has never manifested itself except as self-presence. The privilege granted to consciousness therefore signifies the privilege granted to the present ... *This privilege is the ether of metaphysics, the element of our thought that is caught in the language of metaphysics.*

In her Introduction to Derrida's *Glas*, Kamuf seeks to link the "logocentric baggage" that organizes "the metaphysics of presence" to "the phallus as a mark of presence," as the "sexual scene behind or before ... the scene of philosophy." This is related to what might be interpreted as Derrida's effort to out-psychoanalyze psychoanalysis, to deconstruct and disclose the (alleged) "Sexual Difference in Philosophy," albeit a "feminist-inspired" attempt that Husserl, Brentano, and Frege might have diagnosed as excessive psychologism:[119]

> With the term logocentrism, introduced in *Of Grammatology*, Derrida introduces in an economical manner the set of traits organizing the metaphysics of presence around a center ... logocentric baggage ...: the privilege accorded the phallus as a mark of presence ... a continuity of phallus in logos, and thereby it indicates a certain sexual scene behind or before—but always within—the scene of philosophy.

With these terminological differences (without distinctions?) in place, we now move to Derrida's somewhat scattered earlier remarks on worldliness, the world, and consciousness, to be followed by an exposition and analysis—or deconstructive destruction?—of the world, and its fictional and/or literal end(ings), as disclosed in Derrida's last seminars.

The Origin, and Play, of the Real and the Ideal World

In *Speech and Phenomena*, Derrida links "the world" to "the movement of temporalization," of which "the present alone," it's "presence of the present" is "the ultimate form of being ... by which I transgress empirical existence ... worldliness, etc.":[120]

> The present alone is all there is and every will be ... The relation to the presence of the present as the ultimate form of being and of ideality is the move by which I transgress empirical existence, factuality, contingency, worldliness, etc.—*first of all my own empirical existence, factuality, contingency, worldliness, etc ... the "world" is originally implied by the movement of temporalization.*

These seemingly Heideggerian "moments" are in Derrida's work *Glas* complemented by statements, recalling, perhaps inadvertently, those of the great Swiss psychologist Jean Piaget (1896–1980)[121] about the origins of a child's consciousness in the "culture ... informed by the knowledge of the parents," and the child's consequent education in "relieving" the "contradiction between the real world and the ideal world":[122]

> The child's consciousness does not come to the world as to a material and inorganic exteriority. *The world is already elaborated when education begins, is a culture penetrated, permeated, informed by the "knowledge of the parents"... So the child raised itself in(to) the "contradiction between the real world and the ideal world.* The process of education consists in relieving this contradiction."

Just as the child's sense of the "origin" of the world is co-extensive with the origin of the child's world, as "penetrated, permeated, and informed" by the parents' knowledge and the cultural education in which families are embedded, so, according to Derrida, is *the origin of the world for me that in which "I have lived a unique story."* And my "story" as an individual human being is played out not merely in a contemporary globalized world of inequality and violence, but only against the backdrop of a non-world and the "worldless" environment of animals and other creatures and objects with whom "we share a world." We also live in a specific, historical world, which is for those in the West, the European world, which, according to Derrida, following Husserl, "was born out of the ideas of Kant" as a "regulative, metaphysical world," and that may presage the possibility of "many different worlds to come."

Derrida also argues, plausibly, that the world (for us) cannot be "confused with the world as the globe or the earth," and also that "the thought of the world to come" will be one that is "divisible, unforeseeable, and without condition."[123] As Sean Gaston notes, Derrida, again to some extent following Husserl, also emphasizes:

> that the concept of world can be thought of as both conditioned and unconditioned at the same time ... The uncontained (the universe) has always been in the midst of the contained that contains (the world) and the contained (beings and things). The concept of a world without horizon ... The uncontained with the contained ... The traditional concept of world: the ceaseless attempt ... to establish a vantage point beyond the world ... a transcendental perspective or god-like platform with which to view the world as a whole.[124]

Gaston also stresses the importance for Derrida of language in general and of "a phonocentric" tradition in particular, as the "origin" and "transcendental possibility of the world."[125]

In his discussion of language and the world, Derrida calls phoncentrism and logocentrism "the idealization of language" and describes this linguistic "erasure of the world" as an instance of "the invisible reality of a logos which hears itself speak." Following Kant and the phenomenological tradition, he also makes the distinction between what is "in the world" and what is "outside the world without being in another world."

According to Gaston, Derrida is thereby subverting the Western philosophical tradition's utilization of the concept of world to establish a unique vantage point for the subject:

> The voice takes what is in the world (sound) and produces the ideality of a phenomenological world that stands "outside the world." This voice becomes the possibility of an "idealized world." Language is one more form of the idealization of the world in a long history of idealization.

For Derrida, proceeding from Husserl, "the world" is the "residuum, the remainder that resists a disruption of the phenomenological relation between consciousness and the world." This leads to "the possibilities of the 'end of the world,' the 'threat of its own destruction,'" a theme to which I shall soon return.[126]

Moving in a somewhat different philosophical and grammatological direction, Derrida also shows his indebtedness to Nietzsche in his reflections on the "origin" of "the world" and "the game, or play, of the world." According to Gaston, "the game, or play, of the world [*le jeu du monde*] is to think of writing which is neither in the world nor in 'another world' of world-writing that marks 'the absence of the here-and-now of another transcendent present, of another origin of the world.'"[127]

According to Derrida, Nietzsche, like Heraclitus, celebrates that "joyous affirmation of the play of the world and of the innocence of becoming," and *Nietzsche also describes the world as a "game" or a "play of forces" in which the world is in a continual process becoming, passing way, and destruction.* For Derrida:[128]

> Nietzsche's concept of world—a dynamic relation of forces that are in the process of at once becoming and being destroyed—from Heraclitus … a concept of world no longer tied to gods or to cosmos as the ordering of world. Heraclitus compares the play of the world to the innocent play of a child. Nietzsche contrasts the play of the world, "a becoming and passing away without any moralistic calculations" to the erroneous history of the so called "true world" as a world that is truly available and unavailable.

So What Is "World" According to Derrida (and Heidegger)?

In *The Animal that I am* (TAT), Derrida glosses Heidegger's three "comparative theses" on the world, namely that "… the stone is worldless [*weltlos*], the animal is poor in world [*weltarm*], [and] man is world-forming [*weltbildend*]… it is a matter of knowing what world is …"[129]

Derrida reiterates and amplifies these reflections in *The Beast and the Sovereign* (BS2), where he poses the question regarding, *not only what world is, but also "what world means," leading to "what is now called globalization* [French*: mondialisation], but also the world of* phenomenological and ontological meditations …"[130]

Following Heidegger, Derrida also links these questions to "finitude and loneliness, isolation and solitude," and appears to return to the question launching these metaphysical inquiries: What is world? Seemingly to "answer" this question, Derrida again paraphrases Heidegger in stating:[131]

> in "On the Essence of Ground," but also and especially in *Being and Time*, which is a book on the world, on being-in-the-world, he [Heidegger] had proposed a radically new approach to the question of world: We begin with the first of our three questions: *What is world?* And he distinguishes three possibilities … three "paths" for dealing with the question: The "*first path* toward an initial clarification" is that of historiography, "the *history of the word* 'world' …" But "the history of the word provides only the exterior"; one has therefore to go further, into the "historical development of the concept it contains." That is the path he attempted to pursue … On that path, which begins with the Greek *cosmos*, the Christian conception of world is particularly important.

Derrida continues his inquiry into "world" by citing Heidegger:[132]

> The most familiar aspect of the problem reveals itself in the distinction between God and world. The world is the totality of beings outside of and other than God … And man in turn is also a part of the world understood in this sense. Yet man is not simply regarded as part of the world within which he appears and which he makes up in part. *Man also stands over against the world. This standing-over-against is a "having" of world man, therefore, has the world, whereas … the animal … that it doesn't have it, rather, that it has it without having it … Thus man is, first, a part of the world, and second, as this part he is at once both master and servant of the world.*

According to Derrida, Heidegger brackets or eventually may even reject the two "paths" (of historiography and conceptual history) taken in *Being and Time* in favor of a *"third path," the "the path of comparative examination," in order develop "a new problematic of the world," one in which the "worldless" stones and plants, and the "poor-in-world" animals are compared (invidiously, some might say) to humans, who "do not inhabit the same world" as everything else on the Earth:*[133]

> He [Heidegger] says nothing against the treatment of world in *Being and Time*, against the analysis of "being-in-the-world [*In-der-Welt*

> *sein*]," but what interests him here is a "third path." If one therefore wants to take seriously the theses on the animal, it is necessary to know that Heidegger develops them within a new problematic of world, which is neither that attempted in "The Essence of Ground" nor that of *Being and Time*; he wants a third path—the path of a *comparative examination.*

While I think these three paths are not mutually exclusive—in fact, I have tried to "take" all three at various times in this book—the path seemingly taken by Derrida, following in Heidegger's *Pathmarks, arrives at a destination in which we humans are "homeless" and therefore "also without this world*":[134]

> That, then, *is what the world is, namely the whole in so far as we are this path on the way toward it, but toward it insofar as the path traces itself in it, breaks itself in it, opens itself in it, inscribes itself in it ... what we also call the entirety, the whole and the everywhere, the everywhere of the world, is where we are not at home. Are we not then justified in saying this, namely that we are also without this world, or poor in world, like the stone or the animal ...*?

Perhaps, but later in BS2, Derrida summarizes Heidegger's (and his own?) concept of the world as a world of "co-habitation" among its creatures, even as a "common world," or a" world community":[135]

> the one (Heidegger's hypothesis) does not have the world *as such ... nameable,* or is poor in world, even if the other is ... *weltbildend* [world-forming], the question is indeed that of a community of the world that they share and co- habit. This co- of the cohabitant presupposes a habitat, a place of common habitat, whether one calls it the earth (including sky and sea) or else the world as world of life- death. *The common world is the world in which one- lives- one- dies, whether one be a beast or a human sovereign, a world in which both suffer, suffer death, even a thousand deaths* ... For no one will seriously deny the animal the possibility of inhabiting the world ... no one will deny that these living beings, that we call the beast *and* the sovereign, inhabit a world, what one calls the world, and in a certain sense, the *same* world. There is a habitat of the animal as there is a habitat of plants, as there is a habitat for every living being. *The word "world" has at least as a minimal sense the designation of that within which all these living beings are carried (in a belly or in an egg), they are born, they live, they inhabit and they die.*

So there we are, within "the world," whose meaning has at least this minimal sense. Or are we? For, in a possibly stunning move, Derrida, again glossing Heidegger, also, perhaps inconsistently or at least paradoxically, also seems in his TAT to conclude that:[136]

> *We don't finally know what world is! At bottom it is a very obscure concept!* At the point where he advances like an army, armed with theses, solid, positive theses, it buckles, and he says in the end: decidedly, this concept of world is obscure ... at bottom, we don't know what world is.

Indeed, the very beliefs that there is a/the/this/our/my world, that "the world" has a "meaning," and that we can understand or even "know" "the world" and its "meaning," might be nothing more than comforting fictions, "scientific" or otherwise.[137] Toward the end of BS2, Derrida raises this and other discomforting ideas regarding "no world" and the "end of the world," both in a literary or metaphorical sense, and in the quite literal sense of the "annihilation" of the world, due to war or some other world-destroying threat to life on earth.

The Annihilation of the World and the End of the World—For Derrida—and for Us?

The theme of the "end of the world," or "this fiction of the end of the world," is introduced by Husserl in *Ideas I* (p. 110), and recast by Derrida as the disappearance, the end, the annihilation, and the destruction of the world. Derrida, in his *Rams*, asks us to:[138]

> *Imagine there is No world; the end of the world (French: la fin du monde) marks the reality, the possibility, and the fiction of a destruction that is at once unthinkably total and entirely individual. The end of the world as a whole is linked to the death of the other ... Death is nothing less than an end of the world. Not only one end ... the end of a life or a living being. Death marks ... The absolute end of the one and only world ... the end of the unique world ...* the end of the totality of what is or can be presented as the origin of the world for any unique living being, be it human or not. The survivor ... feels responsible ... without world (German: *weltlos*), in a world without world, as if without earth beyond, the end of the world ... The death of the other [is] linked to the end of the world in totality ... *The possibility of the total destruction of the world, each time there is a death of the other ... Death is nothing else than the end of the world.*

Death—the end of the other, if it is someone else who has died, or my own death if I cease to exist in this life-world—must, according to Heidegger and Derrida, imply the end of *my "unique"* world. But the "possibility of the total destruction of the world," the cessation of life for *all* humans, plants, animals, and organic terrestrial entities and the destruction of the life-world for everyone and everything on Earth—has become more than a mere "possibility" since the dawn of the nuclear age, the so-far uncontrolled increase in global warming, and the possible uncontrollability as such pandemics as severe acute respiratory syndrome coronavirus 2 (SARS-CoV-2), the virus that causes coronavirus disease 2019 (COVID-19).

While Derrida obviously could not analyze the more recent existential risks to our and other species, given his departure from this world in 2004, in his last seminars, he did respond in some detail to Heidegger's ontological question in *Being and Time—Was ist Welt*? (What is World?)[139] He also reflected on the possible annihilation and destruction of the world, especially in the post-World-War II political context.

In discourse reminiscent of the German *Existenz*-philosopher Karl Jaspers as well as the political writings of Albert Einstein, Sigmund Freud, and Bertrand Russell, toward the end of BA2, Derrida ruminates, in a strikingly unambiguous style, on the threat posed to this world by war:[140]

> We know today that, for at least a century now, all wars are *world* wars, worldwide and planetary. *What is at stake in what we call war is more than ever worldwide, i.e. the institution and the appropriation of the world, the world order, no less*. The establishment of a sovereign power and right over the world, of which, still more than the UN, the Security Council would be the emblematic figure, the shadow theater, with all that this collusion of the strongest and the biggest and *the winners of this world* (i.e. of the last world war) can contain of comedy, of strength and weakness, of phantasm and necessity … *And in every war, at stake from now on is an end of the world, both in the sense of ends and purposes that some want to impose by force, force without right or the force of right, force of law … on the interpretation and future of the totality of beings, of the world and the living beings that inhabit it; and at stake is an end of the world, in the sense that what is threatened is not only this always infinite death of each and every one … that individual death … was each time the end of the world, the end.*

And, so it's fitting to conclude this chapter with the destruction of "the world, of any possible world," or as Derrida proclaims:

> the whole end of *the* world (German: *Die Welt ist fort,* English: "the world is gone"), not a particular end of this or that world, of the world of so and so, of this one or that one, male or female, of this civilian, this man, this woman, this child, *but the end of the world in general, the absolute end of the world—at stake is the end of the world … in the sense that what is threatened, in this or that world war, is therefore the end of the world, the destruction of the world, of any possible world, or of what is supposed to make of the world a cosmos, an arrangement, an order, an order of ends, a juridical, moral, political order, an international order resistant to the non- world of death and barbarity* … The armed word of politicians, priests and soldiers is more than ever incompetent, unable to measure up to the very thing it is speaking and deciding about, and

> that remains to be thought, that trembles in the name "world," or even in saying good- bye to the world.

And so, on that uplifting note, I bid "Adieu" to M. Derrida's words and worlds and say "Salut" to the Conclusion of this book.

Notes

1 For a recent "ranking" of 20th-century philosophers by mainly Anglophonic analytically-oriented contributors, who, predictably, put Wittgenstein, Russell, and, perhaps surprisingly, the "possible world" theorist David Lewis, in the first three spots, and such much better-known "continental" philosophers as Heidegger, Sartre, and Merleau-Ponty well down, or even completely off the list (as is the case for Husserl and Herbert Marcuse), see https://leiterreports.typepad.com/blog/2009/03/lets-settle-this-once-and-for-all-who-really-was-the-greatest-philosopher-of-the-20thcentury.html.

2 As previously noted, because of the COVID-19 pandemic, lockdowns, and related travel and library research restrictions, I have had limited access to Wittgenstein's works. The hard copies I do have and to which I will refer are Ludwig Wittgenstein, *Major Works Selected Philosophical Writings* (New York: HarperCollins, 2009, abbreviation: PW), which contains Wittgenstein's *Tractatus Logico-Philosophicus* (abbreviation: TLP), *Studies for Philosophical Investigations* (*The Blue Book* and *The Brown Book*, Abbreviations: BB1 and BB2, respectively), and *On Certainty* (Abbreviation: OC); and Anthony Kenny, ed., *The Wittgenstein Reader* (Oxford: Blackwell Publishers, 2002, Abbreviation: WR), which contains excerpts from various works. I have also had online access to: Wittgenstein's *Philosophical Investigations* (abbreviation: PI), Second Edition, trans. G.E.M. Anscombe (Oxford: Basil Blackwell, 1986), https://static1.squarespace.com/static/54889e73e4b0a2c1f9891289/t/564b61a4e4b04eca59c4d232/1447780772744/Ludwig.Wittgenstein.-.Philosophical.Investigations.pdf; Project Gutenberg's Wittgenstein, *Tractatus Logico-Philosophicus* (abbreviation: TLP), contributor: Bertrand Russell, and trans. C.K. Ogden, www.gutenberg.org/files/5740/5740-pdf.pdf; and Ludwig Wittgenstein, *On Certainty* ed. G.E.M. Anscombe and G.H. von Wright, trans. Denis Paul and G.E.M. Anscombe (Basil Blackwell, Oxford, 1969–75), https://prawfsblawg.blogs.com/files/wittgenstein-on-certainty.pdf.

Other published works by Wittgenstein include: *The Blue and Brown Books* (Oxford: Basil Blackwell, 1969); *Culture and Value*, trans. Peter Winch (Oxford: Basil Blackwell, 1980); and *Lectures and Conversations on Aesthetics, Psychology and Religious Belief*, ed. Cyril Barrett (Oxford: Basil Blackwell, 1966); *Last Writings on the Philosophy of Psychology*, vol. 1, 1982, vol. 2, 1992, G.H. von Wright and H. Nyman (eds.), trans. C.G. Luckhardt and M.A.E. Aue (trans.) (Oxford: Blackwell, 1982 and 1992, respectively); *The Wittgenstein: Conversations, 1949–1951*, O.K. Bouwsma; J.L. Kraft and R.H. Hustwit (eds.) (Indianapolis, IN: Hackett, 1986).

Succinct overviews of Wittgenstein's life and work include: Anat Biletski and Anat Matar, "Ludwig Wittgenstein," *The Stanford Encyclopedia of Philosophy* (Spring 2020 Edition. Abbreviation: Biletski SEP), Edward N. Zalta, ed., https://plato.stanford.edu/archives/spr2020/entries/wittgenstein/; and Duncan

J. Richter, *Ludwig Wittgenstein, International Encyclopedia of Philosophy* (abbreviation: Richter IEP), https://iep.utm.edu/wittgens/.

Biographies of Wittgenstein include, most notably, Ray Monk, *Ludwig Wittgenstein: The Duty of Genius* (London: Jonathan Cape, 1990), as well as Norman Malcolm. *Ludwig Wittgenstein: A Memoir* (Oxford and New York: Oxford University Press, 1984).

The secondary literature on Wittgenstein is, predictably, immense, and still growing. Some salient works include: G.E.M. Anscombe, *An Introduction to Wittgenstein's Tractatus* (Philadelphia, PA: University of Pennsylvania Press, 1971); G.P. Baker, *Wittgenstein's Method: Neglected Aspects*, ed. and introd. Katherine J. Morris (Oxford: Blackwell, 2004); G.P. Baker, and P.M.S. Hacker, *Wittgenstein: Understanding and Meaning, Volume 1 of an Analytical Commentary on the Philosophical Investigations*, 2nd edition (Oxford: Blackwell, 2005)—*Wittgenstein: Rules, Grammar and Necessity, Volume 2 of an Analytical Commentary on the Philosophical Investigations*, 2nd edition (Oxford: Blackwell, 2009); Anat Biletzki, *(Over)Interpreting Wittgenstein*, Leiden: Kluwer, 2003): Max Black, *A Companion to Wittgenstein's Tractatus* (Ithaca, NY: Cornell University Press, 1967); Alice Crary and Rupert Read (eds.), *The New Wittgenstein* (London: Routledge, 2000): Hans-Johann Glock, A *Wittgenstein Dictionary* (Oxford: Blackwell, 1996); P.M.S. Hacker, *Wittgenstein: Meaning and Mind, Volume 3 of an Analytical Commentary on the Philosophical Investigations* (Oxford: Blackwell, 1990)—*Wittgenstein: Mind and Will, Volume 4 of an Analytical Commentary on the Philosophical Investigations* (Oxford: Blackwell, 1996)—*Wittgenstein: Connections and Controversies* (Oxford: Oxford University Press, 2001)—*Comparisons and Context* (Oxford: Oxford University Press, 2013); Andy Hamilton, *Routledge Philosophy Guidebook to Wittgenstein and On Certainty* (London: Routledge, 2014); M.B. Hintikka and J. Hintikka, *Investigating Wittgenstein* (Oxford: Blackwell, 1986); Paul Horwich, *Wittgenstein's Metaphilosophy* (Oxford: Oxford University Press, 2012); Alan Janik and Stephen Toulmin, *Wittgenstein's Vienna* (New York: Simon and Schuster, 1973).

Toulmin (1922–2009), who was on my Berkeley doctoral dissertation committee, once rhapsodized to me about having been in one of Wittgenstein's last Cambridge seminars. See https://en.wikipedia.org/wiki/Stephen_Toulmin.

Other secondary sources include A. Kenny, *Wittgenstein* (Cambridge, MA: Harvard University Press, 1973); James C. Klagge, *Wittgenstein: Biography and Philosophy* (Cambridge: Cambridge University Press, 2001)—*Wittgenstein in Exile* (Cambridge, MA: MIT Press, 2010)—*Taking Wittgenstein at His Word: A Textual Study* (Princeton, NJ: Princeton University Press, 2009); Saul Kripke, *Wittgenstein on Rules and Private Language* (Cambridge, MA: Harvard University Press, 1982); Oskari Kuusela, *The Struggle against Dogmatism: Wittgenstein and the Concept of Philosophy*, Cambridge, MA: Harvard University Press, 2008); Norman Malcolm, *Wittgenstein: Nothing is Hidden* (Oxford: Basil Blackwell, 1986); Colin McGinn, *Wittgenstein on Meaning* (Oxford: Blackwell, 1984); Marie McGinn, *Routledge Philosophy Guidebook to Wittgenstein and the Philosophical Investigations*, 2nd edition (London: Routledge, 2013)—*Elucidating the Tractatus*, Oxford: Oxford University Press, 2009); Thomas McNally, *Wittgenstein and the Philosophy of Language: The Legacy of the Philosophical Investigations* (Cambridge: Cambridge University Press, 2017); Michael Morris, *Routledge Philosophy Guidebook to Wittgenstein and the Tractatus* (London: Routledge, 2008); Hannah Pitkin, *Wittgenstein and Justice: On the Significance of Ludwig Wittgenstein for Social and Political Thought* (Berkeley, CA: University of California Press, 1972); Rupert Reed, *Wittgenstein's Liberatory Philosophy Thinking Through His Philosophical Investigations* (London: Routledge, 2021); Hans D. Sluga and David G. Stern (eds.), *The Cambridge Companion to Wittgenstein* (Cambridge: Cambridge University Press, 1996); Peter Sullivan and Michael Potter (eds.), *Wittgenstein's Tractatus: History and Interpretation* (Oxford: Oxford University Press, 2013); and David G. Stern, *Wittgenstein's Philosophical Investigations: An Introduction* (Cambridge: Cambridge University Press, 2004).

3 See the section on Husserl in this book. For Frege, see Edward N., Zalta "Gottlob Frege," *The Stanford Encyclopedia of Philosophy* (Fall 2020 Edition), Edward N. Zalta, ed., https://plato.stanford.edu/archives/fall2020/entries/frege/; and Kevin C. Clement, Internet Encyclopedia of Philosophy, https://iep.utm.edu/frege/.

4 Bertrand Russell, *A History of Western Philosophy And Its Connection with Political and Social Circumstances from the Earliest Times to the Present Day* (New York: Simon and Schuster, 1945), available online at: www.ntslibrary.com/PDF%20Books/History%20of%20Western%20Philosophy.pdf. While very long and seemingly encyclopedic in its coverage of ancient and medieval philosophy, including numerous minor and peripheral figures, Russell's tome nonetheless omits, for example, any references to Heidegger, Husserl, and phenomenology, as well as to Marxist-inspired philosophers after Karl Marx, even though their works might have been available to him (at least in German). For overviews of Russell's life and works, see Andrew David Irvine, "Bertrand Russell," *The Stanford Encyclopedia of Philosophy* (Spring 2021 Edition), Edward N. Zalta, ed., forthcoming, https://plato.stanford.edu/archives/spr2021/entries/russell/; Ray Monk, *Bertrand Russell: The Spirit of Solitude* (London: Jonathan Cape, 1996)—*Bertrand Russell: The Ghost of Madness* (London: Jonathan Cape, 2000)—and Anthony Palmer (eds.), *Bertrand Russell and the Origins of Analytic Philosophy* (Bristol: Thoemmes Press, 1996); and Rosalind Carey, "Bertrand Russell: Metaphysics," *Internet Encyclopedia of Philosophy*, https://iep.utm.edu/russ-met/.

5 These and additional biographical details are found in Monk, *op. cit.*, Biletski, *op. cit.*, and Richter, *op. cit.* A fuller account of "Wittgenstein's Vienna" is the eponymous book by Janik and Toulmin, *op. cit.*

6 For Wittgenstein's technical education in and contributions to physics and engineering, see Ian Lemco, "Wittgenstein's aeronautical investigation," *The Royal Society Journal of the History of Science*, December 2006, available at: https://royalsocietypublishing.org/doi/10.1098/rsnr.2006.0163.

7 See Biletski, *op. cit.*, and Monk, *op. cit.*, 41.

8 See Hans Sluga, *Wittgensteins Welt*, which recently appeared in English and appears in Christoph Limbeck-Lilienau and Friedrich Stadler, *The Philosophy of Perception Proceedings of the 40th International Ludwig Wittgenstein Symposium*, in: *Publications of the Austrian Ludwig Wittgenstein Society – New Series,* 26 (Berlin: De Gruyter, 2019), 399–417. Sluga contrasts Schopenhauer's *Welt* with Wittgenstein's and claims to find some striking similarities.

I am not entirely persuaded of the degree of this resemblance—in part because it's unclear to me how much of Schopenhauer's work Wittgenstein had actually read—except for the facts the both philosophers begin and end their early masterpieces, the TLP for Wittgenstein and *The World as Will and Representation* for Schopenhauer, with stunning statements about *die Welt* and that certain passages in Schopenhauer, in isolation, seem eerily prescient of similar-sounding ones in Wittgenstein. See the relevant passages on Schopenhauer earlier in this book.

9 The Vienna Circle (German: *Wiener Kreis*) was a group of scientifically-trained philosophers and philosophically-interested scientists who met from 1924 to 1936 for often-weekly discussions under the (nominal) leadership of Moritz Schlick and the continuing participation of Otto Neurath and Rudolf Carnap, among many distinguished thinkers. While they varied in their approaches to scientific knowledge, the participants were linked by a "positivist" reliance on empirical methods and logical proof, and an aversion to theories and methodologies they deemed "metaphysical," "un-" or "pseudo-scientific," "dogmatic," and/or "nonsensical," a stance similar to the one taken in Wittgenstein's *Tractatus* but later abandoned to a considerable degree. See Thomas Uebel, "Vienna Circle", *The Stanford Encyclopedia of Philosophy* (Summer 2020 Edition), Edward N. Zalta, ed., https://plato.stanford.edu/archives/sum2020/entries/vienna-circle/.

10 Wittgenstein, in Monk, *op. cit.*, 579.

11 For an overview of Popper's life and work, see Stephen Thornton, "Karl Popper," *The Stanford Encyclopedia of Philosophy* (Winter 2019 Edition), Edward N. Zalta, ed., https://plato.stanford.edu/archives/win2019/entries/popper/. For the distinction between the "context of discovery" and "context of justification" regarding scientific hypotheses in particular, see Minwoo Seo and Hasok Chang, "Context of Discovery and Context of Justification," in the *Encyclopedia of Science Education*, 2015 Edition, ed. Richard Gunstone, https://link.springer.com/referenceworkentry/10.1007%2F978-94-007-2150-0_239.

12 Sluga, "Wittgensteins Welt," *op. cit.*, and comments in a private email message in 2020.

13 The German edition of Wittgenstein's "Secret Diaries," including his notes from the wars, is: *Geheime Tagebücher* (abbreviation: GT) (Vienna: Turia & Kant and C.G. Luckhardt, 1979).

14 Sluga emphasizes this point and refers in particular to statements made by Wittgenstein during the war and printed in his "Secret Diaries" (GT):

> in the spirit of Tolstoy [Wittgenstein writes] "Man is powerless in the flesh but free by the spirit." (GT 1991: 21) The individual will cannot change the world, just as Schopenhauer taught him. Wittgenstein notes: "I can die in an hour. I can die in a month or in a few years. I cannot know and cannot do for or against it. This is life." And so he asks himself: "So how do I have to live in order to exist in every moment?" (GT 1991: 28) And shortly before that he gave himself an answer: "Man is not allowed to depend on chance. Neither from the favorable nor from the unfavorable" (GT 1991: 27) That also means no longer depending on the world, insofar as everything that happens and being in it is accidental … the *Tractatus* is ultimately and above all (also) a book of war. *The world of experience that the text strives to leave behind is present again in its absence. The randomness of the world, what it means to see it correctly, and what the ethical implications that result from it, form the basic structure of the entire work.*
>
> (Sluga, *op. cit.*, 475–80)

15 Maurer, *op. cit.*, 142.

16 *Ibid.*

17 Wittgenstein, TLP, online edition, *op. cit.*, 89–90.

18 Mauer, *ibid.*

19 Sluga alludes to Wittgenstein's possible struggle with his own feelings in general, and as well in all likelihood with his sexual feelings (Wittgenstein is often considered to have been gay, although the extent to which, if any, he acted on these feelings is less clear):

> only a single sentence in the Tractatus indicates human feelings: "The world of the happy is different from that of the unhappy." (TLP: 6.43) We may ask: Is the world of the unhappy for Wittgenstein perhaps the normal world of experience and that of the happy man, the world seen correctly? Among the feelings that move Wittgenstein are sexual ones. They worry him because they don't allow fulfillment.
>
> (Sluga, *ibid.*, 470)

20 Wittgenstein, "Letter to Malcolm dated 16-11-44," in Norman Malcolm, *Ludwig Wittgenstein. A Memoir* (Oxford: Oxford University Press, 1984), 93–4. For the importance of Wittgenstein's experiences during, and perceptions of, World Wars I and II, see Nicolás Sánchez Durá, "Wittgenstein on War and Peace," in Luigi Perissinotto, ed., *The Darkness of this Time, Ethics, Politics, and Religion in Wittgenstein* (Milan: Mimesis Press, 2013), 161–81.

21 See the Wikipedia article on the *Tractatus*, especially for its mention of the "new, resolute" reading and interpretation of that book in particular and of Wittgenstein's work ore generally: https://en.wikipedia.org/wiki/Tractatus_Logico-Philosophicus.

22 Wittgenstein, TLP, *op. cit.*, 25.

23 German: "*Die Welt ist, alles was der Fall ist.*"

24 Wittgenstein's own definition(s) of propositions may be culled from such statements in his TLP as: "The proposition is a picture of reality …" (4.01–4.023) *Ibid.*, 39–41.

25 For Clinton's impeachment hearings and the related scandal see https://en.wikipedia.org/wiki/Impeachment_of_Bill_Clinton.

26 Sluga, *op. cit.*, 435–38.

27 *Ibid.*

28 Wittgenstein, TLP, *op. cit.*, 28.

29 Sluga, *ibid.*, 439–42.

30 Sluga's formulation of this "dilemma" is:

> The *Tractatus* begins … with a nonsensical sentence. But … why this beginning? Probably to help … "to see the world correctly …" *But what can it mean to see the world correctly if the expression "the world" has no meaning?*
>
> (*Ibid.*, 444–5)

Sluga continues:

> the first sentence of the *Tractatus* says not only that the world consists of facts which are individually accidental, *but also that the world itself in its entirety is accidental.* And this … makes it understandable why Wittgenstein speaks of astonishment in his lecture on ethics that there is something at all and not nothing, that the world exists. And … this astonishment found its expression … in the idea that God created the world out of nothing.
>
> (Sluga, *ibid.*, 447–8)

This has profound implications, not only for what "the world" may or may not be, but also for ethics, which, implicitly, can only lie "outside the world." Sluga argues that for Wittgenstein:

> *"the meaning of the world must lie outside … it." And this randomness of the world shows … that there can be no values in the world.* Random values … are not values … What is not accidental cannot be part of the world, because otherwise it would be accidental again. And … there can be no "propositions of ethics," no ethical theory, and no ethical rules.
>
> (Sluga, *ibid.*, 449–50)

31 TLP, *op. cit.*, 39.
32 *Ibid.*, 44.
33 *Ibid.*, 44–5.
34 *Ibid.*, 27–8.
35 *Ibid.*, 56 and 71–2.
36 *Ibid.*, 74–5.
37 For an analysis and critique of "eliminative materialism," see Charles Webel and Anthony Stigliano, "'Beyond Good and Evil?' Radical Psychological Materialism and the 'Cure' for Evil," *op. cit.*
38 *TLP*, *op. cit.*, 90–1.
39 *Ibid.*, 80–1.
40 *Ibid.*, 82.
41 *Ibid.*, 87.
42 *Ibid.*
43 *Ibid.*, 88.
44 *Ibid.*, 88–9.
45 *Ibid.* 89 and 161.
46 *Ibid.* 90 and 162,
47 In the last sentence of *The Brown Book* of 1934–35, Wittgenstein explicitly declares his account to be descriptive, not explanatory: "Our method is purely descriptive; the descriptions we give are not hints of explanations." Ludwig Wittgenstein, *Major Works*, *op. cit.*, 241.
48 Wittgenstein, *The Blue Book*, in *Major Works*, *op. cit.*, 104–5, 119–20, and 143.
49 *Ibid.*, 155.
50 See Webel and Stigliano, *op. cit.*
51 Wittgenstein, PI (online edition), *op. cit.*, #96, p. 44.
52 *Ibid.*, #97, p. 45.
53 *Ibid.*, #92–3, p. 43.
54 Wittgenstein, *On Certainty*, online edition, *op. cit.*, #20, p. 3.
55 *Ibid.*, #30, p. 4.
56 *Ibid.*, #31, p. 4.
57 *Ibid.*, #65, p. 6.
58 *Ibid.*, #94–5, p. 9.
59 *Ibid.*, #209, p. 18.
60 *Ibid.*, #219–20, 325–7, and 336, pp. 18 and 26. For a more extensive discussion of "reasonable" and "reasonableness," see Webel, *Politics of Rationality*, *op. cit.*, esp. 200–16.
61 For "postanalytic," "postphilosophy," often "pragmatic" developments, often stemming from Richard Rorty, see https://en.wikipedia.org/wiki/Postanalytic_philosophy.

62 Sluga, *ibid.*
63 *Ibid.*, 466–7.
64 Rupert Read, *Wittgenstein's Liberatory Philosophy Thinking Through His Philosophical Investigations* (New York and London: Routledge, 2021), 31.
65 *Ibid.*, 34.
66 Wittgenstein, *TLP*, *op. cit.*, 88 and 161, respectively for the English and German.
67 Wittgenstein, PI, *op. cit.*, #109, p. 46, and #116, p. 48.
68 *Ibid.*, #255, p. 91.
69 Wittgenstein, as cited in Hans Sluga, "Wittgensteins Welt," op. cit., 400. This citation originally appeared in David G. Stern, et al.: *Wittgenstein Lectures, Cambridge 1930–1933. From the Notes of G.E. Moore* (Cambridge: Cambridge University Press, 2016). Sluga was also kind enough to share with me two related unpublished presentations, "Seeing the world aright"—a lecture he gave at an Oxford Wittgenstein conference in August, 2019, and "The darkness of this time"—a lecture he gave in Xi'an, China in November, 2019.
70 Often dubbed the (analytic) "philosopher's philosopher," Kripke is a legendary figure in recent and contemporary Anglophonic philosophical circles. For an overview, see www.britannica.com/biography/Saul-Kripke.

Quine is widely considered to have been the most influential analytic philosopher since Wittgenstein, see: https://plato.stanford.edu/entries/quine/. For Quine's relevance for "possible worlds" theories, see Willard Van Orman Quine, *Ontological Relativity and Other Essays* (New York: Columbia University Press, 1969), and his "Worlds Away," *Journal of Philosophy*, 73 (1976). 859–63; reprinted in Quine, *Theories and Things* (Cambridge, MA: Harvard University Press, 1981).
71 See Leiter Reports, op. cit., https://leiterreports.typepad.com/blog/2009/03/lets-settle-this-once-and-for-all-who-really-was-the-greatest-philosopher-of-the-20thcentury.html. For Leiter's "ranking" of "the most important Western philosophers of all time," see: https://leiterreports.typepad.com/blog/2017/04/the-most-important-western-philosophers-of-all-time. For a critique, within the American Philosophical Association no less, of such rankings (and of the sometimes rank ranker), see https://againstprofphil.org/2019/07/08/why-the-leiter-reports-professional-academic-philosophy-rankings-are-misguided-misleading-and-mistaken/.
72 Brian Weatherson, "David Lewis," *The Stanford Encyclopedia of Philosophy* (Winter 2016 Edition), Edward N. Zalta ed., https://plato.stanford.edu/archives/win2016/entries/david-lewis/.
73 Scott Dixon, "David Lewis," *Internet Encyclopedia of Philosophy*, https://iep.utm.edu/d-lewis/.
74 See James Garson, James, "Modal Logic," *The Stanford Encyclopedia of Philosophy* (Fall 2018 Edition), Edward N. Zalta, ed., https://plato.stanford.edu/archives/fall2018/entries/logic-modal/.
75 See Saul Kripke, *Naming and Necessity* (Cambridge, MA: Harvard University Press, 1980), and Christopher Menzel, "Possible Worlds," *The Stanford Encyclopedia of Philosophy* (Winter 2017 Edition), Edward N. Zalta, ed., https://plato.stanford.edu/archives/win2017/entries/possible-worlds/.
76 David Lewis, *On the Plurality of Worlds* (Oxford: Blackwell, 1986, abbreviation: OPW), vii.
77 David Lewis, *Counterfactuals* (Cambridge, MA: Harvard University Press), 84.

78 Lewis, OPW, *op. cit.*, 1–2.

79 In Aristotelian metaphysics, entelechies are the complete realization and final form of some potential concept or function; the conditions under which a potential thing becomes actualized.

80 *Ibid.*, vii.

81 *Ibid.*, 2.

82 *Ibid.*, 3.

83 *Ibid.*

84 *Ibid.*, 69–71. According to Achille Varzi, "Mereology," *The Stanford Encyclopedia of Philosophy* (Spring 2019 Edition), Edward N. Zalta, ed., https://plato.stanford.edu/archives/spr2019/entries/mereology/:

> Mereology (from the Greek *meros,* "part") is the theory of parthood relations: of the relations of part to whole and the relations of part to part within a whole. Its roots can be traced back to the early days of philosophy, beginning with the Presocratics and continuing throughout the writings of Plato, Aristotle (especially the *Metaphysics*) and Boethius … Mereology occupies a prominent role also in the writings of medieval ontologists and scholastic philosophers … Leibniz's *Monadology* (1714), and Kant's early writings … As a formal theory of parthood relations … mereology made its way into our times mainly through the work of Franz Brentano and of his pupils, especially Husserl's third *Logical Investigation* (1901) … It was not until Leśniewski's *Foundations of the General Theory of Sets* (1916) and his *Foundations of Mathematics* (1927–1931) that a pure theory of part-relations was given an exact formulation … it is only with the publication of Leonard and Goodman's *The Calculus of Individuals* (1940), partly under the influence of Whitehead, that mereology has become a chapter of central interest for modern ontologists and metaphysicians.

85 Lewis, OPW, *ibid.*

86 Lewis, *ibid.*, 92–3.

87 *Ibid.*, 96.

88 Also see David Lewis, *Philosophical Papers*, (two volumes) (Oxford: Oxford University Press, 1983 and 1986), and Alvin Plantinga, 'Transworld Identity or Worldbound Individuals?' in *Logic and Ontology*, ed. Milton Munitz (New York: New York University Press, 1973).

89 Lewis, OPW, *Ibid.*, 106.

90 *Ibid.*, 136–7.

91 *Ibid.*, 142 and 145.

92 *Ibid.*, 205, 230.

93 *Ibid.*, 211.

94 *Ibid.*, 230.

95 For Carnap, see *Der Logische Aufbau der Welt* (Leipzig: Felix Meiner Verlag, 1928), trans. Rolf A. George, *The Logical Structure of the World* (Berkeley, CA: University of California Press, 1967); and Hannes Leitgeb and André Carus, "Rudolf Carnap," *The Stanford Encyclopedia of Philosophy* (Fall 2020 Edition), Edward N. Zalta, ed., https://plato.stanford.edu/archives/fall2020/entries/carnap/.

For Popper, see Stephen Thornton, "Karl Popper," *The Stanford Encyclopedia of Philosophy* (Winter 2019 Edition), Edward N. Zalta (ed.), https://plato.stanford.edu/archives/win2019/entries/popper/. Popper was lauded by many thinkers in the analytic tradition and was the teacher of numerous distinguished philosophers of science, including Paul Feyerabend, who was on my doctoral dissertation committee, as was Stephen Toulmin, Popper also influenced the Hungarian–American billionaire and philanthropist George Soros, whose "Open Society Foundation" promotes the kind of "liberalism" articulated in Popper's political writings.

96 According to Amy Kind in *The Internet Encyclopedia of Philosophy*:

> The term "qualia" (singular: *quale*) was introduced into the philosophical literature in its contemporary sense in 1929 by C.I. Lewis in a discussion of sense-data theory … In contemporary usage, the term has been broadened to refer more generally to properties of experience. Paradigm examples of experiences with qualia are perceptual experiences (including experiences like hallucinations) and bodily sensations (such as pain, hunger, and itching). Emotions (like anger, envy, or fear) and moods (like euphoria, ennui, or anxiety) are also usually taken to have qualitative aspects. Qualia are often referred to as the *phenomenal* properties of experience, and experiences that have qualia are referred to as being *phenomenally conscious*.
>
> (https://iep.utm.edu/qualia/)

97 Chalmers, CW, *op. cit.*, 2–9, and 20.

98 In his two-volume *The Theory of Communicative Action*, Habermas has lengthy analyses of the works of such American sociologists as George Herbert Mead (1863–1931) and Talcott Parsons (1902–79), as well as numerous references to Wittgenstein, Carnap, and the American pragmatists, as well, of course, in-depth discussions of "the world." Habermas's recent two-volume *magnum opus*, *Auch eine Geschichte der Philosophie* [*Also a History of Philosophy*] (Berlin: Suhrkamp Verlag, 2019), is, inter alia, a historical genealogy of the dominant forms of Occidental post-metaphysical thinking, as developed philosophically through its symbiosis with religion and as successively transformed in science, law, politics, and society.

99 Habermas's recent books also have not-infrequent references to Wittgenstein, as well as to John Rawls (1921–2002), John Searle (1932–), and Noam Chomsky (1928–), and to many other recent and contemporary Anglophonic speakers. Curiously, however, they also contain only a few references to Michel Foucault, and none to such major French thinkers as Jacques Derrida, Jean-Luc Nancy, and Alain Badiou, or to such logicians and philosophers of mind as Lewis and Chalmers.

100 Derrida, *Voice and Phenomenon*, 12, cited in Sean Gaston, *The Concept of World from Kant to Derrida*, *op. cit.*, 125 and 198.

101 Derrida, *Of Grammatology*, 47 and 65, cited in Gaston, *ibid.*, 125 and 198.

102 For an overview of Žižek's works and life, see: https://en.wikipedia.org/wiki/Slavoj_%C5%BDi%C5%BEek. I happened to attend a "lecture" by Žižek about 20 years ago at Columbia University. While it was a heady mixture of terms and trends in Lacanian psychoanalytic theory, Hegelian-Marxism, and whatever else came to Žižek's feverish mind at the time, I found it, and the few published writings of Žižek I have perused, to be imaginative and wide-ranging, but often superficial and faddish, an impression shared, perhaps appropriately in this case, by many analytic philosophers, but also, unfortunately, often unjustly extended by them to Critical Theory and continental philosophy more generally.

103 For succinct overviews of Derrida's life and works, see Leonard Lawlor, "Jacques Derrida," *The Stanford Encyclopedia of Philosophy* (Fall 2019 Edition), Edward N. Zalta, ed., https://plato.stanford.edu/archives/fall2019/entries/derrida/; and Jack Reynolds, International Encyclopedia of Philosophy, https://iep.utm.edu/derrida/.

English translations of Derrida's main books include: *Acts of Religion*, ed., Gil Anidjar (London: Routledge, 2002); *Adieu to Emmanuel Levinas*, trans., Michael Naas and Pascalle-Anne Brault (Stanford, CA: Stanford University Press, 1999); *The Animal that Therefore I am* (abbreviation: ATI), ed. Marie-Loiuse Mallet, trans., David Wills (New York: Fordham University Press, 2008); *The Beast and the Sovereign* (Volume 1) (abbreviation: BS 1), trans. Geoffrey Bennigton (Chicago, IL: University of Chicago Press, 2009); *The Beast and the Sovereign* (Volume 2) (Abbreviation: BS 2), trans. Geoffrey Bennington (Chicago, IL: University of Chicago, 2011); *The Death Penalty* (Volume 1), trans. Peggy Kamuf (Chicago, IL: University of Chicago Press, 2014); *The Death Penalty* (Volume 2), trans. Elizabeth Rottenberg (Chicago, IL: University of Chicago Press, 2017); *Dissemination*, trans., Barbara Johnson (Chicago, IL: University of Chicago Press, 1981); *The Ear of the Other: Otobiography, Transference, Translation*, trans., Peggy Kamuf (New York: Schocken, 1985); *Edmund Husserl's Origin of Geometry: An Introduction*, trans., John P. Leavey, Jr. (Lincoln, NE: University of Nebraska Press, 1989); *The Gift of Death and Literature in Secret. Second Edition*, trans., David Wills (Chicago, IL: University of Chicago Press, 2008); *Glas*, abbreviation: G), trans., John P. Leavey, Jr. and Richard Rand (Lincoln, NE: University of Nebraska Press, 1986); *Heidegger: The Question of Being and History*, trans., Geoffrey Bennington (Chicago, IL: University of Chicago Press, 2013); *Limited Inc*, trans., Samuel Weber, (Evanston, IL: Northwestern University Press, 1988); *Margins of Philosophy*, (abbreviation: MP), trans., Alan Bass (Chicago, IL: University of Chicago Press, 1982); *Of Grammatology* (abbreviation: OG), trans., Gayatri Spivak (Baltimore, MD: The Johns Hopkins University Press, 1974); *Of Spirit*, trans., Rachel Bowlby (Chicago, IL: University of Chicago, 1989); *Philosophy in the Time of Terror: Dialogues with Jürgen Habermas and Jacques Derrida*, ed., Giovanna Borradori (Chicago, IL: University of Chicago Press, 2003); *Points … Interviews, 1974–1994*, trans., Peggy Kamuf and others (Stanford, CA: Stanford University Press, 1995); *The Problem of Genesis in Husserl's Philosophy*, trans., Marion Hobson (Chicago, IL: University of Chicago Press, 2003); *Religion*, trans., Samuel Weber (Stanford, CA: Stanford University Press, 1998); *Rogues*, trans., Pascale-Anne Brault and Michael Naas (Stanford, CA: Stanford University Press, 2005); "Signature Event Context," in *Glyph: Johns Hopkins Textual Studies*, 1977, 172–97; *Specters of Marx*, abbreviation: SM), trans., Peggy Kamuf (New York: Routledge, 1994); *Theory and Practice*, trans. David Wills (Chicago, IL: University of Chicago Press, 2019); *Speech and Phenomena*, trans., David B. Allison (Evanston, IL: Northwestern University Press, 1973, abbreviation: SP); *Voice and Phenomenon*, trans., Leonard Lawlor (Evanston, IL: Northwestern University Press, 2011, abbreviation: VP); *The Work of Mourning*, eds., Pascale-Anne Brault and Michael Naas (Chicago, IL: University of Chicago Press, 2001); *Writing and Difference*, trans., Alan Bass (Chicago, IL: University of Chicago, 1978).

Selected Secondary Literature on Derrida includes: Pheng, and Suzanne Guerlac, eds., *Derrida and the Time of the Political* (Durham, NC: University of North Carolina Press, 2009); Hélène Cixous and Jacques Derrida, *Veils*, trans., Geoff Bennington (Stanford, CA: Stanford University Press, 2001); Andrew Cutrofello, *Continental Philosophy: A Contemporary Introduction* (New York and London: Routledge, 2005); Gilles Deleuze and Claire Parnet, *Dialogues*, trans. Hugh Tomlinson and Barbara Habberjam (New York: Columbia University Press, 1987); Vincent Descombes, *Modern French Philosophy*, trans., L. Scott-Fox and J.M. Harding (New York: Cambridge University Press, 1980); Gary Gutting, *French Philosophy in the Twentieth Century* (New York: Cambridge University Press, 2001); Samir Haddad, Samir, *Derrida and the Inheritance of Democracy* (Bloomington, IN: Indiana University Press, 2013); David Farrell Krell, *Derrida and our Animal Others: Derrida's Final Seminar, "The Beast and the Sovereign,"* (Bloomington, IN: Indiana University Press, 2013); Leonard Lawlor, *Derrida and Husserl: The Basic Problem of Phenomenology* (Bloomington, IN: Indiana University Press, 2002) and—*This is not Sufficient: An Essay on Animality and Human Nature in Derrida* (New York: Columbia University Press, 2007); W.J.T. Mitchell and Arnold Davidson, eds., *The Late Derrida* (Chicago, IL: University of Chicago Press, 2007); Jitendranath Mohanty, *Taking on the Tradition: Jacques Derrida and the Legacies of Deconstruction* (Stanford, CA: Stanford University Press, 1997); Michael Naas, *The End of the World and other Teachable Moments: Jacques Derrida's Final Seminar* (Bronx, NY: Fordham University Press, 2015); Benoist Peeters, *Derrida: A Biography*, trans., Andrew Brown (Cambridge: Polity Press, 2013); Jason Powell, *Derrida: A Biography* (London: Continuum, 2006); Nicolas Royle, *Deconstruction: A User's Guide* (London: Palgrave Macmillan, 2000); John Sallis, ed., *Deconstruction and Philosophy* (Chicago, IL: University of Chicago Press, 1987); Alan Schrift, *Twentieth Century French Philosophy: Key Themes and Thinkers* (Oxford: Blackwell Publishing, 2006); John Searle, "Reiterating the Differences: A Reply to Derrida," in *Glyph: Johns Hopkins Textual Studies*, 1977, 198–208; David Wood, ed., *Derrida: A Critical Reader* (Cambridge, MA: Blackwell, 1994), and—ed., *Of Derrida, Heidegger, and Spirit* (Evanston, IL: Northwestern University Press, 1993).

104 For Deleuze, see Daniel Smith and John Protevi, "Gilles Deleuze," *The Stanford Encyclopedia of Philosophy* (Spring 2020 Edition), Edward N. Zalta, ed., https://plato.stanford.edu/archives/spr2020/entries/deleuze/.

For Althusser, see William Lewis, "Louis Althusser," *The Stanford Encyclopedia of Philosophy* (Spring 2018 Edition), Edward N. Zalta, ed., https://plato.stanford/edu/archives/spr2018/entries/althusser/.

For Lyotard, see Peter Gratton, "Jean François Lyotard," *The Stanford Encyclopedia of Philosophy* (Winter 2018 Edition), Edward N. Zalta, ed., https://plato.stanford.edu/archives/win2018/entries/lyotard/.

For Barthes, see https://en.wikipedia.org/wiki/Roland_Barthes.

105 In 1961, Foucault published what many consider his greatest work, *History of Madness* (*Madness and Civilization*). Derrida participated in a seminar taught by Foucault and wrote a critique of that work, "Cogito and the History of Madness" (1963), which severely criticized *History of Madness* and that resulted in a dispute between Derrida and Foucault that was never totally resolved. It did not help that, in 1983, the Berkeley philosopher of language

John Searle told the very influential US periodical *The New York Review of Books* of a remark about Derrida allegedly made by Foucault in a private conversation with Searle himself. Derrida later decried Searle's gesture as gossip, and also objected to the use of a supposedly mass circulation magazine to fight an academic debate. According to Searle (*op. cit.*), Foucault called Derrida's prose style "terrorist obscurantism," and stated:

Michel Foucault once characterized Derrida's prose style to me as "*obcurantisme terroriste*." The text is written so obscurely that you can't figure out exactly what the thesis is (hence "*obscurantisme*") and when one criticizes it, the author says, "*Vous m'avez mal compris; vous e tes idiot*" ("*You have poorly understood me—you idiot" hence* " terrorist").

See Lawlor, "Jacques Derrida," *op. cit.*; and https://en.wikipedia.org/wiki/Limited_Inc.

I attended Foucault's seminar during the early 1980s at Berkeley on "The History of Sexuality" and attempted to discuss Derrida with Foucault, who was perhaps the strangest person I've ever met and who insisted on discussing his body, which at the time was being ravaged by the as-yet-unclassified malady later to be called HIV/AIDS—and who, remarkably, stated "Who is that?" in reply to my queries about Derrida …

During that time, I also briefly met Derrida, Ricoeur, and Hans-Georg Gadamer (1900–2002), who were, perhaps, the three liveliest interlocutors I have known and the embodiments of philosophical humanists.

For Gadamer, see Jeff Malpas, "Hans-Georg Gadamer," *The Stanford Encyclopedia of Philosophy* (Fall 2018 Edition), Edward N. Zalta, ed., https://plato.stanford.edu/archives/fall2018/entries/gadamer/.

That said, some readers may be puzzled at the absence of a section on Foucault in my book. The reasons are that in the few works of Foucault available to me during the pandemic I could find very few references to "the world," and also that Foucault's methods and theories are more suited to my ensuing book on *The Reality of the World* than to *The Idea of the World.* An overview of Foucault's life and works is provided by Gary Gutting and Johanna Oksala, "Michel Foucault," *The Stanford Encyclopedia of Philosophy* (Spring 2019 Edition), Edward N. Zalta, ed., https://plato.stanford.edu/archives/spr2019/entries/foucault/.

For Lévi-Strauss, see https://en.wikipedia.org/wiki/Claude_L%C3%A9vi-Strauss.

For Lacan, see Adrian Johnston, "Jacques Lacan," *The Stanford Encyclopedia of Philosophy* (Fall 2018 Edition), Edward N. Zalta, ed., https://plato.stanford.edu/archives/fall2018/entries/lacan/.

For Levinas, see Bettina Bergo, "Emmanuel Levinas," *The Stanford Encyclopedia of Philosophy* (Fall 2019 Edition), Edward N. Zalta, ed., https://plato.stanford.edu/archives/fall2019/entries/levinas/.

For Ricoeur, see David Pellauer and Bernard Dauenhauer, "Paul Ricoeur," *The Stanford Encyclopedia of Philosophy* (Spring 2021 Edition), Edward N. Zalta, ed., forthcoming, https://plato.stanford.edu/archives/spr2021/entries/ricoeur/.

106 For Hyppolite, see https://en.wikipedia.org/wiki/Jean_Hyppolite.

107 For example, in 1977, Sam Weber, the editor of the then-new highbrow journal *Glyph* invited John Searle to write a response to Derrida's article "Signature Event History," which, inter alia, discusses the "speech act"

theory of one of Searle's Oxford mentors, the "ordinary-language" philosopher J.L. Austin (1911–60). Searle's response, a ten-page essay called "Reiterating the Differences: A Reply to Derrida," is a very critical assessment of Derrida's essay. Derrida responded to Searle's "Reply" with his own 90-page critique of "Sarl" called "Limited Inc. a b c." See Jacques Derrida, "Signature Event Context," in *Limited Inc.*, trans. Samuel Weber and Jeffrey Mehlman (Evanston, IL: Northwestern University Press, 1988); and https://en.wikipedia.org/wiki/Limited_Inc.

Fifteen years later, in 1992, Derrida was offered an honorary degree by Cambridge University. A group of analytic philosophers, including Quine, wrote an open letter to the *Times of London* and the university in which they strongly objected to Derrida receiving this honorary degree. The letter asserted that "Derrida's work does not meet accepted standards of clarity and rigor," and

> Academic status based on what seems to us to be little more than semi-intelligible attacks upon the values of reason, truth, and scholarship is not, we submit, sufficient grounds for the awarding of an honorary degree in a distinguished university.

Despite the letter, Cambridge University awarded Derrida the degree ... See Lawlor, *op. cit.*

108 The only texts by and on Derrida to which I have had direct access during the pandemic are *A Derrida Reader Between the Blinds* (abbreviation: DR), ed. Peggy Kamuf (New York: Columbia University Press, 1991), which contains substantial excerpts from OG, SP, and G, as well as Derrida's "Letter to a Japanese Friend," and short selections from many of Derrida's other works; BS2; TAT, and Gaston, *op. cit.*, esp. 99–64.

109 For "Deconstruction," see https://en.wikipedia.org/wiki/Deconstruction.

110 Derrida, Signature Event Context," in DR, *op. cit.*, 108.

111 Derrida, "Letter to a Japanese Friend," in DR, *op. cit.*, 272–5.

112 Peggy Kamuf, Introduction to "Differance" in *Margins of Philosophy*, in DR, *op. cit.*, 59.

113 *Ibid.*, 61 and 67.

114 Derrida, *Speech and Phenomena*, in DR, *op. cit.*, 24.

115 Derrida, OG, in DR, *op. cit.*, 40–3.

116 *Ibid.*, 31–2, 34, and 46.

117 Derrida, *Speech and Phenomena*, in DR, *op. cit.*, 10–11.

118 Derrida, "Differance" in *Margins of Philosophy*, in DR, *op. cit.*, 68.

119 Kamuf, Introduction to "The Sexual Difference in Philosophy," in DR, *op. cit.*, 313.

120 Derrida, *Speech and Phenomena*, in DR, *op. cit.*, 13 and 27.

121 There is, of course, an enormous literature by and on Piaget. For a brief summary, see https://en.wikipedia.org/wiki/Jean_Piaget.

122 Derrida, *Glas*, in DR., *op. cit.*, 347.

123 Derrida, as cited and paraphrased by Gaston, in Gaston, *op. cit.*, 115–16, 118–19, 143–4, and Derrida, *The World of the Enlightenment to Come*, 155, in Gaston, *ibid.*, 195. I am quoting or paraphrasing Gaston somewhat extensively because he provides an excellent summary of Derrida's thoughts on

"the world" and also had access to texts by Derrida to which I have not had access during the pandemic.

124 Gaston, *op. cit.*, 123.

125 Derrida, *Voice and Phenomenon*, 12, as cited in Gaston, *op. cit.*, 125 and 198.

126 Gaston, paraphrasing and citing Husserl, *Ideas* 1, and Derrida, "No Apocalypse," in *Psyche: Inventions of the Other.* Gaston, *op. cit*, 102–3 and 192–3.

127 Derrida, OG, 47 and 65, cited in Gaston, *op. cit.*, 125.

128 Nietzsche, *Twilight of the Idols*, and Derrida, *The Beast and the Sovereign*, in Gaston, 127 and 199. "World-play" is associated by Heidegger with human finitude and death as the "yet unthought standard measure of the unfathomable," Heidegger, The *Principle of Reason*, trans. Reginald Lilly (Bloomington, IN: Indiana University Press), 112–13, cited in Gaston, *op. cit.*, 124–7 and 198–9.

129 Derrida, TAT, *op. cit.*, 151–2.

130 Derrida, BA2, *op. cit.*, 11.

131 Derrida, TAT, *op. cit.*, 152–3.

132 Heidegger "On the Essence of Ground," trans. William McNeill, in Heidegger, *Pathmarks*, ed. William McNeill (Cambridge: Cambridge University Press), cited in Derrida, TAT, *op. cit.*, 152.

133 Derrida, TAT, *op. cit.*, 153 and 156.

134 Derrida, BS2, *op. cit.*, 101 and 104.

135 *Ibid.*, 264–7.

136 Derrida, TAT, *op. cit.*, 151.

137 See Derrida, BS2, op. cit., 267–8, where he says:

> the vague comforting feeling of understanding each other, of speaking among ourselves the same language, and sharing an intelligible language, in a consensual communicative action, for example in the use of the words "world" (*Welt, world, mundo*), "our common world," the unity of the world, etc., *that does not suffice for it to be true … just about the same meaning in the same useful function to similar vocables or signs, etc. For example the word "world" as totality of what is, etc. That no one has ever come across, right? Have you ever come across the world as such? There seems to be in this refined utilitarian nominalism.*

As Gaston says, "*There is always the possibility that there is no world, but the world remains a necessary fiction.*" Gaston, *op. cit.*, 133.

138 Derrida, "Rams, 2003" in Gaston, *op. cit.*, 105–6, and 140.

139 Derrida, TAT, *op. cit.*, and Derrida, BS2, *op. cit.*

140 Derrida, BS2, *op. cit.*, 259–60. Also see Karl Jaspers, *The Future of Mankind*, trans. E.B. Ashton (Chicago, IL: University of Chicago Press, 1961). For Albert Einstein, see *Einstein on Politics: His Private Thoughts and Public Stands on Nationalism, Zionism, War, Peace, and the Bomb*, ed. David E. Rowe and Robert Schulmann (Princeton, NJ: Princeton University Press, 2013). For the Freud/Einstein correspondence "Why War?" see Charles Webel and Jorgen Johansen, ed., *Peace and Conflict Studies: A Reader* (Routledge: London, 2012), Part III, as well as the "Russell-Einstein Manifesto," in Part II of the same book.

A Conclusion Without an End, or an End Without a Conclusion?

David and Ann Willmore of Bookends of Fowey, UK

> I think it is true that humanity will triumph eventually, only I fear that at the same time the world will become a large hospital and each will become the other's humane nurse.
>
> Johann Wolfgang von Goethe[1]

> If you want to get pleasure out of life you must attach value to the world.[2]
>
> Johann Wolfgang von Goethe

DOI: 10.4324/9781315795171-6

Most books end with a Conclusion, which is, well, a conclusion, that is a summary and restatement of the principal theses and findings presented in the preceding chapters, a perhaps novel take on those ideas and results, and suggestions for possible future inquiries and research. As you might have already detected, *The World as Idea* is not like most books ... And this Conclusion will not be like most conclusions, either ...

But, so as not totally to disappoint your possible expectations, while I will have virtually nothing to say in this volume about "principal theses and findings"—thereby running the risk of disappointing some readers—I will have a bit more to propose in terms of my take, or "theses," on "The Idea of the World," and, so not to disappoint all readers, will have somewhat more to say about "How This World Might End," a semi-conclusion to this book, and a preview of forthcoming attractions, and extractions, for the sequel to this book, *The Reality of the World.*

Some Ideas About the "World as Idea"

Given the number, range, breath, and occasional depth (principally with European philosophers from Kant through Merleau-Ponty) of approaches to conceptualizing the world, there is not much I wish to add, or subtract (Lewis excepted), especially since a lengthier personal summation of the intellectual and political landscapes surveilled will be a major component of this book's successor volume. However, it might be remiss of me entirely to defer my own analysis for another time. So, here are some ideas-in-progress about the world as idea:

1 The "world" is an idea, an intellectual construct used in philosophy, religion, science, mass communications, and elsewhere. What the term denotes depends on the historical context and stylistic constructions of the users of the term.

   In this work, which focuses principally on the history of the idea of the world, the English substantive "world" is prefaced, usually, by either the definite article "the" or by "this." This is neither arbitrary nor insignificant. For "the" world is a term frequently used to refer to some external, physical entity, either the Earth or what might be called the "container" for what's on our planet. Whereas "this" world is used in this book and its companion volume to signify the totality of cultures and civilizations that have fashioned "a" world on the Earth, and not, yet, on other astrophysical entities, although such possible worlds are not to be excluded sometime in the future. In any event, the world, this world, a world, and possible worlds, are all part of the history of ideas. And this book is an intellectual history of the world; my next volume on *The Fate of This World* will focus on *The Reality of the World* from a more "materialist" perspective.

2 My own ideas of *the idea of the world* are syncretic and theoretical, drawing mainly on Aristotle, modern science, Kant, Nietzsche,

> Schopenhauer, Wittgenstein, and Merleau-Ponty, and not primarily on theologians and mainstream religious traditions, classical empiricists, skeptics, mainstream analytic philosophers, postmodernists, and nihilists. Obviously, the ideas of the world presented in this book have overwhelmingly been those of dead white Western males of privilege, although some important contributions to our understanding of the world by non-Western thinkers and scientists, as well as by dissident or marginalized Occidental writers, are also included. While some female and non-white authors are mentioned, they constitute a small minority. This is not deliberately "sexist," "racist," or any other "-ist" or "-ism." It is a reflection of the consensus in the Western intellectual, scientific, and religious traditions, at least through the early 21st century, as to whose and which ideas of the world have sufficiently "mattered" as to have been acknowledged and cited as "influential."
>
> I am aware of this selection bias, and this book has not escaped it. In the next, companion volume to *The World as Idea,* which will focus on "The World as Reality," principally as elaborated in the social and natural sciences since, roughly, the mid-19th century, I hope to be more inclusive of voices absent from this book. *The Reality of the World* will do its best also to include more ideas and findings of living, as well as dead, thinkers, scientists, and social researchers, including biologists, physicists, cosmologists, sociologists, anthropologists, psychologists, economists, political scientists and political theorists, and feminists, as well as environmental and political activists, and futurists, both Western and non-Western. My next book should also end with a Conclusion that is one in the normal sense of the word, but not with an end or ending to the world both as idea and as reality.

Regarding "the end of the world as we now know it," the COVID-19 pandemic provides an unwelcome but still instructive occasion for "thinking the unthinkable" so that we might collectively manage to ensure, at least for the short-term in geo- and astrophysical terms, a future that is not post-apocalyptic. In this connection, it is also useful if not sobering to consider the long-term future of our planetary habitat.

The Long-Term Future of the Earth

Our world's physical habitat, the Earth, has an expected long-term future that is tied to that of the Sun. Unless current predictions are wrong, over the next 1.1 billion years, solar luminosity will increase by 10%, and over the next 3.5 billion years, by 40%. Earth's increasing surface temperature will accelerate the inorganic carbon cycle, reducing CO_2 concentration to levels lethally low for plants in approximately 100–900 million years. The lack of vegetation will result in the loss of oxygen in the atmosphere, making animal life impossible.

Planet Earth's ultimate fate is unclear. If there is no omnicidal extinction-level event, Earth is expected to be habitable until the end of photosynthesis about 500 million years from now. Some scientific models forecast that about a billion years from now, all surface water on Earth will have disappeared and the mean global temperature will reach 70 °C (158°F). Even if the Sun were eternal and stable—which it is not—27% of the water in the oceans as they are today will descend to the mantle of the Earth in one billion years, due to reduced steam venting from mid-ocean ridges. But if nitrogen is removed from the atmosphere, life may continue until a runaway greenhouse effect occurs 2.3 billion years from now.

Scientific models estimate that about 2.8 billion years from now, the surface temperature of the Earth may reach 149 °C (300°F), even at the poles. At this point, any remaining life will be extinguished due to the extreme conditions. If all of the water on Earth has evaporated by this point, the planet will stay in the same conditions, with a steady increase in the surface temperature, until the Sun becomes a red giant. If not, then in about 3–4 billion years, the amount of water vapor in the lower atmosphere will rise to 40% and a "moist greenhouse" effect will commence once the luminosity from the Sun reaches 35–40% more than its present-day value. A "runaway greenhouse" effect will ensue, causing the atmosphere to heat up and raising the surface temperature to around 1,330 °C (2,420°F). This is sufficient to melt the surface of the planet. However, most of the atmosphere will be retained until the Sun has entered the red giant stage.

As a red giant, the Sun is predicted to lose about 30% of its mass, so, without tidal effects, Earth will move to an orbit 1.7 Astronomical Units (AUs), which is 250 million km (160 million miles) from the Sun, when that star reaches its maximum radius. Most, if not all, remaining life will be destroyed by the Sun's increased luminosity (peaking at about 5,000 times its present level). With the extinction of any remaining life about 2.8 billion years from now, it is also expected that Earth's biosignatures will disappear,[3] to be replaced by signatures caused by non-biological processes.

The most rapid part of the Sun's expansion into a red giant will occur during its final stages, when the Sun will be more than 7.5 billion years old. It is then likely to expand to swallow both Mercury and Venus, reaching a maximum radius of 1.2 AU. The Earth will interact tidally with the Sun's outer atmosphere, which would serve to decrease Earth's orbital radius. Drag from the chromosphere of the Sun would also reduce the Earth's orbit. A simulation indicates that Earth's orbit will eventually decay due to tidal effects and drag, causing it to enter the Sun's atmosphere and be vaporized.

But even if the Earth is not destroyed by the expanding red giant Sun, models predict that the remaining planets in the solar system will be ejected from it. If this does not happen to the Earth, the ultimate fate of the planet may be that it collides with the black dwarf Sun, due to the decay of its orbit via gravitational radiation, in about 10^{20} (100 trillion) years. There are at least a half-dozen other scenarios that forecast the end of our planet.[4]

The End of This World as We Know It?

Human Extinction?

Long before the Earth is vaporized, the human race may come to an end, with either a bang or a whimper. Human extinction is the hypothetical complete end of the human species. This may result either from natural causes or due to anthropogenic (human) causes, but the risks of extinction through natural disaster, such as a meteorite impact or the explosions of super-volcanoes (including Yellowstone), are generally considered to be comparatively low.

Many possible scenarios of anthropogenic extinction have been proposed, such as radical climate change due to runaway global warming, nuclear annihilation, biological warfare—including the release of a pandemic-causing agent—and ecological collapse. Some scenarios center on emerging technologies, such as advanced malign artificial intelligence, biotechnology, or self-replicating nanobots, which are machines or robots whose components are at or near the scale of a nanometer (one-billionth of a meter) and carry out specific tasks. The probability of anthropogenic human extinction within the next hundred years is the topic of an active debate. Collectively, these real and hypothetical threats to human existence are called Existential and Global Catastrophic Risks.

Existential and Global Catastrophic Risks

"Existential risks" are risks that threaten the entire future of humanity, whether by causing human extinction or by otherwise permanently crippling human civilization by leaving the survivors of a near extinction-level-event (ELE) without sufficient means to rebuild society to current standards of living. A global catastrophic risk is a hypothetical future event that could damage human well-being on a global scale, endangering or even destroying modern civilization, but not necessarily leading to our extinction.

Potential global catastrophic risks include *anthropogenic* risks, i.e., those caused by humans (including malign technology and global governance, as well as omnicidal climate change), and *non-anthropogenic*, or external, risks. Examples of technology risks are hostile artificial intelligence and destructive biotechnology or nanotechnology. Insufficient or malign global governance creates risks in the social and political domain, such as a global war, including nuclear holocaust; bioterrorism using genetically modified organisms; cyberterrorism destroying such critical infrastructures as the electrical grid and worldwide web, or the failure to manage a natural pandemic. Problems and risks in the domain of Earth system governance include global warming; environmental degradation—including extinction of species; famine as a result of non-equitable resource distribution; human overpopulation; crop failures, and non-sustainable agriculture. Examples of non-anthropogenic risks are an asteroid impact event; a supervolcanic eruption; a lethal gamma-ray burst; a geomagnetic storm destroying electronic equipment; natural long-term climate change; hostile extraterrestrial life, and, in the very-distant future, the Sun transforming into a red giant star engulfing the Earth.

Until relatively recently, most existential and global catastrophic risks were natural, such as the supervolcanoes and asteroid impacts that led to mass extinctions millions of years ago. The technological advances of the last century, while responsible for great progress and achievements, have also exposed humanity in particular and life on Earth in general to new existential risks, putting this world as a whole in peril. Cumulatively, these risks have created what the philosopher Nick Bostrom has called "The Vulnerable World" hypothesis.[5]

A nuclear holocaust, whose possibility dates to the creation and detonation by the United States of the first atomic bombs in 1945, is the first human-made existential risk.[6] A global thermonuclear war, between the US and China or between, the US and Russia, for example, could kill a large percentage of the human population, and even a local ("theater") nuclear conflict, between India and Pakistan, for example, could kill hundreds of millions. As more research into nuclear threats was conducted, scientists realized that the resulting "nuclear winter" could be even deadlier than the war itself, potentially killing most people and possibly endangering life on Earth.

Biotechnology and genetics often inspire as much fear as excitement, as people worry about the possibly negative effects of cloning, gene splicing, gene drives, and a host of other genetics-related advancements. While biotechnology provides a great opportunity to save and improve lives, it also increases existential risks associated with manufactured pandemics and the loss of genetic diversity.

Artificial intelligence (AI) has long been associated with science fiction, but it's a field that's made significant strides in recent years. As with biotechnology, there is great opportunity to improve lives with AI, but if the technology is not developed safely, there is also the chance that someone could accidentally or intentionally unleash a malign and seemingly omnipotent AI system that might ultimately cause the elimination of humanity.

Climate change, whether anthropogenic, natural, or a combination of the two, is an increasing threat that many people, governments, and nongovernmental organizations are trying to address. As the global average temperature rises, droughts, floods, fires, extreme storms, and other "natural" catastrophes are becoming the norm. The resulting food, water, and housing shortages could trigger economic instabilities and more regional wars. While climate change itself is unlikely to be an existential risk, at least in the short run, the havoc it wreaks could increase the likelihood of nuclear war or other anthropogenic catastrophes. In the long run, of course, runaway global warming could end, or at least significantly harm, human civilization as it is known today. And then there are, of course, global pandemics.

Global Pandemics

Throughout history, nothing has killed more human beings than infectious disease. The COVID-19 (SARS-CoV-2) pandemic demonstrates how vulnerable we remain, and also how we might prevent and mitigate similar pandemics in the future.

A pandemic[7] (from the Greek *pan*, "all," and *demos*, "people") is an epidemic of an infectious disease that has spread across a large region, for instance, across multiple continents or even globally. An epidemic denotes something, usually infection- or drug-related, that's out of control, such as "the opioid epidemic." An epidemic is often localized to a region, but the number of those infected in that region is significantly higher than normal. For example, when COVID-19 was limited to Wuhan, China, it was an epidemic. The geographical spread of COVID-19 turned it into a pandemic.

Endemics, on the other hand, are a constant presence in a specific location. Malaria is endemic to parts of Africa; ice is endemic to Antarctica (at least for now). A widespread endemic disease, such as the seasonal flu, with a stable number of infected people is not a pandemic, since its outbreaks occur simultaneously in specific regions of the globe rather than being spread worldwide.

Throughout history, there have been a number of pandemics of such diseases as smallpox and tuberculosis. The most fatal pandemic recorded in human history was the Black Death (also known as "The Plague"), which killed an estimated 75–200 million people in the 14th century. Other notable pandemics include the 1918–20 influenza pandemic ("Spanish flu"), which may have infected 500 million and killed more than 50 million people, and the 2009 influenza pandemic (H1N1 or "Swine Flu"), which killed hundreds of thousands. Current pandemics include HIV/AIDS—the ongoing HIV infection and AIDS for which there is still no vaccine—and the 2019–? coronavirus pandemic. According to the World Health Organization, HIV-related causes have killed about 32 million people out of the 75 million who have been infected (roughly 43%) since it was detected in human blood about 60 years ago, and of the approximately 38 million people living with HIV, more than 770,000 died in 2018.

There are many historical examples of pandemics having had a devastating effect on a large number of people. The present unprecedented scale and speed of human movement make it more difficult than ever to contain an epidemic through local quarantines, and the evolving nature of the risk means that both natural pandemics and very virulent and highly transmissible bioengineered pathogens may pose an omnicidal threat to human civilization.

Is Omnicide a "Blessing" or a "Curse?"

Omnicide (the death of everyone) is human extinction as a result of human action. Most commonly, it refers to extinction through nuclear warfare or biological warfare, but it can also apply to extinction due to a global anthropogenic ecological or some other catastrophe.

Some philosophers, animal rights activists, anarchists, and scientists are "antinatalists," who assign a negative value to human birth.[8] Antinatalists argue that people should abstain from procreation because it is allegedly morally wrong and argue that human extinction would be a positive thing for the other organisms on the planet, and for the Earth itself, arguing, for example, that human civilization is in itself omnicidal.

Most scholars and scientists who have addressed the issue of human extinction, however, argue that the potential "benefits" of omnicide are outweighed by the disadvantages—especially to us and to all possible future generations of humans!—of the termination of our species due to one or more global existential risks. These "pronatalists" also claim that because of the inconceivably large number of potential future lives that are at stake, even small reductions of existential risk have great value.[9]

Some Possible Future Scenarios

Planetary management and respecting planetary boundaries have been proposed as approaches to preventing ecological and other global catastrophes. One such approach is geoengineering, which includes the deliberate large-scale engineering and manipulation of the planetary environment to combat or counteract anthropogenic changes in atmospheric chemistry. Food storage has been proposed globally, but the monetary cost would be high.

For example, some "survivalists" stock their survival retreats with multiple-year food supplies. The Svalbard Global Seed Vault in Norway is buried 400 feet (120 meters) inside a mountain on an island in the Arctic. It is designed to hold 2.5 billion seeds from more than 100 countries as a precaution to preserve the world's crops. The surrounding rock is −6 °C (21°F), but the vault is kept at −18 °C (0°F) by refrigerators powered by locally-sourced coal.

Also on the Svalbard archipelago is the Arctic World Archive, including a data storage facility containing information—a veritable world library—stored offline on digital film that has a purported lifetime of at least 500 years and is ostensibly preserved from global threats by the climatic conditions present on the Svalbard archipelago. The facility is located approximately 490 feet (150 meters) below ground inside an abandoned coal mine in a mountain on the only permanently populated Svalbard island—Spitsbergen. Because of this island's permafrost, even if the power to the facility failed, the temperature inside the vault should remain below freezing—meant to preserve the vault's contents for centuries. The Svalbard archipelago is declared demilitarized by 42 nations and is claimed to be one of the most geopolitically secure places in the world. The vault is situated deep enough that it can ostensibly avoid damage from nuclear and electromagnetic pulse weapons.[10]

More speculatively, if society continues to function in some form, and if the biosphere remains habitable, calorie needs for the present human population might in theory be met during an extended absence of sunlight—due to a nuclear winter, for example—given sufficient advance planning. Conjectured solutions include growing mushrooms on the dead plant biomass left in the wake of the catastrophe, converting cellulose to sugar, or feeding natural gas to methane-digesting bacteria.

Other possible future "solutions" to the problems raised by existential risks include space colonization as a way to improve the odds of some humans surviving an extinction-level event. Stephen Hawking and others have advocated colonizing other planets within the solar system once technology progresses sufficiently, in order to improve the chance of human survival from such planet-wide events as global thermonuclear war.

Some social scientists propose the establishment on Earth of one or more self-sufficient, remote, permanently occupied settlements, specifically created for the purpose of surviving a global disaster. Economist Robin Hanson, for example, has argued that a secure refuge permanently housing as few as 100 people would significantly improve the chances of human survival during a range of global catastrophes.[11]

Other scenarios envision that humans could use genetic or technological modifications to split into normal humans and a new species—"posthumans." Such a species could be fundamentally different from any previous life form on Earth, e.g., by merging humans with technological systems, possibly including "thinking" machines. Such scenarios also assess the risk that the "old" human species will be outcompeted and driven to extinction by the new, posthuman entity.

And then, of course, there is the distinct possibility that humanity will not survive a global existential calamity. But much or at least some of the "natural world" might. What would the Earth look like without this world called human civilization?[12]

Out of This World?

Humanity's first space explorer was Soviet cosmonaut Yuri Gagarin, who orbited around the Earth in April, 1961. Since then, more than 550 people have left the Earth for space, of whom about a tenth have been women. Although the Soviet Union initially pulled ahead in the international space race with the first space walks, the US was the first to reach the Moon when its Apollo 11 touched down on the lunar surface in July, 1969. In the next few years, 12 men, all Americans, walked on the Moon, but no one has been back there since 1972. For the past few decades, the only humans in space have been those on the International Space Station [ISS], which has been continuously occupied since late 2000.

The US and Russia have been giving way to new space players. In 2003, China became the third country to put a person into orbit and India plans to follow in 2022. And while it is looking increasingly likely that the ISS will be defunded in the next decade by the nations currently supporting it, several private ventures are considering either taking over the ISS or building their own space stations.

In what is being called the "billionaire's space race," Elon Musk (the founder of Tesla electric cars), former Amazon CEO Jeff Bezos, and Virgin boss Richard Branson all want to send private citizens to space. Their companies, SpaceX, Blue Origin, and Virgin Galactic, respectively, are bent on making human space travel cheaper. They join a handful of commercial space flight companies that already work as contractors for national space agencies. Aerospace industry companies Boeing and Lockheed Martin send heavy launchers into space, but those cost several times more than SpaceX's rocket system, which reduces costs through deploying reusable spacecraft.

Leaving Earth?

One response to the bleak fate of this world is to leave it. No longer in the realm of science fiction, the idea of putting people on Mars has been taken seriously in recent years, especially by multibillionaires. While governmental agencies prioritize returning to the Moon, some private firms are looking to send humans to Mars. Elon Musk has said his life goal is to create a thriving Mars colony as a fail-safe for humanity in case of a catastrophic event on Earth, such as a nuclear war or Terminator-style artificial intelligence coup. For this, SpaceX is developing the Big Falcon Rocket (BFR), which Musk claims could send crewed flights to the red planet by the mid-2020s.

In one of his last interviews before his death in 2018, the acclaimed physicist Stephen Hawking emphasized what he saw as the great challenges and existential threats for humanity in coming decades. Hawking was troubled that humanity was putting all its eggs in one basket—that basket being Earth. For decades, Hawking had been calling for humans to begin the process of permanently settling other planets. Hawking's rationale was that humankind would eventually fall victim to an extinction-level catastrophe—perhaps sooner rather than later. What worried him were so-called low-probability, high impact events—for example, a large asteroid striking our planet or an invasion by hostile extraterrestrials. But Hawking perceived a host of other potential threats, including AI, climate change, genetically modified viruses (perhaps SARS/COVID-related?), and nuclear war, to name a few. In 2016, he told the BBC: "Although the chance of a disaster to planet Earth in a given year may be quite low, it adds up over time, and becomes a near certainty in the next 1,000 or 10,000 years." Hawking was confident that humans would spread out into the cosmos by that time (given the chance), but added: "We will not establish self-sustaining colonies in space for at least the next hundred years, so we have to be very careful in this period."[13]

Hawking's views resemble those of Elon Musk. In 2013, Musk told a conference: "Either we spread Earth to other planets, or we risk going extinct. An extinction event is inevitable and we're increasingly doing ourselves in." These warnings are also similar to those of such futurists as Ray Kurzweil and Max Tegmark, as well as the theoretical physicist Michio Kaku, author of *The Future of Humanity.*

In this book, Kaku forecasts a future for humanity in which dramatic advances in technology and science will force us to rethink our position in the universe. He foresees a time when Mars will become our new home, when intelligent life won't be limited to us (if it ever was …), when immortality will no longer be a fanciful obstacle, and when, ultimately we will expand into colonizing other solar systems. However, according to Kaku, humanity's first step is to leave what will soon be an uninhabitable planet Earth.

I submit that an exit from a future (which may not so far away as one might think …) uninhabitable Earth should be a *last, not first step. Furthermore, if and when the time comes seriously to consider engaging in the colonization and/or creation of other worlds, this should only be done if and when this world, the one we have created on Earth, has been pacified to such an extent that whatever version of the human, or posthuman, species has prevailed, and for survival reasons feels compelled to depart this planet, it does not carry its bellicose baggage, and weaponry, with it, thus potentially condemning not only it but also extraterrestrial life forms to the catastrophic fate of this world.*

The World—a Good, or Bad, Idea?

Ludwig Wittgenstein was insightful when he declared that "the world of the happy man is a different one from that of the unhappy man." This seems to imply that how one feels in an existential sense deeply affects how one perceives what in ordinary language is referred to as "the world," namely the cultural and physical realm of persons and objects within which one lives. Accordingly, the world is "good" if it pleases one and "bad" if it does not.

But what if this world is the chief source of one's distress? And what to do if the world is indeed "a crazy place," or even "evil," especially if one is not a religious believer in "another, better world," to which the righteous will ascend after death or following the tribulations of the "end-time?" And, perhaps from a god's or a more advanced extraterrestrial civilization point of view, the world fabricated by humans on Earth is a threat not merely to the existence of life on this planet, but potentially to organisms and intellects in other worlds as well? In these cases, perhaps the understandable human desire to survive by any means necessary, including the colonization of other potentially habitable worlds is not merely anthropocentric but is also an existential risk to other worlds in ways similar to the existential threats posed by European colonists to the indigenous peoples they infected and enslaved during the periods of colonial expansion. Perhaps, then, the lifeworld of homo sapiens is the real or potential death world not only to itself but also to the lifeworlds of extraterrestrial species, as it has been for countless terrestrial species it has driven to extinction.

Consequently, "the world" may be a good idea in theory, as a (mere) "idea." And, this book has explored the world as idea through an exposition and examination of some of its notable framers, mostly but not exclusively dead white Western males of privilege. That is a strength of this approach, insofar as the ideas of the world generated by these people are frequently powerful and perceptive. It is also a limitation in that this book has generally omitted all "the other voices" of people who did not publish in the venues accessible to this author, or did so but were left out due to my own preferences, limitations, and capabilities. The sequel to this book will, circumstances and good fortune permitting, attempt to address such limitations.

In conclusion, this world may end with a big nuclear bang, or with an overheated whimper, or, simply, in silence. We may not be around to witness the end of the world, even though we may be its cause. The fate of the world is in any case existentially tied to that of the human lifeworld. To a considerable degree, we are the forgers of this world's fate. How we act during the decades to come will largely determine if and how that fate comes to pass.

Notes

1 Johann Wolfgang von Goethe, letter to Frau von Stein, June 8, 1787.
2 Johann Wolfgang von Goethe, allegedly in an album-entry for Schopenhauer.
3 A biosignature, sometimes called a chemical fossil or a molecular fossil, is any substance—such as an element, isotope, or a molecule—that provides scientific evidence of past or present life. Measurable attributes of life include its complex physical or chemical structures and its use of free energy and the production of biomass and wastes. A biosignature can provide evidence for living organisms outside the Earth and can be directly or indirectly detected by searching for their unique byproducts. See https://en.wikipedia.org/wiki/Biosignature.
4 See www.sciencealert.com/8-terrifying-ways-the-world-could-actually-end.
5 According to Nick Bostrom, scientific and technological progress might change people's capabilities or incentives in ways that would destabilize civilization. For example, advances in do-it-yourself biohacking tools might make it easy for anybody with basic training in biology to kill millions; novel military technologies could trigger arms races in which whoever strikes first has a decisive advantage; or some economically advantageous process may be invented that produces disastrous negative global externalities that are hard to regulate. This leads to his concept of a "vulnerable world," roughly, one in which there is some level of technological development at which civilization almost certainly gets devastated by default, i.e., unless it has exited the "semi-anarchic default condition." Nick Bostrom, "The Vulnerable World Hypothesis," *Global Policy*, Volume 10: Issue 4. November 2019, 455–76.

6 See Karl Jaspers. *The Future of Mankind* (Chicago, IL: University of Chicago Press, 1959).

7 The English noun pandemic comes from the Greek words *pan*, meaning "all," and *dêmos*, meaning "people."

8 Antinatalism is in opposition to natalism or pronatalism, and was used probably for the first time as the name of the position of Belgian philosopher Théophile de Giraud in his book *The Art of Guillotining the Procreators: Anti-Natalist Manifesto (L'art de guillotiner les procréateurs: Manifeste anti-nataliste,* 2006). Antitnatalist proponents include philosopher David Benatar, animal rights activist Steven Best, anarchist Todd May, and evolutionary biologist David Barash, who has argued that "we might consider doing the living world a favor by reproducing below the natural replacement rate, so that our species gradually disappeared, not with a nuclear bang but a whimper." David Barash, posted Feb 27, 2020: www.psychologytoday.com/us/blog/peace-and-war/202002/can-the-coronavirus-help-us-see-ourselves-differently.

9 See Nick Bostrom and Milan M. Cirkovic, eds. *Global Catastrophic Risks* (Oxford: Oxford University Press, 2008); Martin Rees. *On the Future: Prospects for Humanity* (Princeton, NJ: Princeton University Press, 2018); https://futureoflife.org/background/existential-risk/Wikipedial; and https://en.wikipedia.org/wiki/Global_catastrophic_risk. Relevant research institutes include the Centre for the Study of Existential Risk, University of Cambridge: www.cser.ac.uk/; and the Future of Humanity Institute, University of Oxford: www.fhi.ox.ac.uk/.

"Pronatalists" include the late astrophysicist Carl Sagan, who wrote in 1983:

> If we are required to calibrate extinction in numerical terms, I would be sure to include the number of people in future generations who would not be born … (By one calculation), the stakes are one million times greater for extinction than for the more modest nuclear wars that kill 'only' hundreds of millions of people. There are many other possible measures of the potential loss—including culture and science, the evolutionary history of the planet, and the significance of the lives of all of our ancestors who contributed to the future of their descendants. Extinction is the undoing of the human enterprise.

The late philosopher Derek Parfit made an anthropocentric utilitarian argument that, because all human lives have roughly equal intrinsic value no matter where in time or space they are born, the large number of lives potentially saved in the future should be multiplied by the percentage chance that an action will save them, yielding a large net benefit for even tiny reductions in existential risk. Another philosopher, Robert Adams, rejects Parfit's "impersonal" views, but speaks instead of a moral imperative for loyalty and commitment to

> the future of humanity as a vast project … The aspiration for a better society—more just, more rewarding, and more peaceful … our interest in the lives of our children and grandchildren, and the hopes that they will be able, in turn, to have the lives of their children and grandchildren as projects.

And more recently, Bostrom has argued that preference-satisfactionist, democratic, custodial, and intuitionist arguments all converge on the commonsense view that preventing existential risk is a high moral priority, even if the exact "degree of badness" of human extinction varies among these philosophies. Parfit claimed that the size of the "cosmic endowment" can be calculated from the following argument: if Earth remains habitable for a billion more years and can sustainably support a population of more than a billion humans, then there is a potential for 10^{16} (or 10,000,000,000,000,000) human lives of normal duration. Bostrom has gone further, stating that if the universe is empty, then the accessible universe can support at least 10^{34} biological human life-years; and, if some humans were uploaded onto computers, could even support the equivalent of 10^{54} cybernetic human life-years. See Bostrom, Nick. "Existential risk prevention as global priority". Global Policy 4.1 (2013): 15–31. S*agan, Carl* (1983). "Nuclear war and climatic catastrophe: Some policy implications". *Foreign Affairs*. 62 (2): 257–92*;* Derek Parfit, *Reasons and Persons* (Oxford: Clarendon Press, 1984); Robert Merrihew Adams *"Should Ethics Be More Impersonal? A Critical Notice of Derek Parfit, Reasons and Persons,"* The Philosophical Review, October 1989, 98 (4): 439–84*;* S. Schubert, L. Caviola, & N.S. Faber, "The Psychology of Existential Risk: Moral Judgments about Human Extinction." *Sci Rep* 9, 15100 (2019), available at: https://doi.org/10.1038/s41598-019-50145-9; Toby Ord, *The Precipice: Existential Risk and the Future of Humanity* (London: Bloomsbury, 2021), and Jim Holt, "The Power of Catastrophic Thinking Should we value human lives in the distant future as much as present ones?" *New York Review of Books*, February 25, 2021, review of Toby Ord, *The Precipice: Existential Risk and the Future of Humanity,* www.nybooks.com/articles/2021/02/25/power-catastrophic-thinking-toby-ord-precipice/?lp_txn_id=1023913.

10 This is according to a documentary on Deutsche Welle, May 10, 2020, and also at https://en.wikipedia.org/wiki/Arctic_World_Archive.

11 See Robin Hanson, *The Age of Em: Work, Love and Life When Robots Rule the Earth* (New York: Oxford University Press, 2016); and www.overcomingbias.com/2017/08/forager-v-farmer-elaborated.html.

12 One such vision of a possible future of the Earth without humans is Alan Weisman. *The World Without Us* (London: Viking, 2008).

13 Stephen Hawking's interview is available at: www.smithsonianchannel.com/videos/stephen-hawkings-stark-warning-for-humans-to-leave-earth/61143; and www.bbc.com/news/science-environment-43408961.

Index

D

E

M

N

Q

R

T

X

Y

Z

www.ingramcontent.com/pod-product-compliance
Ingram Content Group UK Ltd.
Pitfield, Milton Keynes, MK11 3LW, UK
UKHW021450080625
459435UK00012B/440

* 9 7 8 1 0 3 2 1 1 5 6 6 5 *